AF361809

Genetics, Genomics, and Breeding of Legume Crops

Genetics, Genomics, and Breeding of Legume Crops

Editor

Guo-Liang Jiang

Basel • Beijing • Wuhan • Barcelona • Belgrade • Novi Sad • Cluj • Manchester

Editor
Guo-Liang Jiang
Agricultural Research Station
Virginia State University
Petersburg
United States

Editorial Office
MDPI
St. Alban-Anlage 66
4052 Basel, Switzerland

This is a reprint of articles from the Special Issue published online in the open access journal *Agronomy* (ISSN 2073-4395) (available at: www.mdpi.com/journal/agronomy/special_issues/genetics-legume).

For citation purposes, cite each article independently as indicated on the article page online and as indicated below:

Lastname, A.A.; Lastname, B.B. Article Title. *Journal Name* **Year**, *Volume Number*, Page Range.

ISBN 978-3-0365-9916-8 (Hbk)
ISBN 978-3-0365-9915-1 (PDF)
doi.org/10.3390/books978-3-0365-9915-1

Contents

About the Editor

Guo-Liang Jiang

Dr. Guo-Liang Jiang obtained a PhD in Plant Breeding and Genetics at Nanjing Agricultural University, China. Dr. Jiang has many years of applied and basic research in plant breeding and genetics. Dr. Jiang's research interests include cultivar development, germplasm enhancement, breeding principles and methodologies, QTL mapping, molecular marker applications, genome-wide association study (GWAS), genetic analysis and characterization of traits of importance (with emphasis on quality, yield, and resistance/tolerance), gene exploiting and discovery, and trait integration. Dr. Jiang has developed and released over 30 crop cultivars and germplasm lines and published approximately 80 articles in peer-reviewed journals and over 30 book chapters and proceeding papers. Dr. Jiang has accomplished many research projects and received many awards and grants. Dr. Jiang is currently a plant breeder and professor, leading the soybean breeding and genetics program at Virginia State University. Dr. Jiang serves as an editorial board member of six referred journals.

Editorial

Special Issue "Genetics, Genomics, and Breeding of Legume Crops"

Guo-Liang Jiang

Agricultural Research Station, Virginia State University, P.O. Box 9061, Petersburg, VA 23806, USA;
gjiang@vsu.edu or gljiang99@yahoo.com; Tel.: +1-804-524-5953

Citation: Jiang, G.-L. Special Issue "Genetics, Genomics, and Breeding of Legume Crops". *Agronomy* **2021**, *11*, 475. https://doi.org/10.3390/agronomy11030475

Received: 5 February 2021
Accepted: 8 February 2021
Published: 4 March 2021

Publisher's Note: MDPI stays neutral with regard to jurisdictional claims in published maps and institutional affiliations.

Legume crops are grown in the world primarily for their grain seeds that are widely used for human and animal consumption or for the production of oils for industrial uses. Broadly, legumes include well-known crops: soybean, common bean or dry bean, peanut, peas, chickpeas, cowpea, lentils, alfalfa, clover, etc. From a nutritional standpoint, legume crops are a significant source of protein, oil, dietary fiber, carbohydrates and dietary minerals. Economically, they also play an important role in international trade. Genetic improvement has been key to the growth of crop production and will continuously contribute to sustainable agriculture and food security. For the present, plant genetic improvement is in the middle of its evolution from field-based traditional breeding to a new era of application of multiple novel techniques such as marker-assisted selection, genomic prediction and gene-editing, which will be integrated with conventional methods in practical breeding. Research has involved all the traits of importance, including yield, quality, resistance to pests/diseases and abiotic stresses. In this Special Issue, fifteen articles are included, addressing a wide range of research topics in legume crops.

Abiotic stress like drought and salinity is a major limiting factor in crop productivity worldwide. Evaluation and genetic understanding of stress tolerance is essential for development of varieties adapted to the abiotic stresses. In light of insufficient studies of drought response in large, seeded genotypes of Andean origin, Sedlar et al. (2020) constructed a novel Andean intra-gene pool genetic linkage map for quantitative trait locus (QTL) mapping of drought-responsive traits using a recombinant inbred line population derived from a cross of two cultivars differing in their response to drought [1]. They detected 49 QTLs for physiology, phenology, and yield-associated traits under control and/or drought conditions, and validated the QTLs by projection on common bean consensus linkage map. These results confirmed the potential of Andean germplasm in improving drought tolerance in common bean. Tani et al. (2018) evaluated the seedling response of *Medicago. sativa*, *M. arborea*, and their hybrid (*Alborea*) to salt shock and salt stress treatments [2]. They concluded that different components of salt tolerance mechanisms were regulated in the populations of *M. sativa* and *M. arborea*. It appears that the knowledge of different parental mechanisms of salt tolerance could be helpful for incorporating both mechanisms in Alborea populations.

In soybean, seed composition or nutritional quality is of importance to uses and market values. Bulatova et al. (2019) studied a soybean germplasm collection to identify accessions with low trypsin inhibitor content in seeds [3]. They selected and analyzed twenty-nine accessions, parental plants, and two hybrid populations using genetic markers for alleles of the *Ti3* locus, encoding Kunitz trypsin inhibitor (KTI). By marker-assisted selection with Satt228, they obtained some prospective homozygous *ti3/ti3* lines which might be further used in the breeding program for soybean quality improvement. To investigate the variation in seed fatty acid composition of soybeans of different origins, Abdelghany et al. (2020) evaluated a diverse germplasm set of 633 soybean accessions originated from China, United States of America (USA), Japan, and Russia [4]. Their results indicated that the unique accessions identified could be used in the soybean breeding programs to fit various human nutrition patterns across the world.

Short petiole is a valuable trait for the improvement of plant canopy of ideotypes with high yield [5]. Liu et al. (2019) identified a unique soybean mutant line called derived short petiole (*dsp*) with an extremely short petiole, which is also obviously different from most short-petiole lines identified previously. Genetic analysis with 941 F_2 individuals and subsequent segregation analysis of 184 $F_{2:3}$ and 172 $F_{3:4}$ families revealed that the *dsp* mutant was controlled by two recessive genes, named as *dsp1* and *dsp2*. These two recessive genes were located on two nonhomologous regions of chromosome 07 and chromosome 11. A total of 36 and 33 gene models were located within the physical genomic interval of *dsp1* and *dsp2* loci, respectively. They concluded that the identified markers linked with genomic regions responsible for the short-petiole phenotype of soybean could be effectively used in developing ideal soybean cultivars through marker-assisted breeding [5].

Pleiotropy is considered to have a significant impact on multi-trait evolution, but its roles in the evolution of domestication-related traits in crop species remain unclear [6]. In soybean, several maturity loci or quantitative trait loci (QTL) controlling maturity are known to have major effects on both flowering and maturity in a highly correlated pleiotropic manner. Sedivy et al. (2020) conducted a QTL mapping experiment by creating a population derived from a cross between domesticated soybean *G. max* and its wild ancestor *G. soja* that underwent stringent selection for non-pleiotropy in flowering and maturity [6]. Their results suggested that pleiotropy in flowering and maturity could be genetically separated, although pleiotropic loci such as *E* loci that control both flowering and maturity have been preferred by artificial selection during soybean domestication and diversification. The non-pleiotropic loci identified in the study may be helpful to improve soybean yield potential against diverse environments and cultivation systems.

Large seed size in the kabuli chickpea (*Cicer arietinum* L.) is preferred in the market for its high price and superior seedling vigor as well [7]. Comparatively, the double-podded chickpea has advantages in yield and stability over the single-podded chickpea. Kivrak et al. (2020) presented a study with the aim of integrating extra-large-seeded and double-podded traits in the kabuli chickpea, and increasing variation by transgressive segregations [7]. Heritability of important agro-morphological traits were also estimated in their study using F_2 and F_3 populations.

Cooking characteristics in food legume crops have been rarely investigated, although they are a central factor for consumer's choice. Interestingly, Harouna et al. (2019) explored the cooking time and the water absorption capacity upon soaking using 84 accessions of wild *Vigna* legumes [8]. For the first time, they elucidated the relationships between the cooking time and water absorbed during soaking.

Peanut is widely grown around the world, predominantly in China followed by India, Nigeria, the USA and Sudan in terms of total production. Konate et al. (2020) presented a review of the progress in the peanut breeding in Burkina Faso, a relatively small peanut producer with unique climate conditions and cropping systems [9]. They also provided crucial information about opportunities and challenges for peanut research in Burkina Faso, and particularly emphasized the need for institutional attention to genetic improvement of the crop.

Lupinus mutabilis (tarwi) is a species of Andean origin with high protein and oil content, and regarded as a potential crop in Europe [10]. The knowledge of intra-specific genetic variability of the collections helps successful introduction of this crop and establishment of breeding and conservation programs. Guilengue et al. (2020) assessed genetic and genomic diversity of tarwi accessions under Mediterranean conditions using morphological traits, inter-simple sequence repeat (ISSR) markers and genome size [10]. Their results revealed important levels of diversity, which is unrelated to phenotypic diversity, and reflected the recent domestication of the crop.

A unique nature of legume crops is the nitrogen-fixing function. Most legume crops have symbiotic nitrogen-fixing bacteria in structures called root nodules. Therefore, in addition to use of grains for human food and animal feeds, some legume species are grown for cover crop or livestock forage. Crimson clover (*Trifolium incarnatum* L.) is the most common legume cover crop in the United States [11]. However, limited genetic variation for crimson clover was previously found within the National Plant Germplasm System (NPGS) collection. Focusing on traits important for cover crop performance, Moore et al. (2020) assessed the phenotypic and nodule microbial diversity within the NPGS crimson clover collection [11]. They discovered that accession effect was significant for the traits of fall emergence, winter survival, flowering time, biomass per plant, nitrogen (N) content in aboveground biomass, and proportion of plant N from biological nitrogen fixation (BNF). The information generated should be useful for cover crop breeding and production.

Mutation is a powerful tool in creating genetic variability, and mutation breeding has been successfully used in the development of new varieties with unique traits in plants. Based on phenotypic traits, Kim et al. (2020) chose 208 soybean mutants as a mutant diversity pool (MDP), and then investigated the genetic diversity and inter-relationships of these MDP lines using target region amplification polymorphism (TRAP) markers [12]. They suggested that the MDP would have great potential for soybean germplasm enhancement and that TRAP markers are useful for the selection of mutants in soybean mutation breeding. In order to enhance peanut genetic variability, Chen et al. (2020) treated two widely cultivated peanut genotypes, using different concentrations of the mutagen ethyl methyl sulfonate (EMS) for different durations [13]. They found that mutants induced by EMS differed in various phenotypic traits, such as plant height, number of branches, leaf characteristics, and yield and quality in plants of the M_2 generation. They also identified some potentially useful mutants among individuals of the M_2 generation, which were associated with dwarfism, leaf color and shape, high oil or protein content, seed size and coat color. Mutations were stably inherited in M_3-generation individuals.

As a new powerful tool, genomic selection (GS) has attracted increasing attention since it was proposed. An aim of GS is to incorporate molecular information directly into the prediction of individual genetic merit by predicting genomic estimated breeding value (GEBV). Using regularized quantile regression (RQR), Nascimento et al. (2019) predicted the individual genetic merits of the traits associated with flowering time (DFF—days to first flower; DTF—days to flower) in the common bean [14]. They also compared the predictive abilities in predicting the genetic merit between RQR and other methods such as random regression best linear unbiased predictor (RR-BLUP), Bayesian LASSO (BLASSO) and BayesB.

A trend of the present and future plant breeding is the application of integrated multiple methods including state-of-the-art technologies and extensive collaboration. Saxena et al. (2020) reported such an attempt. In order to overcome the productivity barrier for pigeonpea production, a translational pigeonpea genomics consortium (TPGC) was established across multiple states in India [15]. The team has been engaged in deploying modern genomics approaches in breeding and popularizing modern varieties in farmers' fields to augment pigeonpea productivity and production. Through the collaborative effort including farmer's participation, new genetic stock has been developed for trait mapping and molecular breeding initiated for improving resistance to fusarium wilt and sterility mosaic disease [15]. Meanwhile, genomic segments associated with various traits have been identified and participatory varietal selection trials involving a total of 303 farmers have been conducted. It is expected that further progress can be achieved in the near future.

Funding: This research received no external funding.

Acknowledgments: I thank the USDA-NIFA Evans-Allen Research Program for funding in support of this work. I appreciate the great efforts that the authors have made in contributing to the collection of the special topic. I especially thank all the reviewers for their time and dedications to review the manuscripts received. It would not have succeeded without their prompt, thoughtful and constructive reviews. I would also thank the editors in the Editorial Office, especially Aileen Song, the managing editor of this Special Issue, for their excellent in support.

Conflicts of Interest: The author declares no conflict of interest.

References

1. Sedlar, A.; Zupin, M.; Maras, M.; Razinger, J.; Šuštar-Vozli, J.; Pipan, B.; Megli, V. QTL mapping for drought-responsive agronomic traits associated with physiology, phenology, and yield in an Andean intra-gene pool common bean population. *Agronomy* **2020**, *10*, 225. [CrossRef]
2. Tani, E.; Sarri, E.; Goufa, M.; Asimakopoulou, G.; Psychogiou, M.; Bingham, E.; Skaracis, G.N.; Abraham, E.M. Seedling growth and transcriptional responses to salt shock and stress in *Medicago sativa* L., *Medicago arborea* L., and their hybrid (Alborea). *Agronomy* **2018**, *8*, 231. [CrossRef]
3. Bulatova, K.; Mazkirat, S.; Didorenko, S.; Babissekova, D.; Kudaibergenov, M.; Alchinbayeva, P.; Khalbayeva, S.; Shavrukov, Y. Trypsin inhibitor assessment with biochemical and molecular markers in a soybean germplasm collection and hybrid populations for seed quality improvement. *Agronomy* **2019**, *9*, 76. [CrossRef]
4. Abdelghany, A.M.; Zhang, S.; Azam, M.; Shaibu, A.S.; Feng, Y.; Qi, J.; Li, Y.; Tian, Y.; Hong, H.; Li, B.; et al. Natural variation in fatty acid composition of diverse world soybean germplasms grown in China. *Agronomy* **2020**, *10*, 24. [CrossRef]
5. Liu, M.; Wang, Y.; Gai, J.; Bhat, J.A.; Li, Y.; Kong, J.; Zhao, T. Genetic analysis and gene mapping for a short-petiole mutant in soybean (Glycine max (L.) Merr.). *Agronomy* **2019**, *9*, 709. [CrossRef]
6. Sedivy, E.J.; Akpertey, A.; Vela, A.; Abadir, S.; Khan, A.; Hanzawa, Y. Identification of non-pleiotropic loci in flowering and maturity control in soybean. *Agronomy* **2020**, *10*, 1204. [CrossRef]
7. Kivrak, K.G.; Eker, T.; Sari, H.; Sari, D.; Akan, K.; Aydinoglu, B.; Catal, M.; Toker, C. Integration of extra-large-seeded and double-podded traits in chickpea (Cicer arietinum L.). *Agronomy* **2020**, *10*, 901. [CrossRef]
8. Harouna, D.V.; Venkataramana, P.B.; Matemu, A.O.; Ndakidemi, P.A. Assessment of water absorption capacity and cooking time of wild under-exploited Vigna species towards their domestication. *Agronomy* **2019**, *9*, 509. [CrossRef]
9. Konate, M.; Sanou, J.; Miningou, A.; Okello, D.K.; Desmae, H.; Janila, P.; Mumm, R.H. Past, present and future perspectives on groundnut breeding in Burkina Faso. *Agronomy* **2020**, *10*, 704. [CrossRef]
10. Guilengue, N.; Alves, S.; Talhinhas, P.; Neves-Martins, J. Genetic and genomic diversity in a tarwi (Lupinus mutabilis Sweet) germplasm collection and adaptability to Mediterranean climate conditions. *Agronomy* **2020**, *10*, 21. [CrossRef]
11. Moore, V.; Davis, B.; Poskaitis, M.; Maul, J.E.; Kucek, L.K.; Mirsky, S. Phenotypic and nodule microbial diversity among crimson clover (*Trifolium incarnatum* L.) accessions. *Agronomy* **2020**, *10*, 1434. [CrossRef]
12. Kim, D.-G.; Lyu, J.I.; Lee, M.-K.; Kim, J.M.; Hung, N.N.; Hong, M.J.; Kim, J.-B.; Bae, C.-H.; Kwon, S.-J. Construction of soybean mutant diversity pool (MDP) lines and an analysis of their genetic relationships and associations using TRAP markers. *Agronomy* **2020**, *10*, 253. [CrossRef]
13. Chen, T.; Huang, L.; Wang, M.; Huang, Y.; Zeng, R.; Wang, X.; Wang, L.; Wan, S.; Zhang, L. Ethyl methyl sulfonate-induced mutagenesis and its effects on peanut agronomic, yield and quality traits. *Agronomy* **2020**, *10*, 655. [CrossRef]
14. Nascimento, A.C.; Nascimento, M.; Azevedo, C.; Silva, F.; Barili, L.; Vale, N.; Carneiro, J.E.; Cruz, C.; Carneiro, P.C.; Serão, N. Quantile regression applied to genome-enabled prediction of traits related to flowering time in the common bean. *Agronomy* **2019**, *9*, 796. [CrossRef]
15. Saxena, R.K.; Hake, A.; Hingane, A.J.; Kumar, C.V.S.; Bohra, A.; Sonnappa, M.; Rathore, A.; Kumar, A.V.; Mishra, A.; Tikle, A.N.; et al. Translational Pigeonpea Genomics Consortium for accelerating genetic gains in pigeonpea (*Cajanus cajan* L.). *Agronomy* **2020**, *10*, 1289. [CrossRef]

agronomy

Article

Phenotypic and Nodule Microbial Diversity among Crimson Clover (*Trifolium incarnatum* L.) Accessions

Virginia Moore [1,2,*], Brian Davis [3], Megan Poskaitis [4], Jude E. Maul [2], Lisa Kissing Kucek [5] and Steven Mirsky [2]

[1] Department of Crop and Soil Sciences, North Carolina State University (NCSU), Raleigh, NC 27695, USA
[2] Beltsville Agricultural Research Center (BARC), United States Department of Agriculture (USDA), Agricultural Research Service (ARS), Beltsville, MD 20705, USA; jude.maul@usda.gov (J.E.M.); steven.mirsky@usda.gov (S.M.)
[3] Department of Environmental Science and Technology, University of Maryland (UMD), College Park, MD 20742, USA; brianwdavis@gmail.com
[4] School of Plant and Environmental Sciences, Virginia Polytechnic Institute and State University (VT), Blacksburg, VA 24061, USA; megankposkaitis@vt.edu
[5] Dairy Forage Research Center (DFRC), United States Department of Agriculture (USDA), Agricultural Research Service (ARS), Madison, WI 53706, USA; lisa.kucek@usda.gov
* Correspondence: vmmoore3@ncsu.edu; Tel.: +1-301-504-0069

Received: 13 August 2020; Accepted: 8 September 2020; Published: 21 September 2020

Abstract: Crimson clover (*Trifolium incarnatum* L.) is the most common legume cover crop in the United States. Previous research found limited genetic variation for crimson clover within the National Plant Germplasm System (NPGS) collection. The aim of this study was to assess the phenotypic and nodule microbial diversity within the NPGS crimson clover collection, focusing on traits important for cover crop performance. Experiments were conducted at the Beltsville Agricultural Research Center (Maryland, USA) across three growing seasons (2012–2013, 2013–2014, 2014–2015) to evaluate 37 crimson clover accessions for six phenotypic traits: fall emergence, winter survival, flowering time, biomass per plant, nitrogen (N) content in aboveground biomass, and proportion of plant N from biological nitrogen fixation (BNF). Accession effect was significant across all six traits. Fall emergence of plant introductions (PIs) ranged from 16.0% to 70.5%, winter survival ranged from 52.8% to 82.0%, and growing degree days (GDD) to 25% maturity ranged from 1470 GDD to 1910 GDD. Biomass per plant ranged from 1.52 to 6.51 g, N content ranged from 1.87% to 2.24%, and proportion of plant N from BNF ranged from 50.2% to 85.6%. Accessions showed particularly clear differences for fall emergence and flowering time, indicating greater diversity and potential for selection in cover crop breeding programs. Fall emergence and winter survival were positively correlated, and both were negatively correlated with biomass per plant and plant N from BNF. A few promising lines performed well across multiple key traits, and are of particular interest as parents in future breeding efforts, including PIs 369045, 418900, 561943, 561944, and 655006. In 2014–2015, accessions were also assessed for nodule microbiome diversity, and 11 genera were identified across the sampled nodules. There was large variation among accessions in terms of species diversity, but this diversity was not associated with observed plant traits, and the functional implications of nodule microbiome diversity remain unclear.

Keywords: cover crop; crimson clover; germplasm diversity; legume

---

1. Introduction

Cover crops provide a wide range of ecosystem services including nutrient retention, nitrogen fixation, erosion control, weed suppression, improved soil structure and ecology, improved water quality, and beneficial insect conservation [1–7]. Despite these widely acknowledged benefits, adoption has remained limited [8–10]. This is in part due to limited improvements in cover crop genetics.

To increase farmer adoption, there is a need for germplasm improvement in cover crops. A lack of adapted cover crop cultivars (e.g., with enhanced biomass and winter hardiness for various climates) presents a significant barrier to cover crop adoption [11,12]. Historically, plant breeding has focused resources and effort on breeding for increased productivity in cash crops, with remarkable success [13–19]. By comparison, relatively little breeding has focused on ecosystem services (e.g., improving crops for use as cover crops) [20]. Shifting resources towards cover crop breeding could result in significant improvement in cover crop germplasm and ultimately in increased adoption on the landscape. Many species used as cover crops have been bred as cash crops (e.g., for forage production). However, to initiate breeding for use as cover crops, it is necessary to evaluate existing variation for traits of interest specific to cover crop production. Previous studies have assessed the genetic and phenotypic variation available through the National Plant Germplasm System (NPGS) for cover crop species, notably hairy vetch (*Vicia villosa* L.) [21,22].

Crimson clover (*Trifolium incarnatum* L.) is the most common legume cover crop in the United States [10]. Valued for its biomass production and nitrogen contribution, crimson clover is used across many regions; it is commonly used as a forage in the Southeast US and can be grown as a winter or summer annually depending on the climate [23]. A previous screening of the genetic variation in the NPGS crimson clover collection found limited genetic diversity and a substantial overlap in pedigrees [24]. Modern US cultivars show especially low genetic variation, with nearly all cultivars derived from 'Dixie.' However, an assessment of phenotypic variation in the NPGS crimson clover collection has not been conducted for cover crop traits of importance to farmers: nitrogen fixation, biomass production, winter hardiness, and flowering time [12]. Further, identifying which microbes colonize nodules may provide insights into variation in biological nitrogen fixation among crimson clover genotypes. In other legume species, nodule occupancy has been observed to differ both diversity and richness in rhizobia species [25,26]. To improve crimson clover germplasm for cover cropping, the phenotypic variations in existing collections needs to be assessed for traits of interest.

In this study, we assessed phenotypic variation within the NPGS crimson clover collection, and also explored the genetic diversity of nodule microbial populations regulating biological nitrogen fixation (BNF). We evaluated the NPGS crimson clover collection for emergence, winter survival, flowering time, biomass per plant, nitrogen (N) content, and proportion of plant N from BNF. We also evaluated the diversity of nodule microbial populations among accessions, with the understanding that ultimately BNF will be a result of interactions between traits of the resident soil/inoculum population and plant traits that affect nodule entry selectivity, nodule number, and activity.

2. Materials and Methods

2.1. Plant Material

Germplasm consisted of 37 accessions of crimson clover, representing all plant introduction (PI) lines of crimson clover in the NPGS with seed available for distribution. However, seven of these accessions were not available from NPGS after the 2012–2013 season, so the remaining 30 accessions were evaluated in 2013–2014 and 2014–2015 (Table S1).

2.2. Field Evaluation

Phenotypic evaluation of crimson clover accessions was conducted over three growing seasons, with crimson clover direct-seeded by hand in late September 2012, 2013, and 2014 in field plots in Beltsville, MD, USA (39°02′ N, 76°56′ W) on sandy loam soils (Russett Christiana complex). The field design was a randomized complete block design (RCBD) with four replications in each year, except for five accessions planted in 2015, which only had three replications due to limited seed availability.

Each plot was a single row of 0.6 m in length with 1.5 m between plots. Between 37 and 45 seeds were planted per plot, depending on seed availability in each year.

Fall emergence was evaluated in late October of each year by counting the total number of plants in each plot. Winter survival was determined by counting the total number of plants per plot in late April divided by the total number of plants counted in the fall. Flowering time was evaluated by recording percent flowering on a per-plot basis on a scale from 0% (no flower buds present) to 100% (all flowers dried up entire length of head). Flowering evaluations took place periodically between late April and early June. In 2013, evaluation took place on six dates: 23 April, 9 May, 15 May, 24 May, 30 May, and 4 June. In 2014, evaluation took place on five dates: 28 April, 6 May, 13 May, 19 May, and 27 May. In 2015, evaluation took place on eight dates: 25 April, 29 April, 4 May, 7 May, 11 May, 14 May, 18 May, and 21 May. The frequency of evaluations and total duration of evaluation period varied from year-to-year primarily due to the effects of year-to-year weather variation (Table 1) on the rate of growth and development.

Table 1. Total accumulated growing-degree days (GDD), freezing-degree days (FDD), days below freezing without snow cover, total precipitation (mm), and minimum low temperature ($^\circ$C) in Beltsville, MD, from crimson clover planting to final data collection. Weather data were acquired from an on-site weather station at the Beltsville Agricultural Research Center (BARC).

	Growing Season		
Measurement	**2012–2013**	**2013–2014**	**2014–2015**
Growing degree days (GDD)	2208.9	2102.8	2198.2
Freezing degree days (FDD)	84	115	102
Days below freezing without snow cover	80	86	72
Total precipitation (mm)	596.0	802.8	616.2
Minimum low temperature ($^\circ$C)	−11.0	−16.5	−17.0

Once an accession was rated at 50% or greater for flowering, biomass was collected. All plants in the plot were pulled up with roots attached. Plants were counted and the roots were clipped. All plants within a plot were placed in the same brown paper bag and dried. Dry weight was recorded and plants were ground for laboratory evaluation of nitrogen content, proportion of nitrogen from BNF, and metagenomic analysis.

2.3. Laboratory Evaluation

The crimson clover biomass samples were separated into shoots and roots. Shoots were oven-dried (60 $^\circ$C) for approximately 72 h, weighed, and ground to pass a 1.0-mm screen. Tissue C and N concentrations and ^{15}N natural abundance were determined for the shoot material of each accession using a Thermo Delta V Isotope Ratio Mass Spectrometer (Thermo Scientific, Waltham, MA, USA) and Carlo Erba NC2500 Elemental Analyzer (Carlo Erba, Milan, Italy). Isotopic abundance data were expressed as δ^{15}N in parts per thousand (‰), representing the abundance of plant tissue ^{15}N relative to that of atmospheric N2.

For 2015 samples, fresh roots were prepared for nodule collection by carefully removing soil from the root system, breaking as few roots as possible. Nodules were washed in a beaker of deionized water, placed onto a 53-µm sieve, and washed with deionized water until clean of soil and organic matter. The nodules were surface-sterilized by transferring to Falcon tubes containing 10% (*v/v*) bleach, gently inverting the tubes over a five-minute period, and washing again with deionized water until bleach odor was gone.

For each crimson clover accession, 500 mg of surface sterilized nodules were placed in a 1.5 mL Eppendorf tube and submerged in 1× phosphate buffered solution (PBS). Three to four sterilized nodules from each preparation were reserved before mashing and plated on yeast mannitol broth (YMB) to serve as controls for the surface sterilization procedure (no growth observed). The remaining nodules were mashed using an Eppendorf micropestle. The mashed nodules were centrifuged for two minutes at 14,000 rpm to pellet the residual plant biomass. A total of 100 µL of mashed nodules were added to two mL of yeast mannitol broth YMB in sterile, 4.5 mL culture tubes and incubated in the dark on a shaker at 28 °C and 225 rpm for four d. Then, 100 µL of this culture was plated onto YMB plates and incubated at 28 °C for four d. After four days, a bacterial lawn developed on all plates. To collect the cells for analysis, 300 µL of 1× PBS buffer was added to the surface of each plate using a cell scraper, and the bacterial biomass from the surface of the plate was scraped and transferred into a single Eppendorf tube. The bacterial biomass was pelleted and divided into two aliquots; one was immediately prepared for storage as a glycerol stock. The other half was frozen at −20 °C until it was used for genomic DNA extraction and downstream molecular analysis.

The nodule rhizobacteria from 19 accessions were individually processed for genomic DNA extraction using Qiagen power plant genomic DNA (Qiagen Inc., Germantown, MD, USA). Each metagenomic extraction was individually barcoded using the Illumina 30 barcodes of a 96-index bar code kit following manufacturer's instructions (Illumina, Inc., San Diego, CA, USA). The indexed metagenomic libraries were then sequenced on an Illumina nextSeq 500 following manufacturer's instructions using all channels of a flow cell.

2.4. Data Analysis

Six phenotypic performance metrics were assessed: fall emergence, winter survival, flowering time, biomass per plant, nitrogen content, and proportion of plant nitrogen from BNF. For each response variable, a linear mixed model was fit, with accession, year, and year × accession as fixed effects and replicate within year as random effects. Where appropriate, log-transformation was applied to the response variable to achieve homogeneity of variances. For each model, means and least-significant difference (LSD) estimates were extracted.

Fall emergence was calculated as the count of living plants as a proportion of seeds planted. Winter survival was calculated as the count of living plants in the spring as a percent of fall-emerged plants. When more plants were present in the spring than in the fall (e.g., due to spring emergence), winter survival was set to 100%.

Flowering time was evaluated as the growing-degree days (GDD) required to reach 25% flowering on a per-plot basis. GDD was calculated using a base temperature of 0 °C [27]. For plots that reached 25% flowering before the first rating (17 of 323 plots), GDD to 25% was set to the GDD of the first observation date. Likewise, for plots that never reached 25% flowering (5 of 323 plots), GDD to 25% was set to that of the last observation date. For plots that reached 25% flowering between observation dates, a linear interpolation function was used to estimate the GDD at which 25% flowering occurred.

Biomass for each plot was calculated on a per-plant basis by dividing the total plot biomass by the number of plants present in the plot at biomass harvest. The biomass and plant-tissue %N content data showed a non-normal distribution, so a log-transformation was applied. The proportion of tissue-N derived from BNF was estimated using the ^{15}N natural abundance method [28] and statistically analyzed as above. All analyses were conducted in R version 3.6.1 (R Foundation for Statistical Computing, Vienna, Austria) and used package {nlme} (3.1-143) for model fitting [29]. Scripts for data analysis and raw data for the plant phenotypic analysis are available at https://data.nal.usda.gov/dataset/data-phenotypic-and-nodule-microbial-diversity-among-crimson-clover-trifolium-incarnatum-l-accessions.

All metagenomic analyses were conducted with the free on-line bioinformatics suite Kbase (https://kbase.us/). Paired-end reads were uploaded to the Kbase cloud then joined using the join-paired-ends import tool, adapters and indices were trimmed, and joined reads with a Phred quality score over 30 were kept for further analysis and processed separately from this point forward. Paired end libraries were assembled using KBaseGenomeAnnotations.Assembly-5.0, which employs Megahit 1.1 [30] with customized parameter setting. Parameters were set to min-count multiplicity for filtering (k_min+1)-mers, default 2, k-min kmer size (≤127), default 21, k-max kmer size (≤255), default 141, k-step increment of kmer size of each iteration (≤28), default 12, k-list list of kmer size (range 15–255, increment ≤28), min-contig-len minimum length of contigs to output (≥300), default 300.

Assembled contigs were then annotated using Rapid Annotations and the Subsystems Technology toolkit (RASTtk) [31–33]. Default search and alignment parameters are defined at http://rast.nmpdr.org/. Annotated metagenomic libraries were compared to each other using GenomeComparisonSDK v.0.0.4 (https://kbase.us/), which uses a k-mer approach to compute a protein pangenome. This approach builds on the sequence-based homologous protein families identified in the input Pangenome object by adding to the accumulated protein family annotation by identifying putatively similar functions within or among family groups. In addition, we identified the genes for tRNAs, rRNAs, and transposons. This database was then used to identify all the co-occurring protein families, tRNAs, rRNAs, and transposons among all nodule metagenomes. Scripts for data analysis and raw data for the metagenomic analysis are available at https://narrative.kbase.us/narrative/26966.

3. Results

3.1. Fall Emergence

The effects of accession, year, and accession x year were all significant to at least the 0.05 level. Replicate explained a small amount of variance in the model compared with accession, year, and accession x year interaction (Table 2). Fall emergence averaged across years varied from 16.0% (PI 233812) to 70.5% (PI 556990) of seeds planted (Figure 1), with a grand mean of 48.6% of plants emerged across all genotypes. Emergence differences were likely caused by a combination of genetic and seed source effects on seed coat impermeability and seed viability [34,35]. The effect of year on emergence was inconsistent across genotypes (indicated by the significant interaction effect), but there were several genotypes with consistently high or low emergence. The following nine accessions had a mean fall emergence of 60% or greater, and merit consideration as breeding parents: PIs 369045, 556990, 561942, 561943, 561944, 613042, 613043, 613047, and 655006. If low emergence was due to seed coat impermeability rather than non-viable seed, accessions with low fall emergence may also be valuable in a breeding context, as some producers prefer to use hard-seeded crimson clover as a self-seeding forage in pastures. Conversely, other producers prefer lines with low hard-seed for consistent emergence as a cover crop in rotations with annual cash crops [24,34]. Further investigation is needed to determine the effect of genetic and environmental variance on hard seed and seed viability: accessions that exhibited high and low emergence in this study (Figure 1) could be grown in diverging environments, then harvested seed could be assayed for impermeability. As previous research has found hard seed to be heritable [36], divergent selection for hard seed could be a promising breeding target for crimson clover.

Table 2. Results of the analysis of variance for each of the six phenotypic performance metrics: fall emergence, winter survival, flowering time, biomass per plant, nitrogen content, and plant nitrogen from BNF.

Term	Fall Emergence	Winter Survival	Flowering Time	Biomass per Plant	N Content	Plant N from BNF
			F value			
Accession	49.30 **	30.63 ***	2819.88 ***	22.66 ***	140.30 ***	21.61 ***
Year	4.30 *	144.41 ***	16.58 ***	317.17 ***	92.43 ***	77.34 ***
Accession × Year	1.72 **	1.04	1.81 **	1.18	2.16 ***	2.32 ***
			chi square			
Replicate (Year)	0.20	384.86 ***	1.10	357.61 ***	75.77 ***	443.61 ***
			Variance decomposition			
Fixed effects	0.75	0.33	0.83	0.38	0.43	0.45
Random effects	0.02	0.72	0.03	0.74	0.37	0.87
Coefficient of Variation (CV)	0.20	0.23	0.03	0.34	0.13	0.08

p value significance is denoted at the following levels: * <0.05, ** <0.01, *** <0.001.

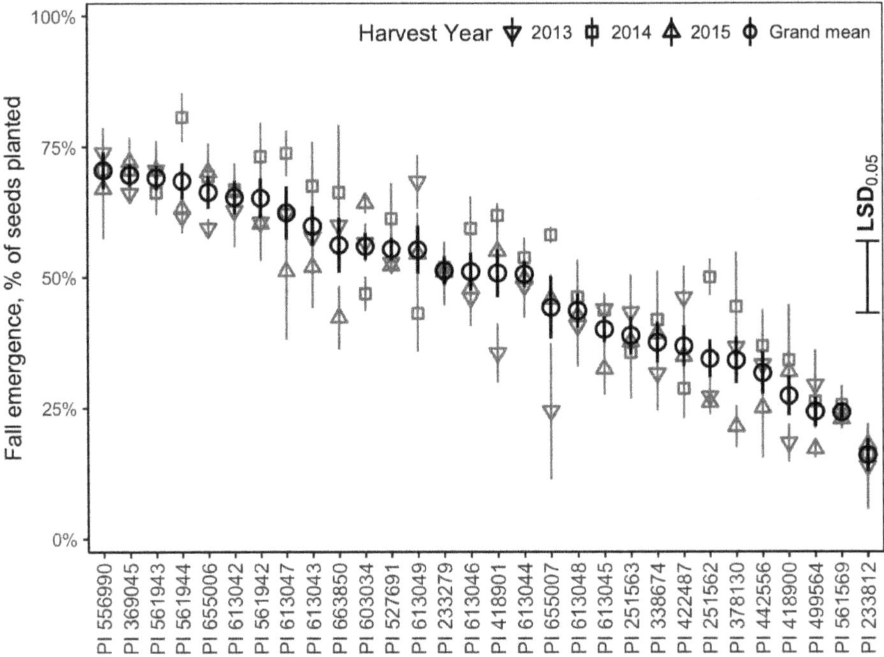

Figure 1. Effect of crimson clover accession on fall emergence in Beltsville, MD, evaluated as emerged plants per seeds planted. Grand mean, harvest year means, and least significant difference (LSD) estimates are displayed.

3.2. Winter Survival

The effects of accession, year, and replicate on winter survival were significant at the 0.001 level. Replicate explained more variance in the model than accession, year, and accession × year interaction combined (Table 2). Winter survival averaged across years varied from 52.8% (PI 418900) to 82.0% (PI 613046) of seeds planted (Figure 2), and a grand mean of 69.6% of emerged plants survived the winter. The 2013–2014 season had lower winter survival than other years (few accessions surpassed a mean of 50% winter survival) but also much greater variability in survival. The reduced winter survival in 2013–2014 is likely explained by the greater number of FDD and days below freezing without snow cover (Table 1). The large contribution of replicate, relative to fixed effects, highlights challenges associated with selection for winter hardiness, especially the complex genetics and physiology of winter hardiness and its high degree of variability even within a single field and growing season [37,38]. However, the significant accession effect and the lack of significant accession × year interaction suggests that agronomically important differences in winter hardiness in crimson clover may be observed even in years with relatively limited winter kill, and that selection can take place even in milder winters.

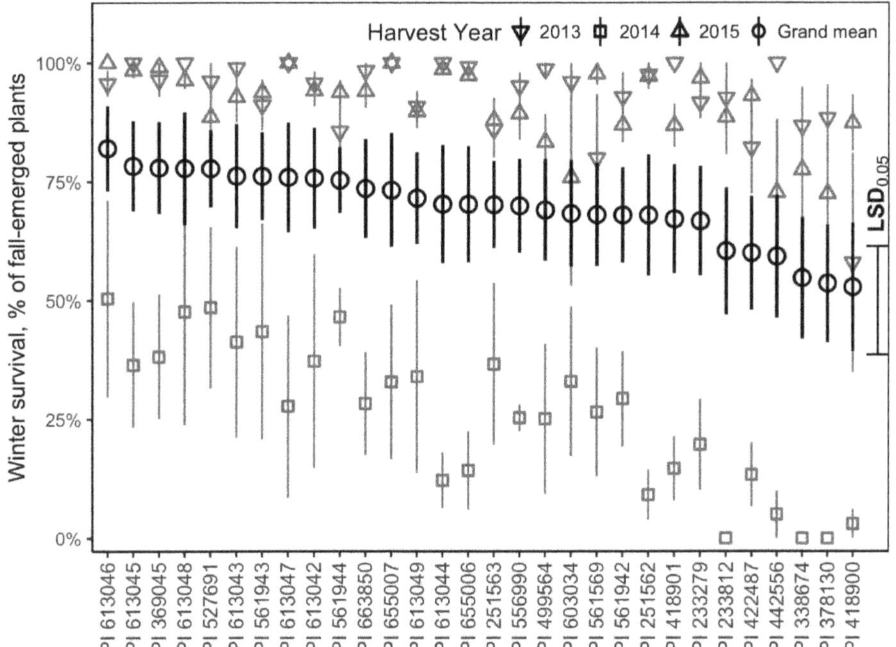

Figure 2. Effect of crimson clover accession on winter survival in Beltsville, MD, evaluated as surviving plants in the spring per plants emerged in the fall. Grand mean, harvest year means, and least significant difference (LSD) estimates are displayed.

3.3. Flowering Time

Effect of accession, year, and accession × year interaction on crimson clover flowering time were significant to at least the 0.01 level. As with fall emergence, the replicate explained a small amount of variance in the model compared with accession, year, and accession × year interaction (Table 2). GDD to reach 25% maturity varied from 1470 (PIs 561943 and 561944) to 1910 (PI 418901) (Figure 3), and accessions reached 25% maturity at a grand mean of 1581 GDD. Although the accession × year interaction was significant, flowering time was relatively consistent across years except for the accessions with the most extreme values. This consistent performance aligns with previous studies in other clover species, which found high heritability for flowering time [39,40]. While the majority of accessions fell within a narrow band of flowering time (around 1600 GDD), three accessions (PIs 561942, 561943, and 561944) originating in the southeast US matured somewhat earlier, with all three reaching 25% maturity prior to 1500 GDD. These accessions represent potential parents for development of an early-maturing population. One accession, PI 418901, was significantly later-maturing than all other accessions; it reached 25% maturity at 1910 GDD, and the next-latest accession (PI 418900) reached 25% at 1690 GDD. This accession, therefore, is an excellent candidate for development of a late-maturing population. Both early- and late-maturing crimson clover may be of interest to farmers, depending on region and cropping system.

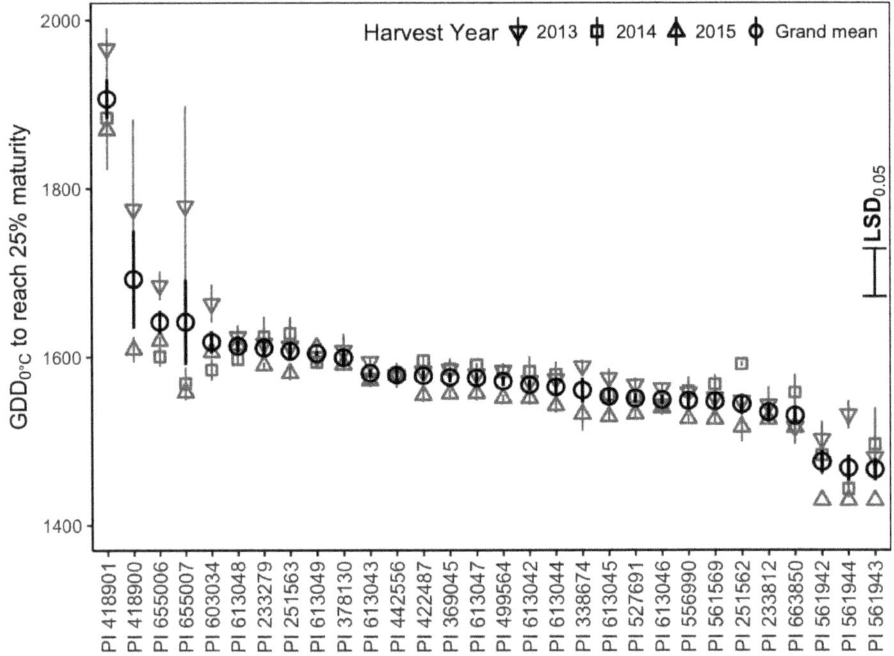

Figure 3. Effect of crimson clover accession on flowering time in Beltsville, MD, evaluated as growing degree days (GDD) to 25% flowering, calculated at a base temperature of 0 °C. Grand mean, harvest year means, and least significant difference (LSD) estimates are displayed.

3.4. Biomass per Plant

Effects of accession, year, and replicate on biomass per plant were significant to at least the 0.01 level. Replicate explained more variance in the model than accession, year, and accession × year interaction combined (Table 2). Biomass per plant varied from 1.52 g plant^{-1} (PI 233279) to 6.51 g plant^{-1} (PI 442556) (Figure 4), and accessions produced a grand mean of 2.97 g plant^{-1}. Across accessions, plants produced less biomass in 2013–2014 than the other years, likely due to the colder winter and/or wetter year (Table 1). Again, the large contribution of replicates within the year highlights the impact of field spatial variability on plant biomass and underscores the challenge of phenotyping and selecting for increased biomass production. While some crimson clover cultivars have been developed with improved plant vigor and forage yields [34], given the limited variance explained by accession within the crimson clover collection, progress in biomass production is likely to be slow in a crimson clover cover crop breeding program as well. However, as for winter survival, accession × year interaction is not significant, suggesting that agronomically important differences in biomass in crimson clover may be observed even in years with low biomass per plant overall.

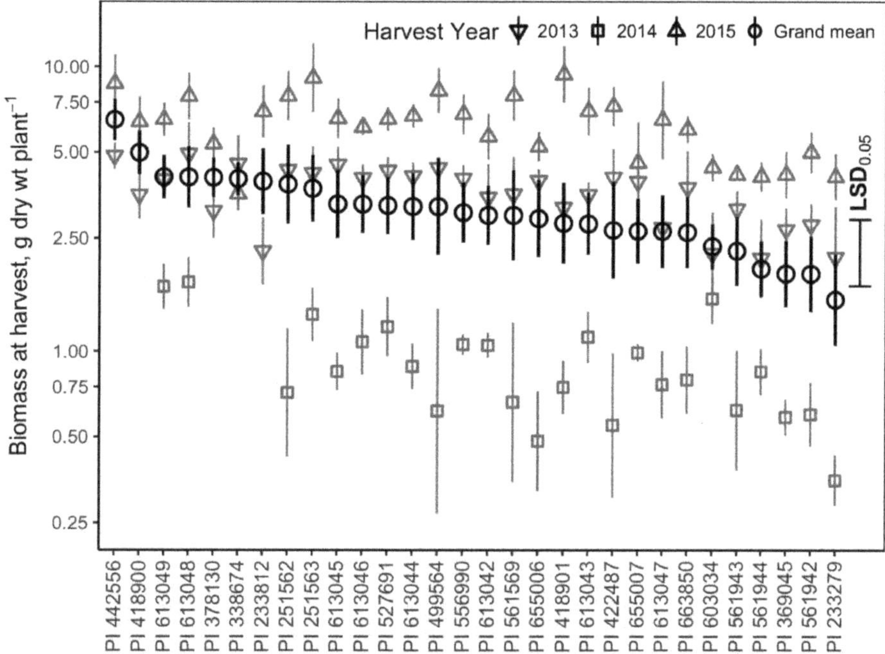

Figure 4. Effect of crimson clover accession on biomass per plant in Beltsville, MD, evaluated as dry weight per plant at harvest. Data are presented on a log scale. Grand mean, harvest year means, and least significant difference (LSD) estimates are displayed.

3.5. Nitrogen Content

The effects of accession, year, accession × year interaction, and replicate on nitrogen content were all significant at the 0.001 level. Replicates explained comparable variance as accession, year, and accession × year interaction combined (Table 2). Plant nitrogen content varied from 1.87% (PI 418901) to 2.24% (PI 233812) (Figure 5), and accessions contained a grand mean of 2.07% nitrogen. In 2013–2014, in which most accessions produced less biomass, many accessions also contained higher nitrogen content. The relationship between yield and crude protein content within cultivars due to seasonal variation has not been sufficiently addressed in the literature and merits further research. However, the range of nitrogen content present in the accessions is relatively narrow, and so in terms of total N contributed by the cover crop, an accession producing more biomass will likely produce more N even if its percent N is on the lower end of the observed spectrum. As with winter survival and biomass, replicate contributed a large portion of the model variance, again indicating that nitrogen content is highly impacted by spatial variability within the field.

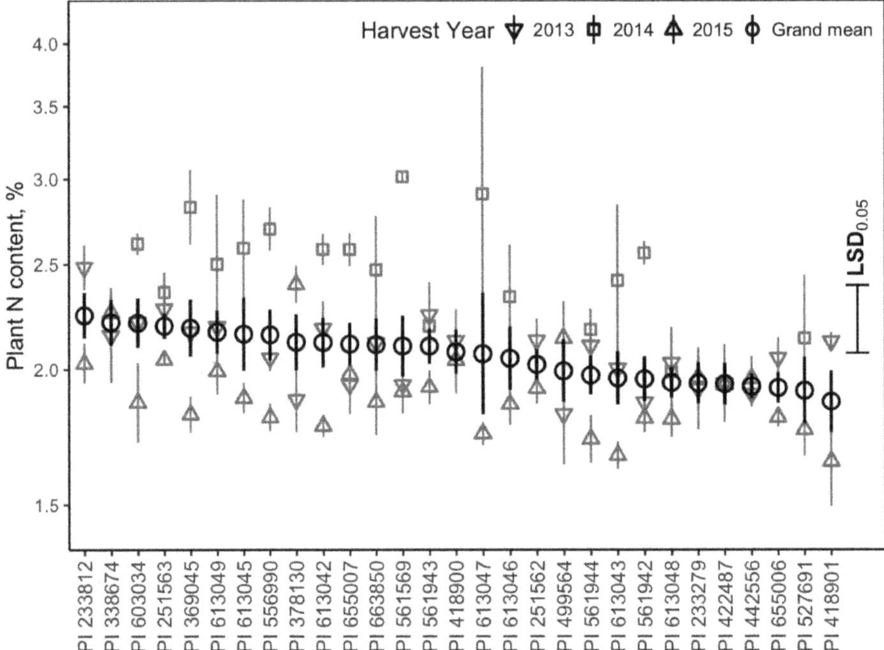

Figure 5. Effect of crimson clover accession on plant nitrogen content in Beltsville, MD, evaluated as percent N of dry weight. Data were log-transformed, but reverse-transformed for display purposes. Grand mean, harvest year means, and least significant difference (LSD) estimates are displayed.

3.6. Proportion of Plant Nitrogen as BNF

Again, the effects of accession, year, accession × year interaction, and replicate on BNF were all significant at the 0.001 level. Again, the replicate explained more variance in the model than accession, year, and accession x year interaction combined (Table 2). The proportion of plant nitrogen as BNF varied from 50.2% (PI 527691) to 85.6% (PI 655006) (Figure 6), and accessions contained a grand mean of 73.1% BNF as a proportion of plant nitrogen. However, the accessions with the highest mean proportion of N as BNF were accessions not included in 2013–2014, a year with lower BNF across all included accessions. Few breeding efforts have focused on nitrogen fixation ability in part due to challenges in implementing efficient screening systems, although differences between clover cultivars have been noted [41]. Again, although significant differences between accessions were observed, the replicate explained a larger portion of the variance than fixed effects, indicating that biological nitrogen fixation is highly impacted by field spatial variability and selection for this trait in a crimson clover cover crop breeding program is likely to be slow.

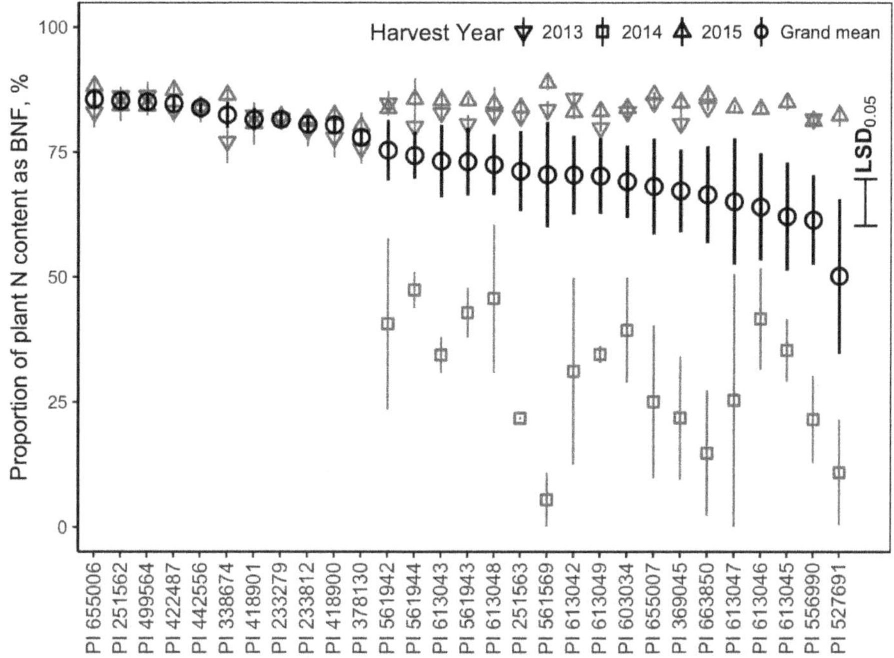

Figure 6. Effect of crimson clover accession on proportion of plant nitrogen content as biologically fixed nitrogen (BNF) in Beltsville, MD, evaluated as percent BNF per total N content. Grand mean, harvest year means, and least significant difference (LSD) estimates are displayed.

3.7. Nodule Metagenome

In total, we constructed nodule metagenomic libraries from 19 of the crimson clover accessions. Each library contained on average 8–10 million reads, equivalent to about 1.51×10^9 bases with an average single read length of 147 bp. Sequence assembly and duplicate sequence removal resulted in metagenomic libraries that ranged in size from 4.6 to 23.7 Mb. The number of contigs generated for each accession ranged in size among the libraries from 300 bp to 190 kb. The total number of contigs ranged from 5693 to 18,765, with a mean of 13,090.

The guanine-cytosine (GC) content of the libraries ranged from 50.4 to 62.8 % GC, which is, on average, lower than the three reference genomes that were used as comparisons (60.6, 60.9, and 61.1 %GC for Rhizobium leguminosarum Rt24.2, Rhizobium leguminosarum WSM1689, Rhizobium leguminosarum WSM2304, respectively) (Figure S1).

Homology to the rRNA operon was used to determine the diversity of bacterial species within each nodule metagenomic library at >99%, identity. Other parts of the genome were identified to the genera or species level by alignment with documented gene fragments in KBASE with a minimum length > 1 kb, and an insertion or deletion gap size of 25 bp. All species identified had a minimum E-value of <0.001. These stringent conditions enabled the identification of many taxa at species level or variants contained within species. Among the nodule metagenomic libraries, the 11 genera identified were Rhizobium, Mesorhizobium, Xanthomonas, Stenotrophomonas, Aquiflexum, Sinorhizobium, Bradyrhizobium, Agrobacetrium, Paenibacillus, Sphingomonas, and Pseudomonas. Bacillus dominated the number of reads that could be identified to a species level with reads specific enough to identify nine distinct Bacillus species in contigs of genomic regions greater than 10 kilobases. Rhizobium was clearly represented but is known to be more challenging to grow on media and therefore may have had a disadvantage on media relative to its importance in the nodule.

We identified 160,000 protein families, tRNA, ORFs, rRNA, and transposon genes; of these, 42% of the protein families were of unknown function. In the 11 bacterial genera identified in large reconstructed genomic contigs, all contained whole or partial complements of the nitrogen fixation complexes (nitrogenase genes, nif, and nod) as well as nitrogen responsive gene operons such as the nor and nos genes. Searches within the nodule metagenomes for other potential plant growth promoting rhizobacterial genes yielded numerous antibiotic genes as well as hydrogenases, which have been associated with increases in efficiency in biological nitrogen fixation.

We used three reference *Rhizobium legumunosarum* strains in our analysis (WSM1689, WSM2304, and Rt24.2 which have 6760, 6700, and 7080 annotated genes, respectively). The reference genomes have about 9500 identified elements that are over 1 kb in size and can be assigned to a specific species. The crimson clover accessions harbored a wide range of diversity within their nodule phytobiomes, from PI 369045, which contained 20,729 orthologous metagenome regions, to PI 561944, which contained 72,930 orthologous metagenome regions. The clover nodule microbiome with the lowest diversity of identifiable genetic regions (PI 369045) has about twice as much genetic diversity as the three reference genomes (Figure S1). The diversity of taxa identified among the nodules range between 12 and 20 different species of bacteria among the clover accessions. This diversity and the sequencing depth of our study (average library size) suggests that, for some accessions, we may not have data covering the full genomes of all genera within the nodule.

Several genes with roles in nitrogen fixation were found across multiple taxa identified within the crimson clover nodules (Table 3). All of the major taxa identified contained the nifH gene, which is a primary gene encoding the nitrogenase enzyme that carries out nitrogen fixation [42]. The narG nitrate reductase gene was found in two of the identified taxa, and has been shown to recycle fixed nitrogen through the dissimilatory nitrate reduction to ammonium (DNRA) pathway [43]. Anaerobic ammonium oxidation has been shown to remove significant amounts of nitrite and ammonium from ecosystems fixing nitrogen [44]. The hlyD gene, which is a multi-drug efflux transporter associated with antimicrobial resistance and survival in toxic environment, was found in three of the identified taxa. These types of excretion systems are likely to be involved with infection and nodule morphology [45]. The hesB gene, which is part of the nitrogenase, nitrogen fixation operon, was also found in three of the identified taxa [46]. Interestingly, the majority of research related to the role of hesB in BNF efficiency and regulation has occurred in algal systems and platonic freshwater cyanobacteria systems [47–49]. The possible roles of these genes in terrestrial leguminous systems has largely been ignored. There is potential to use metagenomic analysis of nodules to understand the efficiency of nitrogen fixation, recycling, and survival of symbionts in the nodule merit.

Table 3. Taxa identified in crimson clover nodules containing genes of interest related to nitrogen fixation. Genes of interest include nifH nitrogenase, narG- respiratory nitrate reductase, hlyD- multi drug efflux transporter, and hesB-, a regulatory gene that is only expressed during nitrogen fixation conditions.

Taxa Identified by 16S Operon	nifH	narG	hlyD	hesB
Acidovorax	X			
Bacillus/Psychrobacillus/Paenibacillus	X	X	X	X
Microbacterium	X			
Pelamonas puraquae	X			
Propionibacterium	X			X
Pseudomonas (fluorescens, koreensis)	X		X	
Rhizobium	X			
Rhizobium leguminosarum	X		X	X
Sphingomonas	X			
Sporosarcina	X	X		

3.8. Correlation among Traits

Associations among plant traits and nodule metagenome features were assessed using Pearson correlation. Among the plant phenotypic traits measured, only fall emergence and winter survival were positively correlated. Negative correlations were observed between fall emergence and biomass per plant, winter survival and biomass per plant, fall emergence and plant N from BNF, and winter survival and plant N from BNF. No correlations with any other traits were observed for flowering time or N content (Table 4).

Table 4. Correlation among plant phenotypic traits and nodule metagenomic library features.

	Fall Emergence	Winter Survival	Biomass per Plant	Flowering Time	N Content	Plant N from BNF	Library Size	Contigs	GC Content
Winter Survival	0.59 ***	1							
Biomass per Plant	−0.60 ***	−0.48 **	1						
Flowering Time	−0.15	−0.23	0.15	1					
N Content	−0.12	−0.09	0.06	−0.25	1				
Plant N from BNF	−0.42 *	−0.65 ***	0.24	0.24	−0.27	1			
Library Size	0.12	0.09	−0.09	0.05	0.07	0.18	1		
Contigs	−0.12	−0.05	−0.04	−0.35	−0.26	−0.06	−0.16	1	
GC Content	−0.10	0.12	−0.04	0.05	−0.26	0.29	0.46 *	−0.03	1
Microbial Diversity	−0.14	−0.20	0.17	−0.05	−0.10	0.25	0.60 **	0.37	0.58 **

p value significance is denoted at the following levels: * <0.05, ** <0.01, *** < 0.001.

Several accessions ranked in the top five accessions for two or three key traits. Of particular interest, PI 369045 was among the top accessions for emergence, survival, and N content. PI 418900 ranked second for both biomass and late flowering. PIs 561943 and 561944 were the earliest-flowering accessions and were both also among the top accessions for emergence. PI 613048 had both high winter survival and biomass per plant, and PI 655006 was among the top accessions for emergence, proportion of N as BNF, and late flowering (Figures 1–6).

Several positive correlations were observed among the nodule metagenome features. However, there were no correlations between any plant phenotypes and any nodule metagenome features.

4. Discussion

Plant breeding is a critical tool to increase cover crop use and efficacy on agricultural landscapes [11,12]. Crimson clover is the most widely used legume cover crop in the United States [10], and there is a need to improve the crimson clover germplasm for important cover cropping traits such as increased biomass and winter hardiness. Modern US cultivars have significant overlap in pedigrees, as most are derived from 'Dixie,' and limited genetic diversity found in a previous study of the NPGS crimson clover collection [24], but this was the first phenotypic assessment of the crimson clover collection for cover crop traits important to farmers.

Despite the limited genetic diversity previously observed, we did find some meaningful phenotypic variation for traits of agronomic importance for cover crop usage. Flowering time showed a large accession effect, and agronomically relevant diversity was observed among accessions. Its lack of correlation with other traits indicates that selection for flowering time may be possible without negative impacts on other agronomically important traits (e.g., due to linkage drag). Both early- and late-maturing populations have utility in different cropping systems. These results suggest that flowering time would be a relatively straightforward trait for selection in a crimson clover breeding program.

Fall emergence also showed a large accession effect, but further work is needed to determine whether this variation is caused by genetics or seed source effects. Low emergence may have been due to hard seededness, in which case accessions with both low and high emergence will be valuable for crimson clover breeding programs, depending on cropping system context [24,34].

Replicate effects explained the most variance in winter survival, biomass per plant, and proportion of plant N as BNF, highlighting the high degree of spatial variability among these traits and the challenges of phenotyping and selection [37,38,41,50,51]. However, unlike other traits, winter survival and biomass per plant did not exhibit accession x year interaction, indicating that despite large annual variation in these traits, agronomically relevant differences in winter hardiness and biomass per plant are observable among crimson clover accessions even in years with mild winters or low biomass overall.

Additional challenges come with the negative correlations among key traits—both fall emergence and winter survival are negatively correlated with both biomass per plant and proportion of plant N as BNF. The presence of these correlations may limit breeders' ability to rapidly integrate all key agronomic traits in a single population. However, several accessions performed among the top five for multiple traits, and should be considered as parents for future breeding populations. All of the nodules contained more genetic diversity than would be expected if a single isolate occupied the nodules; the reference isolates had, on average, 6900 genetic elements [52], whereas the accessions surveyed ranged from 20,000 to 72,000 common genetic elements. This indicates that, in all of the accessions tested, we observed, from 3 to 10 times, the expected diversity of a single bacterial genome (reference genomes), but the range among accessions was quite broad. In the case of PI561944 (~72,000 elements) the genetic diversity is what might be observed if analyzing the genetic diversity of 9–12 genera of bacteria.

Additional analysis of the metagenome data showed that a large proportion of the identified bacterial taxa contained intact or partial nitrogenase operons [42]. There is evidence that genomes of nitrogen-fixing bacteria may have higher GC content then their non-fixing relatives [53]; the metagenomic GC content of the nodule microbiome may be an indicator of nodule populations that harbor a higher or lower proportion of nitrogen-fixing bacteria. Although we did not observe a significant relationship between GC content and BNF, our assessment was not quantitative in regards to extraction of total microbial biomass from the nodules.

Our data suggest that crimson clover accessions contain a wide range of bacteria in their nodules, but transferring these microbes from the nodule to a petri plate likely gave an advantage to the r-selected linages of bacteria (i.e., those with high growth rates) that had successfully infected the nodule [54]. In addition, nodule metagenomic analysis was conducted in only one of the three growing seasons, so we are unable to draw conclusions about the stability of the nodule microbial populations or their impact on annual variation in agronomic traits. We also cannot yet clearly link the genetic diversity of nodule microbiomes of field grown clover to traits important in cover crop production such as biological nitrogen fixation and biomass production.

Although we cannot relate the metagenomic survey to the diversity of phenotypes analyzed for the field work portion of this study, this nodule microbiome dataset gives us a snapshot of the surveyed diversity of plant bacterial symbiosis in planta. Recent studies have shown that the nodule is much more diverse than previously thought; effectively managing the nodule microbiome will require an understanding of interactions among the diversity within the nodule in the context of the plant genotype [55–58]. It is not known whether this observed range of nodule metagenomic diversity is typical among closely related plant varieties, and additional research is needed to determine any salient relationships between microbial diversity and agronomic performance. The degree of annual and spatial variation in N content and BNF especially highlight the need to study environmental and management effects and genotype × environment interactions on crimson clover microbial populations.

5. Conclusions

While the effect of crimson cover accession was significant across all evaluated traits, accessions showed particularly clear differences for fall emergence and flowering time, indicating greater diversity and potential for selection in cover crop breeding programs. By contrast, winter survival, biomass per plant, N content, and plant N from BNF showed greater within- and/or among-year variation. Fall emergence and winter survival were found to be positively correlated, and both of these traits were negatively correlated with both biomass per plant and plant N from BNF. Several accessions were identified as high-performing across several key traits, and are of particular interest as parents in future breeding efforts: PIs 369045, 418900, 561943, 561944, and 655006. We observed larger than expected variation among accessions in terms of the diversity within the nodule microbial population, but did not observe any association between nodule metagenome features and plant phenotypic traits. Therefore, additional research is needed to determine functional implications of nodule microbiome diversity, and thus whether to incorporate such data into crimson clover breeding programs.

Supplementary Materials: The following are available online at http://www.mdpi.com/2073-4395/10/9/1434/s1. Figure S1: Metagenomic analysis of nodule endophytes; Table S1: Crimson clover accessions.

Author Contributions: Conceptualization, S.M. and J.E.M.; methodology, S.M., L.K.K., and J.E.M.; formal analysis, B.D., V.M., and J.E.M.; data curation, B.D., L.K.K., and M.P.; writing—original draft preparation, V.M. and J.E.M.; writing—review and editing, L.K.K. and S.M.; visualization, B.D. and L.K.K.; supervision, S.M.; project administration, V.M. and L.K.K.; All authors have read and agreed to the published version of the manuscript.

Funding: This research received no external funding.

Acknowledgments: Thanks to Victoria Ackroyd for providing edits to the manuscript.

Conflicts of Interest: The authors declare no conflict of interest.

References

1. Drinkwater, L.E.; Wagoner, P.; Sarrantonio, M. Legume-based cropping systems have reduced carbon and nitrogen losses. *Nature* **1998**, *396*, 262–265. [CrossRef]
2. Lu, Y.C.; Watkins, B.; Teasdale, J.; Abdul-Baki, A. Cover crops in sustainable food production. *Food Rev. Int.* **2000**, *16*, 121–157. [CrossRef]
3. Altieri, M.; Nicholls, C. *Biodiversity and Pest Management in Agroecosystems*, 2nd ed.; CRC Press: New York, NY, USA, 2004; ISBN 978-1-56022-923-0.
4. Gardner, J.B.; Drinkwater, L.E. The fate of nitrogen in grain cropping systems: A meta-analysis of 15N field experiments. *Ecol. Appl.* **2009**, *19*, 2167–2184. [CrossRef] [PubMed]
5. Gabriel, J.L.; Garrido, A.; Quemada, M. Cover crops effect on farm benefits and nitrate leaching: Linking economic and environmental analysis. *Agric. Syst.* **2013**, *121*, 23–32. [CrossRef]
6. Schipanski, M.E.; Barbercheck, M.; Douglas, M.R.; Finney, D.M.; Haider, K.; Kaye, J.P.; Kemanian, A.R.; Mortensen, D.A.; Ryan, M.R.; Tooker, J.; et al. A framework for evaluating ecosystem services provided by cover crops in agroecosystems. *Agric. Syst.* **2014**, *125*, 12–22. [CrossRef]
7. Malone, R.W.; Jaynes, D.B.; Kaspar, T.C.; Thorp, K.R.; Kladivko, E.; Ma, L.; James, D.E.; Singer, J.; Morin, X.K.; Searchinger, T. Cover crops in the upper midwestern United States: Simulated effect on nitrate leaching with artificial drainage. *J. Soil Water Conserv.* **2014**, *69*, 292–305. [CrossRef]
8. Singer, J.W.; Nusser, S.M.; Alf, C.J. Are cover crops being used in the US corn belt? *J. Soil Water Conserv.* **2007**, *62*, 353–358.
9. Wallander, S. While Crop Rotations Are Common, Cover Crops Remain Rare. In *Amber Waves: The Economics of Food, Farming, Natural Resources, and Rural America*; United States Department of Agriculture (USDA) Economic Research Service (ERS): Washington, DC, USA, 2013.
10. CTIC. *Report of the 2016–17 National Cover Crop Survey*; Joint publication of the Conservation Technology Information Center (CTIC): West Lafayette, IN, USA; The North Central Region Sustainable Agriculture Research and Education (SARE) Program: St. Paul, MN, USA; The American Seed Trade Association: Alexandria, VA, USA, 2017.
11. Wilke, B.J.; Snapp, S.S. Winter cover crops for local ecosystems: Linking plant traits and ecosystem function. *J. Sci. Food Agric.* **2008**, *88*, 551–557. [CrossRef]
12. Wayman, S.; Kucek, L.K.; Mirsky, S.B.; Ackroyd, V.; Cordeau, S.; Ryan, M.R. Organic and conventional farmers differ in their perspectives on cover crop use and breeding. *Renew. Agric. Food Syst.* **2017**, *32*, 376–385. [CrossRef]
13. Duvick, D.N. Genetic Contributions to Yield Gains of U.S. Hybrid Maize, 1930 to 1980. In *Genetic Contributions to Yield Gains of Five Major Crop Plants*; John Wiley and Sons, Ltd.: Hoboken, NJ, USA, 1984; pp. 15–47. ISBN 978-0-89118-586-4.

14. Duvick, D.N. Plant breeding: Past achievements and expectations for the future. *Econ. Bot.* **1986**, *40*, 289–297. [CrossRef]

15. Meredith, W.R.; Bridge, R.R. Genetic Contributions to Yield Changes in Upland Cotton. In *Genetic Contributions to Yield Gains of Five Major Crop Plants*; John Wiley and Sons, Ltd.: Hoboken, NJ, USA, 1984; pp. 75–87. ISBN 978-0-89118-586-4.

16. Miller, F.R.; Kebede, Y. Genetic Contributions to Yield Gains in Sorghum, 1950 to 1980. In *Genetic Contributions to Yield Gains of Five Major Crop Plants*; John Wiley and Sons, Ltd.: Hoboken, NJ, USA, 1984; pp. 1–14. ISBN 978-0-89118-586-4.

17. Schmidt, J.W. Genetic Contributions to Yield Gains in Wheat. In *Genetic Contributions to Yield Gains of Five Major Crop Plants*; John Wiley and Sons, Ltd.: Hoboken, NJ, USA, 1984; pp. 89–101. ISBN 978-0-89118-586-4.

18. Specht, J.E.; Williams, J.H. Contribution of Genetic Technology to Soybean Productivity—Retrospect and Prospect. In *Genetic Contributions to Yield Gains of Five Major Crop Plants*; John Wiley and Sons, Ltd.: Hoboken, NJ, USA, 1984; pp. 49–74. ISBN 978-0-89118-586-4.

19. Russell, G.E. (Ed.) *Progress in Plant Breeding—1*; Butterworth-Heinemann: Oxford, UK, 1985.

20. Brummer, E.C.; Barber, W.T.; Collier, S.M.; Cox, T.S.; Johnson, R.; Murray, S.C.; Olsen, R.T.; Pratt, R.C.; Thro, A.M. Plant breeding for harmony between agriculture and the environment. *Front. Ecol. Environ.* **2011**, *9*, 561–568. [CrossRef]

21. Yeater, K.M.; Bollero, G.A.; Bullock, D.; Rayburn, A.L.; Rodriguez-Zas, S.L. Assessment of Genetic Variation in Hairy Vetch Using Canonical Discriminant Analysis. *Crop Sci.* **2004**, *44*, 185–189. [CrossRef]

22. Maul, J.; Mirsky, S.; Emche, S.; Devine, T. Evaluating a Germplasm Collection of the Cover Crop Hairy Vetch for Use in Sustainable Farming Systems. *Crop Sci. Madison* **2011**, *51*, 2615–2625. [CrossRef]

23. Clark, A. (Ed.) *Managing Cover Crops Profitably*, 3rd ed.; SARE Outreach: College Park, MD, USA, 2007; ISBN 978-1-888626-12-4.

24. Steiner, J.; Piccioni, E.; Falcinelli, M.; Liston, A. Germplasm Diversity among Cultivars and the NPGS Crimson Clover Collection. *Crop Sci.* **1998**, *38*, 263–271. [CrossRef]

25. Mothapo, N.V.; Grossman, J.M.; Maul, J.E.; Shi, W.; Isleib, T. Genetic diversity of resident soil rhizobia isolated from nodules of distinct hairy vetch (*Vicia villosa* Roth) genotypes. *Appl. Soil Ecol.* **2013**, *64*, 201–213. [CrossRef]

26. Mothapo, N.V.; Grossman, J.M.; Sooksa-nguan, T.; Maul, J.; Bräuer, S.L.; Shi, W. Cropping history affects nodulation and symbiotic efficiency of distinct hairy vetch (*Vicia villosa* Roth.) genotypes with resident soil rhizobia. *Biol. Fertil. Soils* **2013**, *49*, 871–879. [CrossRef]

27. Baxter, L.L.; Grey, T.L.; Tucker, J.J.; Hancock, D.W. Optimizing Temperature Requirements for Clover Seed Germination. *Agrosyst. Geosci. Environ.* **2019**, *2*, 180059. [CrossRef]

28. Shearer, G.; Kohl, D.H. N2-Fixation in Field Settings: Estimations Based on Natural 15N Abundance. *Funct. Plant Biol.* **1986**, *13*, 699–756. [CrossRef]

29. Pinheiro, J.; Bates, D.; DebRoy, S.; Sarkar, D. NLME: Linear and Nonlinear Mixed Effects Models. R Package Version 2019. Available online: https://www.researchgate.net/publication/303803175_Nlme_Linear_and_Nonlinear_Mixed_Effects_Models (accessed on 30 January 2019).

30. Li, D.; Liu, C.-M.; Luo, R.; Sadakane, K.; Lam, T.-W. MEGAHIT: An ultra-fast single-node solution for large and complex metagenomics assembly via succinct de Bruijn graph. *Bioinformatics* **2015**, *31*, 1674–1676. [CrossRef]

31. Aziz, R.K.; Bartels, D.; Best, A.A.; DeJongh, M.; Disz, T.; Edwards, R.A.; Formsma, K.; Gerdes, S.; Glass, E.M.; Kubal, M.; et al. The RAST Server: Rapid Annotations using Subsystems Technology. *BMC Genom.* **2008**, *9*, 75. [CrossRef]

32. Overbeek, R.; Olson, R.; Pusch, G.; Olsen, G.; Davis, J.; Disz, T.; Edwards, R.; Gerdes, S.; Parrello, B.; Shukla, M.; et al. The SEED and the Rapid Annotation of microbial genomes using Subsystems Technology (RAST). *Nucleic Acids Res.* **2013**, *42*, D206–D214. [CrossRef] [PubMed]

33. Brettin, T.; Davis, J.J.; Disz, T.; Edwards, R.A.; Gerdes, S.; Olsen, G.J.; Olson, R.; Overbeek, R.; Parrello, B.; Pusch, G.D.; et al. RASTtk: A modular and extensible implementation of the RAST algorithm for building custom annotation pipelines and annotating batches of genomes. *Sci. Rep.* **2015**, *5*, 8365. [CrossRef] [PubMed]

34. Knight, W.E. Crimson Clover. In *Clover Science and Technology*; John Wiley and Sons, Ltd.: Hoboken, NJ, USA, 1985; pp. 491–502. ISBN 978-0-89118-218-4.

35. Williams, W.A.; Elliott, J.R. Ecological Significance of Seed Coat Impermeability to Moisture in Crimson, Subterreanean and Rose Clovers in a Mediterranean-Type Climate. *Ecology* **1960**, *41*, 733–742. [CrossRef]
36. Bennett, H.W. The Effectiveness of Selection for the Hard Seeded Character in Crimson Clover1. *Agron. J.* **1959**, *51*, 15–16. [CrossRef]
37. Fowler, D. Selection for Winterhardiness in Wheat. II. Variation within Field Trials. *Crop Sci.* **1979**, *19*, 773–775. [CrossRef]
38. Gusta, L.V.; Wisniewski, M. Understanding plant cold hardiness: An opinion. *Physiol. Plant.* **2013**, *147*, 4–14. [CrossRef]
39. Nair, R.M.; Craig, A.D.; Rowe, T.D.; Biggins, S.R.; Hunt, C.H. Genetic variability and heritability estimates for hardseededness and flowering in balansa clover (*Trifolium michelianum* Savi) populations. *Euphytica* **2004**, *138*, 197–203. [CrossRef]
40. Salisbury, P.A.; Aitken, Y.; Halloran, G.M. Genetic control of flowering time and its component processes in subterranean clover (*Trifolium subterraneum* L.). *Euphytica* **1987**, *36*, 887–902. [CrossRef]
41. Abberton, M.T.; Marshall, A.H. Progress in breeding perennial clovers for temperate agriculture. *J. Agric. Sci.* **2005**, *143*, 117–135. [CrossRef]
42. Gaby, J.C.; Buckley, D.H. A Comprehensive Evaluation of PCR Primers to Amplify the nifH Gene of Nitrogenase. *PLoS ONE* **2012**, *7*, e42149. [CrossRef]
43. Smith, C.J.; Nedwell, D.B.; Dong, L.F.; Osborn, A.M. Diversity and Abundance of Nitrate Reductase Genes (narG and napA), Nitrite Reductase Genes (nirS and nrfA), and Their Transcripts in Estuarine Sediments. *Appl. Environ. Microbiol.* **2007**, *73*, 3612–3622. [CrossRef] [PubMed]
44. Yamazaki, M.; Thorne, L.; Mikolajczak, M.; Armentrout, R.W.; Pollock, T.J. Linkage of genes essential for synthesis of a polysaccharide capsule in Sphingomonas strain S88. *J. Bacteriol.* **1996**, *178*, 2676–2687. [CrossRef] [PubMed]
45. Bozsoki, Z.; Cheng, J.; Feng, F.; Gysel, K.; Vinther, M.; Andersen, K.R.; Oldroyd, G.; Blaise, M.; Radutoiu, S.; Stougaard, J. Receptor-mediated chitin perception in legume roots is functionally separable from Nod factor perception. *Proc. Natl. Acad. Sci. USA* **2017**, *114*, E8118–E8127. [CrossRef] [PubMed]
46. Yan, Y.; Yang, J.; Dou, Y.; Chen, M.; Ping, S.; Peng, J.; Lu, W.; Zhang, W.; Yao, Z.; Li, H.; et al. Nitrogen fixation island and rhizosphere competence traits in the genome of root-associated *Pseudomonas stutzeri* A1501. *Proc. Natl. Acad. Sci. USA* **2008**, *105*, 7564–7569. [CrossRef]
47. Di Cesare, A.; Cabello-Yeves, P.J.; Chrismas, N.A.M.; Sánchez-Baracaldo, P.; Salcher, M.M.; Callieri, C. Genome analysis of the freshwater planktonic *Vulcanococcus limneticus* sp. nov. reveals horizontal transfer of nitrogenase operon and alternative pathways of nitrogen utilization. *BMC Genom.* **2018**, *19*, 259. [CrossRef]
48. Huang, T.-C.; Lin, R.-F.; Chu, M.-K.; Chen, H.-M. Organization and expression of nitrogen-fixation genes in the aerobic nitrogen-fixing unicellular cyanobacterium *Synechococcus* sp. strain RF-1. *Microbiol. Read.* **1999**, *145*, 743–753. [CrossRef]
49. Liu, D.; Liberton, M.; Yu, J.; Pakrasi, H.B.; Bhattacharyya-Pakrasi, M. Engineering Nitrogen Fixation Activity in an Oxygenic Phototroph. *mBio* **2018**, *9*. [CrossRef]
50. Riday, H. Correlations between visual biomass scores and forage yield in space planted red clover (*Trifolium pratense* L.) breeding nurseries. *Euphytica Dordr.* **2009**, *170*, 339–345. [CrossRef]
51. Riday, H.; Brummer, E.C. Forage Yield Heterosis in Alfalfa. *Crop Sci.* **2002**, *42*, 716–723. [CrossRef]
52. Terpolilli, J.; Rui, T.; Yates, R.; Howieson, J.; Poole, P.; Munk, C.; Tapia, R.; Han, C.; Markowitz, V.; Tatiparthi, R.; et al. Genome sequence of *Rhizobium leguminosarum* bv trifolii strain WSM1689, the microsymbiont of the one flowered clover Trifolium uniflorum. *Stand. Genom. Sci.* **2014**, *9*, 527. [CrossRef]
53. Mcewan, C.E.A.; Gatherer, D.; Mcewan, N.R. Nitrogen-Fixing Aerobic Bacteria have Higher Genomic GC Content than Non-Fixing Species within the Same Genus. *Hereditas* **1998**, *128*, 173–178. [CrossRef] [PubMed]
54. Provorov, N.A.; Andronov, E.E.; Onishchuk, O.P. Forms of natural selection controlling the genomic evolution in nodule bacteria. *Russ. J. Genet.* **2017**, *53*, 411–419. [CrossRef]
55. Muresu, R.; Porceddu, A.; Sulas, L.; Squartini, A. Nodule-associated microbiome diversity in wild populations of Sulla coronaria reveals clues on the relative importance of culturable rhizobial symbionts and co-infecting endophytes. *Microbiol. Res.* **2019**, *221*, 10–14. [CrossRef]
56. Caetano-Anollés, G.; Gresshoff, P.M. Plant genetic control of nodulation. *Annu. Rev. Microbiol.* **1991**, *45*, 345–382. [CrossRef] [PubMed]

57. Hirsch, S.; Kim, J.; Muñoz, A.; Heckmann, A.B.; Downie, J.A.; Oldroyd, G.E.D. GRAS Proteins Form a DNA Binding Complex to Induce Gene Expression during Nodulation Signaling in *Medicago truncatula*. *Plant Cell* **2009**, *21*, 545–557. [CrossRef] [PubMed]
58. Walker, S.A.; Viprey, V.; Downie, J.A. Dissection of nodulation signaling using pea mutants defective for calcium spiking induced by Nod factors and chitin oligomers. *Proc. Natl. Acad. Sci. USA* **2000**, *97*, 13413–13418. [CrossRef] [PubMed]

2020, 10, 1289

Article

Translational Pigeonpea Genomics Consortium for Accelerating Genetic Gains in Pigeonpea (*Cajanus cajan* L.)

Rachit K. Saxena [1,*], Anil Hake [1], Anupama J. Hingane [1], C. V. Sameer Kumar [2], Abhishek Bohra [3], Muniswamy Sonnappa [4], Abhishek Rathore [1], Anil V. Kumar [1], Anil Mishra [5], A. N. Tikle [5], Chourat Sudhakar [2], S. Rajamani [6], D. K. Patil [7], I. P. Singh [3], N. P. Singh [3] and Rajeev K. Varshney [1,*]

[1] International Crops Research Institute for the Semi-Arid Tropics (ICRISAT), Patancheru 502324, Telangana State, India; H.Anil@cgiar.org (A.H.); H.Anupama@cgiar.org (A.J.H.); a.rathore@cgiar.org (A.R.); anil.kumar@cgiar.org (A.V.K.)

[2] Professor Jayashankar Telangana State Agricultural University (PJTSAU), Rajendranagar, Hyderabad 500030, Telangana State, India; sameerkumar1968@gmail.com (C.V.S.K.); chouratsudhakar2@gmail.com (C.S.)

[3] ICAR—Indian Institute of Pulses Research (IIPR), Kanpur 208024, Uttar Pradesh, India; abhi.omics@gmail.com (A.B.); ipsingh1963@yahoo.com (I.P.S.); npsingh.iipr@gmail.com (N.P.S.)

[4] Agricultural Research Station (ARS)—Kalaburagi, University of Agricultural Sciences (UAS), Raichur 584104, Karnataka, India; muniswamygpb@gmail.com

[5] Rajmata Vijayaraje Scindia Krishi Vishwa Vidyalaya (RVSKVV), RAK College of Agriculture (RAKCA), Sehore 466001, Madhya Pradesh, India; anil1961.mishra@gmail.com (A.M.); antiklepb@gmail.com (A.N.T.)

[6] Regional Agricultural Research Station, Lam Farm, Guntur 5220324, Andhra Pradesh, India; srajamanibreeder@gmail.com

[7] Agricultural Research Station (ARS)—Badnapur, Vasantrao Naik Marathwada Krishi Vidyapeeth (VNMKV), Parbhani 431402, Maharashtra, India; dkpatil.05@gmail.com

[*] Correspondence: r.saxena@cgiar.org (R.K.S.); r.k.varshney@cgiar.org (R.K.V.)

Received: 17 July 2020; Accepted: 21 August 2020; Published: 31 August 2020

Abstract: Pigeonpea is one of the important pulse crops grown in many states of India and plays a major role in sustainable food and nutritional security for the smallholder farmers. In order to overcome the productivity barrier the Translational Pigeonpea Genomics Consortium (TPGC) was established, representing research institutes from six different states (Andhra Pradesh, Karnataka, Madhya Pradesh, Maharashtra, Telangana, and Uttar Pradesh) of India. To enhance pigeonpea productivity and production the team has been engaged in deploying modern genomics approaches in breeding and popularizing modern varieties in farmers' fields. For instance, new genetic stock has been developed for trait mapping and molecular breeding initiated for enhancing resistance to fusarium wilt and sterility mosaic disease in 11 mega varieties of pigeonpea. In parallel, genomic segments associated with cleistogamous flower, shriveled seed, pods per plant, seeds per pod, 100 seed weight, and seed protein content have been identified. Furthermore, 100 improved lines were evaluated for yield and desirable traits in multi-location trials in different states. Furthermore, a total of 303 farmers' participatory varietal selection (FPVS) trials have been conducted in 129 villages from 15 districts of six states with 16 released varieties/hybrids. Additionally, one line (GRG 152 or Bheema) from multi-location trials has been identified by the All India Coordinated Research Project on Pigeonpea (AICRP-Pigeonpea) and released for cultivation by the Central Variety Release Committee (CVRC). In summary, the collaborative efforts of several research groups through TPGC is accelerating genetics gains in breeding plots and is expected to deliver them to pigeonpea farmers to enhance their income and improve livelihood.

Keywords: pigeonpea; genomics; TPGC; FPVS; multi-location trials

---

1. Introduction

Pigeonpea is a pulse crop grown in many countries of the world and plays an important role in sustainable nutritional food security. India ranks first in pigeonpea cultivation area (5.58 mha) and production (4.29 mt) in the world [1]. In the last five years, productivity of pigeonpea in India has shown an increasing trend (11.42%) from 693 (2009–2013) to 774 kg/ha (2014–2018), however, it is lower by ~10% compared to world productivity (761 kg/ha in 2009–2013 and 850 kg/ha in 2014–2018) [1]. Moreover, disproportionate yield gaps between research plots and in farmers' fields of a given variety are also a major concern in India [2]. On the other hand, demand for the pulses is continuously increasing and it has been estimated that 32 million tons of pulses will be required by the year 2030 and 50 million tons of pulses by year 2050 (Vision 2050: Indian Institute of Pulses Research, 2013, www.iipr.res.in). To match these requirements, pulse breeders have been engaged in developing superior varieties but could not achieve the daunting task. In recent times molecular breeding approaches have been successful in developing superior varieties and enhance the production of cereal crops, like rice [3–6], wheat [7–11], sorghum [12–14], maize [15–17], and pearl millet [18,19], and also in legume crops, such as chickpea [20–24] and soybean [25–27]. However, such approaches have not been used until recently in pigeonpea, primarily due to limited information on genes/markers associated with traits. In this direction, the International Initiative on Pigeonpea Genomics (IIPG) decoded and published the genome sequence of pigeonpea in 2012 [28]. As a result of this breakthrough, a significant amount of genomic information has become available [29,30]. However, the availability of the genome sequence or the large-scale of molecular markers alone was not enough to improve crop productivity. These resources can be used as tools to harness the genetic diversity present in the germplasm collection for enhancing the precision and efficiency of crop improvement programs. Therefore, just after the decoding of the pigeonpea genome, a series of consultations with a large number of stakeholders, including the Department of Agriculture Cooperation and Farmers Welfare (DACFW), Indian Council of Agricultural Research (ICAR), several state agricultural universities (SAUs), the private sector, and USAID-India, were conducted to use genome sequence information for translational genomics research for crop improvement. As a continuous effort in translational genomics research for pigeonpea improvement, with funding support from DACFW, the Translational Pigeonpea Genomics Consortium (TPGC) of nine research institutions/agricultural universities representing six different states of India was established in 2017 (Figure 1). During the past three years TPGC has made significant progress (described in the sections below) including: (a) development of new genetic stock for trait mapping, (b) deployment of genomics-assisted breeding in 11 popular varieties of pigeonpea, (c) evaluation of 100 improved lines for their performance in multi-location trials, and (d) demonstration of improved crop varieties in more than 303 farmers' fields across 129 villages from 15 districts of six states (Andhra Pradesh, Karnataka, Madhya Pradesh, Maharashtra, Telangana, and Uttar Pradesh). Several improved lines have also been put in the varietal identification pipeline of Indian Council of Agricultural Research. Molecular markers associated with seed protein content, diseases resistance and yield contributing traits, improved lines with higher yield potential and disease resistance were also identified. In summary, the TPGC has been established with an aim to deploy modern genomics information for pigeonpea improvement, develop/identify new and improved varieties, and to enhance the adoption of superior lines in farmers' fields. The present article reports the significant research achievements of the TPGC as international public goods (IPGs) that will be helping and guiding future pigeonpea improvement programs. Furthermore, this article may also inspire other less-studied crop communities to take similar consortium-based approaches for crop improvement.

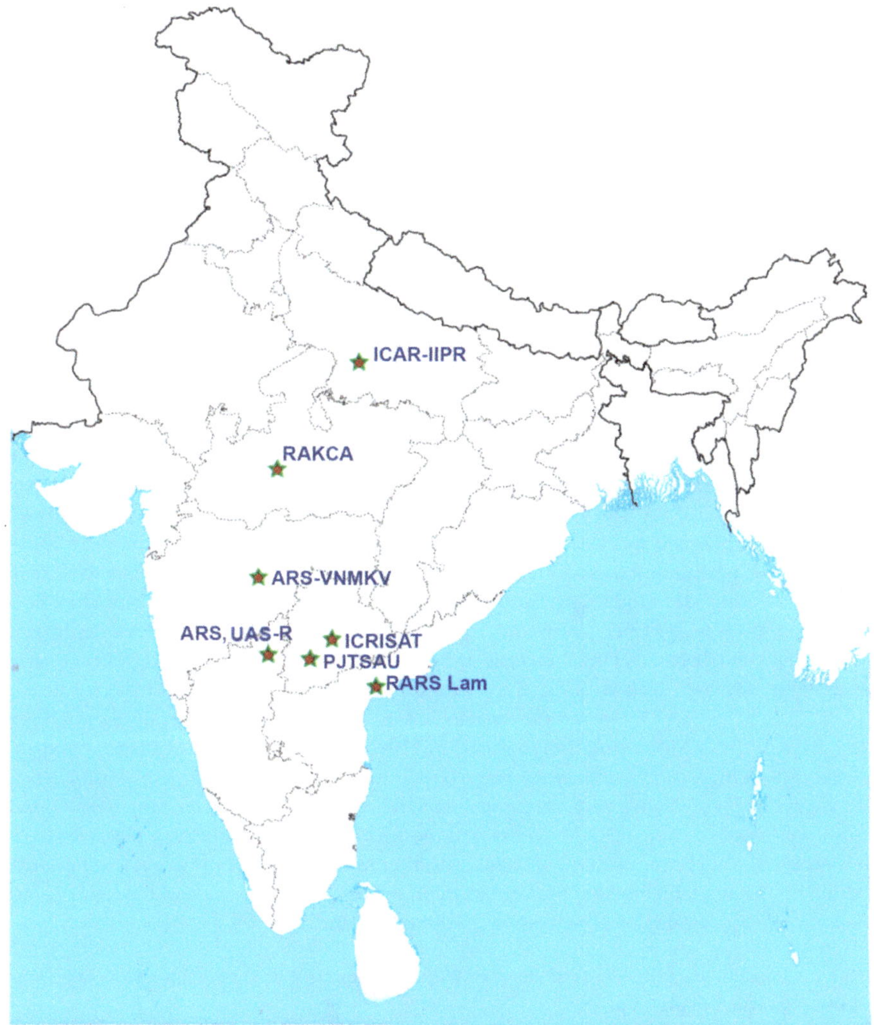

Figure 1. Translational Pigeonpea Genomics Consortium: Marked places on the map shows indicative locations of experimentation. ICAR-IIPR: ICAR- Indian Institute of Pulses Research (IIPR), Kanpur, Uttar Pradesh; RAKCA: RAK College of Agriculture (RAKCA), Sehore, Madhya Pradesh; ARS-VNMKV: Agricultural Research Station (ARS)-Badnapur, Vasantrao Naik Marathwada Krishi Vidyapeeth (VNMKV), Parbhani, Maharashtra; ARS, UAS-R: Agricultural Research Station (ARS)-Kalaburagi, University of Agricultural Sciences (UAS), Raichur, Karnataka; ICRISAT: International Crops Research Institute for the Semi-Arid Tropics (ICRISAT), Patancheru, Telangana; PJTSAU: Professor Jayashankar Telangana State Agricultural University (PJTSAU), Telangana; RARS Lam: Regional Agricultural Research Station, Lam Farm, Guntur, Andhra Pradesh.

2. Advances in Genetics and Genomics

2.1. Novel Breeding and Genetic Materials

A range of genetic and breeding material has been developed in pigeonpea during last few years for their effective use in genomics [30]. This material includes segregating bi-parental populations for target traits [31–35], diverse genetic stocks including reference set [36], core and mini-core [37,38], and back-cross populations [39]. Further, to reap the advantages of family based mapping, the development of multi-parent mapping populations was initiated in the year 2012. During the last two years these family-based mapping populations, especially Nested Association Mapping Population (NAM), have reached the recombinant inbred line (RIL) stage. At present, multi-location evaluation of the NAM population is underway. Another family-based mapping population i.e., Multi-Parent Advanced Generation Inter-Cross (MAGIC), has also been advanced during the last two years and it has reached to RIL stage in the year 2019–2020. The generation of these family-based mapping populations in pigeonpea have provided new genetic combinations for trait discovery and also for genomics applications for new cultivar development. The significant features of these NAM and MAGIC populations have been provided below.

2.2. Nested Association Mapping Population (NAM)

The NAM population consisting of 2224 RILs in pigeonpea was developed by crossing 10 pigeonpea diverse founder lines as female parent to a common pollen parent ICPL 87119 (Table 1, Figure 2). The seeds of 2224 RILs of the NAM population were sown in an augmented block design with spacing of 45 × 30 cm in a single row of 1 m length during the 2018–2019 cropping season. The first set of phenotyping data on the stabilized NAM population (F_6 plants) was collected in cropping season 2018–2019 at ICRISAT. The year 2018–2019 was also used for seed multiplication of the NAM population so that planned multi-location trials could be conducted. During the year 2018–2019, RILs of the NAM population were evaluated for agronomic traits including days to first flower, days to 50% flowering, days to 75% maturity, number of primary and secondary branches, pods per plant, pod and grain weight per plant, and 100 seed weight as per the pigeonpea descriptor [40]. Preliminary analysis of phenotyping data collected on NAM population showed significant variations for the above mentioned traits in RILs. For instance, a range of 60 days to 141 days with a mean value of 95.14 days has been observed for days to 50% flowering across NAM population (Figure 3). Similarly, RILs in the NAM population matured in 110–218 days with a mean value of 146.91 days. In the case of two important yield measuring/contributing traits, i.e., seeds per pod (1.2–6.8) and 100-seed weight (3.37–16.99 g), a wide range of variation has been observed. It is important to mention that multi-location evaluation of NAM over two or more locations per family in six states for two years has been planned. The comprehensive data analysis after multi-location evaluation will provide exact information on the available phenotypic variability in the NAM population and to identify elite lines suitable for major agroecologies of pigeonpea cultivation in India.

Table 1. Characteristic features of parents used in the development of Nested Association Mapping (NAM) population in pigeonpea.

Genotypes	Features
Nested Parent	
Asha (ICPL 87119)	Genome sequence available, leading variety, resistant to FW and SMD
Founder Parents	
HPL 24	High protein content, medium duration, compact, susceptible to FW and resistant to SMD, inter-specific derivative
ICP 7035	Medium duration, SMD resistant, large purple seed, high sugar
ICP 8863	Mid-late, highly resistant to FW and susceptible to SMD, an extensively grown variety in Northern Karnataka and Maharashtra region of India

Table 1. *Cont.*

Genotypes	Features
ICPL 87	Early duration, determinate, short, high combiner
ICPL 88039	Extra early maturity, indeterminate, good yield
ICPL 85063	Medium duration, indeterminate, good yield, more branching
MN 1	Super early, small seeded, determinate
ICP 28	Early maturity, local variety
ICPL 85010	Early maturity, local variety
ICP 7263	Determinate, long podded, white seeded

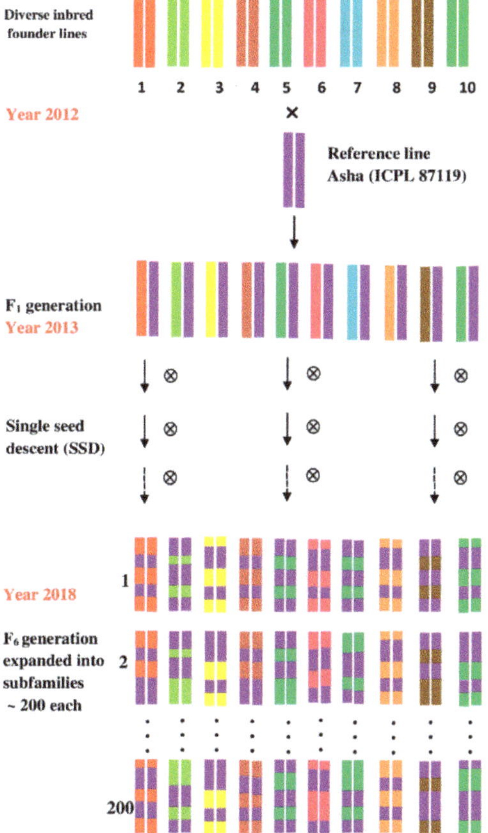

Figure 2. Nested association mapping (NAM) population in pigeonpea. Inbred, Asha (ICPL 87119) was used for crossing with 10 diverse inbred lines separately. The hybrids from each of 10 crosses were selfed to develop F₂s. From these F₂ seeds, at least 200 progenies were generated in each of 10 crosses by single seed decent method to take these lines to F₆ generation.

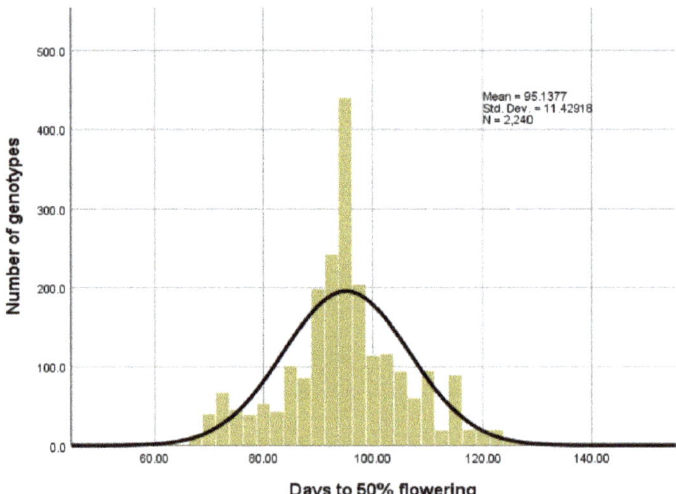

Figure 3. Frequency distribution for days to 50% flowering in NAM population of pigeonpea. A range of 60 days to 141 days with a mean value of 95.14 days has been observed for days to 50% flowering across NAM population.

2.3. Multi-Parent Advanced Generation Inter-Cross (MAGIC)

To bring diversity from landraces and superior varieties, the MAGIC population has been developed as per the crossing scheme of Cavanagh et al. [41] using eight crossing parents (ICP 7426, HPL 24, ICP 11605, ICP 14209, ICP 14486, ICP 5529, ICP 7035, and ICP 8863) by 28 two-ways, 14 four-ways, and seven eight-ways crosses (Table 2, Figure 4). These eight parents have significant variations for agronomic (pod numbers per plant, maturity, days to flowering, grain yield), quality (sugar and protein content), and disease resistance (fusarium wilt and sterility mosaic disease) traits. The F_6 seeds for this population were harvested from the F_5 plants in year 2018–2019 and seed multiplication was undertaken for F_6 plants in cropping season 2019–2020. Homozygous lines (~1300 RILs) obtained from multi-parent crossing approach will be used for high resolution trait mapping that is otherwise not possible by using conventional bi-parental mapping populations. This population also offers new breeding material with enhanced diversity and combined desirable traits (e.g., early maturity, high seed protein content, higher yield, and disease resistance). Some of these lines can be used as parents in future breeding programs or can be put in the varietal release pipelines.

Table 2. Characteristic features of parents used in the development of Multi-Parent Advanced Generation Inter-Cross (MAGIC) population of pigeonpea.

Genotypes	Features
ICP 7426	High pod numbers, medium duration
HPL 24	High protein content, medium duration, compact, susceptible to FW and resistant to SMD, inter-specific derivative
ICP 11605	Early flowering, germplasm line
ICP 14209	High number of pods, germplasm line
ICP 14486	Early flowering, germplasm line
ICP 5529	Medium duration, obcordate leaves, compact plant, modified flower
ICP 7035	Medium duration, SMD resistant, large purple seed, high sugar content
ICP 8863	Mid-late, highly resistant to FW and susceptible to SMD, an extensively grown variety in Northern Karnataka and Maharashtra region of India

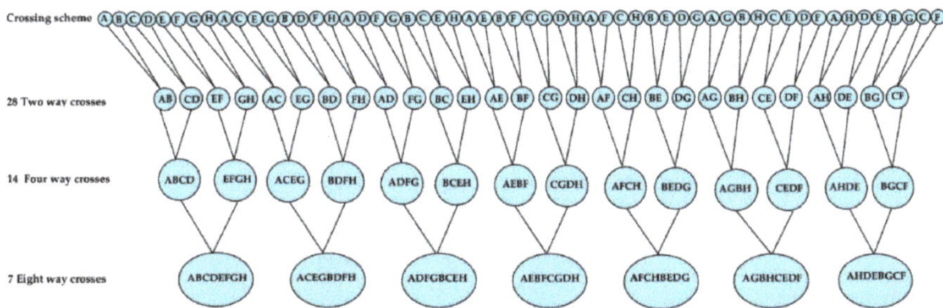

Figure 4. Development of a MAGIC population in pigeonpea. Eight genotypes (ICP 7426, HPL 24, ICP 11605, ICP 14209, ICP 14486, ICP 5529, ICP 7035 and ICP 8863) were used in crossing program as parents to develop MAGIC population. Crossing scheme included, 28 two-ways, 14 four-ways and 7 eight-ways crosses. Further, hybrids coming from 7 eight-way crosses were selfed to generate ~1300 F_6 lines.

3. Genomics Advances and Genomics-Assisted Breeding

3.1. Marker Assisted Back-Crossing for Fusarium Wilt (FW) and Sterility Mosaic Diseases (SMD) Resistance

Fusarium wilt (FW) and sterility mosaic diseases (SMD) are considered as major yield reducing biotic stresses in pigeonpea [42–45]. Pathogen variability have further added to the severity of these diseases [46–48]. Moreover, the known resistant varieties are witnessing the breakdown of resistance. Therefore, it has been planned to introduce, combine, or reconstruct the resistance for FW and SMD in leading pigeonpea varieties from different agro-climatic zones of India following marker-assisted back-crossing (MABC). In this endeavor, backcross populations were developed by crossing 11 mega varieties (recurrent parent, susceptible to FW and/or SMD) with ICPL 20096 (donor parent, resistant to FW and SMD) following two cycles of backcrosses (Table 3) during cropping seasons (rainy) 2017–2018 to 2018–2019. These 11 released varieties were crossed with ICPL 20096 as recipient parents to generate F_1s in respective crossing combinations. True F_1s from respective crosses were identified using molecular markers. Subsequently confirmed F_1s of respective crosses were used to make backcrosses with the recipient or recurrent parent. The backcross seeds (BC_1F_1) from respective crosses were harvested and tested with markers for foreground selection and second backcrossing (Table 4). It is important to note that we have developed two different sets of 10 markers each associated with FW and SMD resistance (unpublished). These markers have been used for foreground selection in BC_1F_1 plants. Those BC_1F_1 plants carrying the highest heterozygosity for 10-marker panel were used for second round of backcrossing with respective recurrent parents for maximum genome recovery. In this way, at present, we have reached the stage of BC_2F_1 seeds. It has been planned to advance and obtain BC_2F_2 seeds from BC_2F_1 plants in upcoming years. Further selfed BC_2F_2 seeds from each cross will be sown in sick plot for evaluation of FW and SMD incidences. Most promising lines identified for FW and SMD resistance will be subjected to yield evaluations in hotspot regions of the country. The MABC-bred improved lines showing higher disease resistance and similar or better yield performance as compared to recurrent parent in both stressed and normal conditions will enter varietal identification and release pipelines.

Table 3. List of pigeonpea mega varieties targeted for introgression of resistance to fusarium wilt (FW) and sterility mosaic disease (SMD) resistance.

Varieties	# Centre	Introgression of the Trait
LRG 41	RARS-Lam	FW and SMD
LRG 52	RARS-Lam	FW and SMD
ICPL 88039	ICRISAT	FW
UPAS 120	ICRISAT	FW
ICP 8863	ARS-Kalaburagi	SMD
TS 3R	ARS-Kalaburagi	SMD
TJT 501	RAKCA-Sehore	FW
JKM 189	RAKCA-Sehore	FW
BDN 711	ARS-Badnapur	FW
PRG 176	PJTSAU	FW and SMD
Bahar	ICAR-IIPR	FW

Centre undertaking marker-assisted backcrossing program in selected preferred varieties in the region.

Table 4. Details and current status of marker assisted backcrossing for improving elite lines for resistance to fusarium wilt and sterility mosaic disease.

S. No.	Recurrent Parent	Donor Parent	F_1 Plants	F_1s Confirmed	No. of BC_1F_1s Grown	No. of BC_1F_1s Confirmed
1	BDN 711	ICPL 20096	17	17	92	85
2	ICP 8863	ICPL 20096	30	21	92	78
3	TS 3R	ICPL 20096	26	22	92	80
4	TJT 501	ICPL 20096	28	27	19	18
5	JKM 189	ICPL 20096	25	7	60	52
6	Bahar	ICPL 20096	30	28	92	89
7	PRG 176	ICPL 20096	11	9	5	4
8	LRG 41	ICPL 20096	11	11	5	5
9	LRG 52	ICPL 20096	4	2	7	6
10	ICPL 88039	ICPL 20096	11	11	132	122
11	UPAS 120	ICPL 20096	28	15	88	65

3.2. Development of Trait-Associated Markers

Availability of molecular markers governing various traits are prerequisites for deploying genomics-assisted breeding in crop improvement. While the MABC approach is being deployed for improving resistance to FW and SMD, molecular mapping for several quality- and yield-related traits was also undertaken by TPGC. These findings have been considered as important milestones for pigeonpea improvement. A summary on the development of molecular markers has been presented below:

A bi-parental trait mapping study by Yadav et al. [49] identified 26 quantitative trait loci (QTLs) on nine CcLGs (except CcLG01 and CcLG05) for 10 traits viz. cleistogamous flower (Cl), shriveled seed (ShS), plant height (PH), number of primary branches (PB), number of secondary branches (SB), pods per plant (PP), seeds per pod (SP), seed count per plant (SC), seed weight (SW), and seed size (100 seed weight) (SS) using RILs of ICPL 99010 × ICP 5529. The phenotypic variance explained (PVE) by each QTL ranged from 9.1% to 50.6% (Table 5, Table S1). Similarly, a total of 17 CAPS/dCAPS markers were identified for seed protein content (SPC) using high (HPL 24, ICP 5529) and low (ICP 11605, ICPL 87119) seed protein content (SPC) containing pigeonpea genotypes by whole genome re-sequencing (WGRS) data [50]. Further, these markers were validated using F_2 population of ICP 5529 × ICP 11605. Out of 17 CAPS/dCAPS markers, four genic-CAPS/dCAPS markers, spc003 (NADH-GOGAT), spc107 (copper transporter), spc017 (protein kinase), and spc100 (BLISTER) revealed co-segregation with SPC (Table S1). Using five F_2 populations, Obala et al. [51] identified 192 QTLs across 10 CcLGs for five traits, namely, seed protein content (SPC) seed weight (SW), seed yield (SY), growth habit (GH), and days to first flowering (DFF) with PVE of 0.7–91.3%. Major effect (PVE ≥ 10%) QTLs included 14 QTLs for SPC, 16 QTLs for SW, 17 QTLs for SY, 19 QTLs for GH, and 24 QTLs for DFF (Table S1).

Table 5. Summary on QTLs/genomic segments identified for target traits in pigeonpea.

Trait	Number of QTLs/Genomic Segments Identified	# PVE Range (%)	Reference
Cleistogamy	5	9.10–50.60	Yadav et al. 2019
Seed shape	3	11.80–37.20	Yadav et al. 2019
Seed size	2	29.50–33.90	Yadav et al. 2019
Seed protein content	19	2.20–23.50	Obala et al. 2019; 2020
100 seed weight	18	10.10–46.60	Obala et al. 2020
Seed yield	18	10.20–53.00	Obala et al. 2020
Growth habit	21	10.90–91.30	Obala et al. 2020
Days to first flowering	28	10.90–47.60	Obala et al. 2020

Phenotypic variation explained.

4. Promising Pigeonpea Lines Identified through Multi-Location Trials

To select high yielding superior lines from multi-location trials, three trials (one each for super-early (>90 days), early (140–160 days), and medium maturity (>160–180 days) group) were conducted during cropping seasons of 2017–2018 (year 1) and 2018–2019 (year 2) for grain yield (Figure 5, Table S2). Thirty genotypes of super-early maturity group were tested in year 1 at five centers, namely, Kalaburagi, Kanpur, Tandur, Lam, and Patancheru, while in year 2, it was tested at three centers, namely, Tandur, Lam, and Kanpur. Likewise, 30 genotypes of the early maturity group were tested in year 1 at Kanpur, Tandur, Lam, Kalaburagi, and Patancheru, while in year 2, these were tested at Kanpur, Tandur, Lam, and Badnapur. The 40 genotypes of the medium-maturity group were tested at Kalaburagi, Tandur, and Lam for two consecutive years, while one time these were tested at Patancheru (year 1), Badnapur (year 2), and Sehore (year 2). In multi-location trials, seeds of each entry were sown in four rows of 3 m length with spacing of 90 × 30 cm, 45 × 20 cm, and 30 × 10 cm for medium/late duration, early and super-early maturity groups, respectively, in a randomized complete block design with three replications. The mean for each genotype in a replication was calculated using the observations recorded from the whole plot. The overall mean for each genotype was calculated using the values from each replication. Minimum and maximum means of the genotypes were considered to record the range. Individual analysis of variance (ANOVA) was carried out in order to partition the variation due to different sources following the method of Panse and Sukhatme [52]. Combined ANOVA was computed using a general linear mixed model using the procglm function of SAS version 9.2 [53]. The stability analysis of selected genotypes for grain yield was done using the data recorded during the rainy season of 2017–2018 and 2018–2019 across various locations. A GGE biplot (site regression analysis) was used to illustrate the genotype plus genotype-by-environment variation using principal component (PC) scores from singular value decomposition (SVD). A GGE biplot with average-environment coordination (AEC) and polygon view was drawn to examine the performance of all genotypes within a specific environment and to simultaneously select genotypes based on stability and mean performance [54]. The model for the GGE based on SVD of the first two PCs is given by:

$$Y_{ij} - \mu - \beta_j = \lambda_1 \xi_{i1} \eta_{j1} + \lambda_2 \xi_{i2} \eta_{j2} + \varepsilon_{ij} Y_{ij} - \mu - \beta = \lambda_1 \xi_{i1} \eta_{j1} + \lambda_2 \xi_{i2} \eta_{j2} + \varepsilon_{ij}$$

where Y_{ij} is the mean performance of genotype i in environment j, μ is the grand mean, β_j is the environment j main effect, λ_1 and λ_2 are the singular values of the first and second PC, ξ_{i1} and ξ_{i2} are the eigenvectors for genotype i, η_{j1} and η_{j2} are the eigenvectors for environment j, and ε_{ij} is the residual effect.

Figure 5. An overview of pigeonpea field with super-early pigeonpea lines at flowering stage and remaining lines at vegetative stage.

4.1. Performance and Stability of Genotype and Environment for Grain Yield

In super-early, early- and medium-maturity groups of multi-location trials, individual (Table S3) as well as combined analysis of variance (ANOVA) (Table 6) revealed significant differences in genotype, environments, and genotype × environment interaction (GEI) effects for grain yield. To know the best test environment and superior genotype (high yield and stable), GGE ((genotype (G) + (genotype (G) × environment (E) interaction)) biplot analysis was conducted with the phenotyping data recorded in multi-location trials. A GGE biplot explained 73.58, 66.53, and 67.85% of total variation of the environment-centered G by E table for grain yield for super-early, early and medium duration trials respectively. GGE biplot analysis revealed two mega environments in each trial. For instance, in super-early, Tandur and rest environments (Kalaburagi_2017, Patancheru_2017, Kanpur and Lam), for early, Patancheru and rest environments (Kanpur, Tandur, Lam, Kalaburagi and Badnapur) and for medium duration trial, Badnapur, Lam_2018 and rest environments (Tandur, Kalaburagi, Patancheru and Lam_2017) were observed for the performance of grain yield (Figure 6). In super-early trials, Lam_2017 and Tandur (2017 and 2018) identified as more discriminating environment while Kalaburagi, Patancheru, Kanpur, and Lam_2018 identified as least discriminating environment for grain yield (Figure 6a). Likewise, in early duration trial, Lam_2017 was identified as the most discriminating, however, Kalaburagi_2017, Kanpur_2017, and Tandur_2017 were identified as average discriminating, while Kanpur_2018, Lam_2018, Badnapur_2017, and Patancheru_2017 were identified as the least discriminating environments for grain yield (Figure 6b). In medium-duration trials, Sehore_2018 is the most discriminating, Tandur is average discriminating, while rest environments Lam, Kalaburagi, Patancheru, and Badnapur were identified as least discriminating environments for grain yield (Figure 6c).

Table 6. Combined analysis of variance for grain yield of super-early, early and medium duration trials during cropping season 2017–2018 and 2018–2019.

Trial	Effects				
		Environment	Rep (Env)	Genotype	Genotype × Environment
Super-early (30 genotypes)	df	7.0	16.0	28.0	196.0
	F	2245.8 **	10.7 **	14.2 **	17.9 **
Early (30 genotypes)	df	7.0	14.0	29.0	203.0
	F	1305.0 *	7.9 **	24.8 **	15.6 **
Medium (40 genotypes)	df	8.0	15.0	34.0	272.0
	F	1093.2 **	11.7 **	22.0 **	14.1 **

*: Significant at 0.05 probability level, **: Significant at 0.01 probability level.

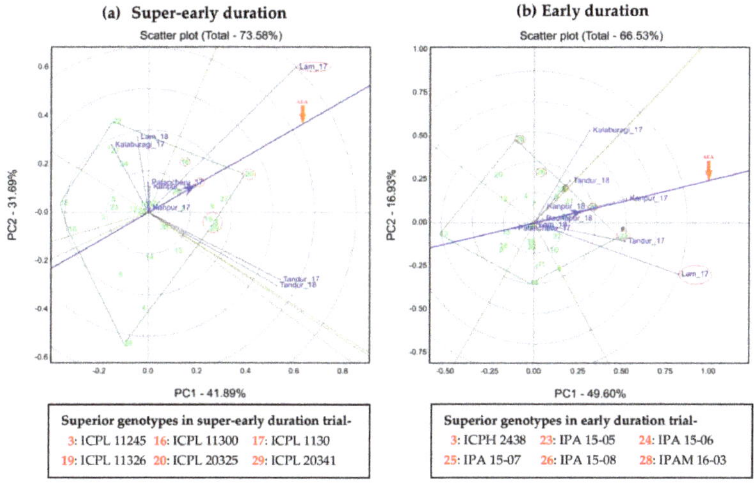

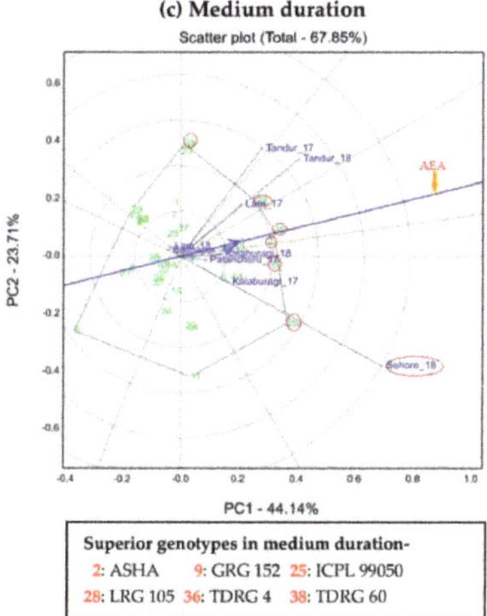

Figure 6. GGE biplot showing ranking of genotypes and environments for mean and stability for grain yield in (**a**) super-early, (**b**) early and (**c**) medium duration trials over locations in India during cropping season 2017–2018 and 2018–2019.

4.2. Promising Lines in Super-Early Duration Trial

Pigeonpea lines, namely, ICPL 20325 at Tandur, ICPL 11300 at Lam and ICPL 11279 at Kanpur exhibited yield advantage (14.8–154.4%) over MN1 (check) across the years (Table S4). Whereas, ICPL 20326 at Kalaburagi and ICPL 11301 at Patancheru have also shown yield advantages of 20.2–109% as compared to check (Table S4). One pigeonpea line, ICPL 20325, has been identified as a superior line with higher grain yield and high stability over the years and across the locations (Figure 6a) with 24.09% yield advantage over MN1 (Table S4).

4.3. Promising Lines in Early Duration Trial

IPA 15-05 at Lam and Tandur and IPA 15-06 at Kanpur revealed yield advantage (33.3–127.9%) over ICPL 88039 (check) across the years (Table S5). Whereas, on the other testing locations, IPAM 16-03 at Kalaburagi, IPA 15-08 at Patancheru and ICPL 92047 at Badnapur exhibited significant yield advantage (4.3–118.2%) over ICPL 88039 (Table S5). Overall, IPA 15-05 exhibited the highest grain yield and gained a yield advantage of 52.29% over ICPL 88039 across the locations over the years with high stability (Table S5, Figure 6b). Importantly, two lines IPA 15-03 and IPA 15-06 had 7.47 and 8.81% yield advantage, respectively, over the best check when evaluated in initial varietal trial (IVT) of AICRP-Pigeonpea for the northwestern plain zone (NWPZ) and central zone (CZ). Furthermore, with an average yield of 1791 kg/ha, IPA 15-06 has shown 11.44% yield advantage over the best check in CZ in advanced varietal trial 1 (AVT 1) of AICRP-Pigeonpea.

4.4. Promising Lines in Medium Duration Trial

ICPL 99050 at Kalaburagi, LRG 105 at Lam, TDRG 60 at Tandur revealed higher grain yield and yield advantage up to 285.9% over the checks across the years (Table S6). At Badnapur, TJT 501 and at Patancheru, AGL 1603-4 revealed a yield advantage (0.6–89.4%) over the checks (Table S6). At Sehore, ICPL 99050 exhibited a yield advantage of 23.7–397.7% over the checks. GRG 152 followed by LRG 105 and ICPL 99050 exhibited higher yield as compared to different checks and stability (Table S6, Figure 6c).

5. Enhancing Varietal Adoption through Farmer Participatory Varietal Selection (FPVS) Trials

In order to enhance the adoption of available varieties/hybrids, during the last two years a total of 303 FPVS trials have been conducted in 129 villages of 15 districts of six states, namely, Andhra Pradesh, Karnataka, Madhya Pradesh, Maharashtra, Telangana, and Uttar Pradesh. FPVS trials were conducted using 4–5 improved cultivars for the evaluation of grain yield during rainy season of 2017–2018 (year 1) and 2018–2019 (year 2) (Table 7). The seeds of each entry for FPVS trial were sown in 1000 square meter plots on farmers' fields. Data analysis for the FPVS trials was conducted following similar methods mentioned in multi-location trials.

Table 7. List of varieties/hybrids used for farmer participatory varietal selection trial in the six states of India during cropping season 2017–2018 and 2018–2019.

SN	State	Districts	Variety/Hybrids [#]	No. of FPVS		Total FPVS
				2017–2018	2018–2019	
1	Andhra Pradesh	Guntur, Prakasam, and Kurnool	LRG 52, **LRG 105, LRG 160,** ICPH 2433, and ICPH 2740	30	30	60
2	Karnataka	Kalaburagi, Bidar and Yadgir	TS 3R, GRG 811, ICPL 332, **BSMR 736** and ICPH 2433	25	30	55
3	Madhya Pradesh	Sehore, Shajapur, and Rajgarh	**ICPH 2671,** ICPH 2443, JKM 189, TJT 501, and ICPL 88039	25	30	55
4	Maharashtra	Aurangabad, Jalna, and Parbhani	**BDN 711,** BDN 716, BSMR 853, and BSMR 736	-	28	28

Table 7. *Cont.*

SN	State	Districts	Variety/Hybrids [#]	No. of FPVS		Total FPVS
				2017–2018	2018–2019	
5	Telangana State	Mahabubnagar, Rangareddy, and Warangal	ICPL 161, ICPL 99050, **ICPL 332 WR**, PRG 176, and RGT 1	30	30	60
6	Uttar Pradesh	Kanpur, Banda, and Chitrakoot	ICPH 2740, IPA 203, JKM 189, and ICPL 88039	25	20	45

[#] Lines in "bold" identified as best performing lines by farmers in respective state.

5.1. Performance and Stability of Genotype and Environment for Grain Yield

The combined analysis of variance (ANOVA) revealed significant differences in genotype but non-significant genotype × environment interaction (GEI) effects for grain yield at locations in Karnataka and Maharashtra states over the years (Table 8), while combined ANOVA revealed significant genotype, environment and GEI effects over the years and locations for grain yield in locations at Telangana, Andhra Pradesh, and Madhya Pradesh (Table 8). Stability analysis for grain yield was conducted using a GGE biplot for FPVS where significant GEI was observed at Madhya Pradesh, Andhra Pradesh, and Telangana. A GGE biplot explained 99.48%, 99.14%, and 98.75% of total variation of the environment-centered G by E table for Madhya Pradesh, Andhra Pradesh, and Telangana, respectively. Due to non-significant GEI, the environment had no effect on the performance of genotypes and the stable performance of genotypes was observed at Karnataka (Table S7, Figure 7a) and Maharashtra (Table S8, Figure 7b). In Madhya Pradesh, Sehore was identified as the most discriminating and representative environment, and significantly differed with Shajapur and Rajgarh for grain yield (Table S9, Figure 8a). In Andhra Pradesh, Guntur region is highly discriminating, more representative, and significantly differed with Prakasam and Kurnool for grain yield (Table S10, Figure 8b). In Telangana, Mahabubnagar was identified as the most discriminating region, and significantly differed with Warangal and Rangareddy for grain yield (Table S11, Figure 8c).

Table 8. Combined analysis of variance for grain yield of farmer participatory varietal selection trials during cropping season 2017–2018 and 2018–2019.

FPVS Trial		Effects		
		Environment	Genotype	Genotype × Environment
Andhra Pradesh	df	2.0	4.0	8.0
	F	145.8 **	23.78 **	6.6 **
Maharashtra	df	2.0	3.0	6.0
	F	0.09 NS	63.9 **	1.36 NS
Madhya Pradesh	df	2.0	4.0	8.0
	F	13.5 **	109.9 **	5.1 **
Karnataka	df	2.0	3.0	6.0
	F	1.3 NS	9.8 **	1.4 NS
Telangana	df	2.0	4.0	8.0
	F	4.11 *	9.56 **	14.58 **

NS: Non-significant at 0.05 probability level, *: Significant at 0.05 probability level, **: Significant at 0.01 probability level.

5.2. Selection of High-Performing Varieties in Different States

In Karnataka over the years, BSMR 736 revealed significantly higher grain yield at Bidar, Kalaburagi, and Yadgir, respectively, with yield advantage of 8.6–29.9% over TS 3R (check). Across the location and over the years, BSMR 736 gained a 22.3% yield advantage over TS 3R (Table S7). In Maharashtra, BDN 711 exhibited the highest grain yield in Aurangabad, Jalna, and Parbhani districts, over BSMR 736 (check), with yield advantage in the range of 18.7–35.2% (Table S8). Across the locations in

Maharashtra, BDN 711 followed by BDN 716 revealed significantly higher grain yield over BSMR 736 with a yield advantage of 23.3–27.5%. In Madhya Pradesh over the years, ICPH 2671 significantly exhibited the highest grain yield at Rajgarh, Shajapur, and Sehore with a yield advantage in the range of 18.3–35.3% over the check variety, ICPL 88039. In Madhya Pradesh, over the years and across the locations, ICPH 2671 revealed superior performance in terms of higher yield, stability, and gained yield advantage of 34.72% over ICPL 88039 (Table S9). In Andhra Pradesh at Guntur, LRG 105 followed by LRG 160 exhibited significantly higher grain yield over check, LRG 52 with yield advantage in the range of 5.1–5.2%. Likewise, at Prakasam, LRG 105 followed by LRG 160 revealed significantly higher grain yield over LRG 52 with yield advantage in the range of 6.9 to 7.2%. At Kurnool, LRG 52 followed by LRG 160 revealed significantly higher grain yield over LRG 52 with yield advantage in the range of 7.1–7.5%. Overall, in Andhra Pradesh, LRG 160 followed by LRG 105 exhibited the highest grain yield with yield advantage in the range of 0.9–4.2% over LRG 52. A GGE biplot indicated LRG 160 was better performing and highly stable across the locations (Table S10). In Telangana, ICPL 332 WR significantly exhibited the highest grain yield at Mahabubnagar, Warrangal, and Rangareddy over check, TS 3R with yield advantage in the range of 6.5–17.8%. Across the location in Telangana ICPL 332 WR revealed the highest grain yield over TS 3R with yield advantage of 14.6% (Table S11).

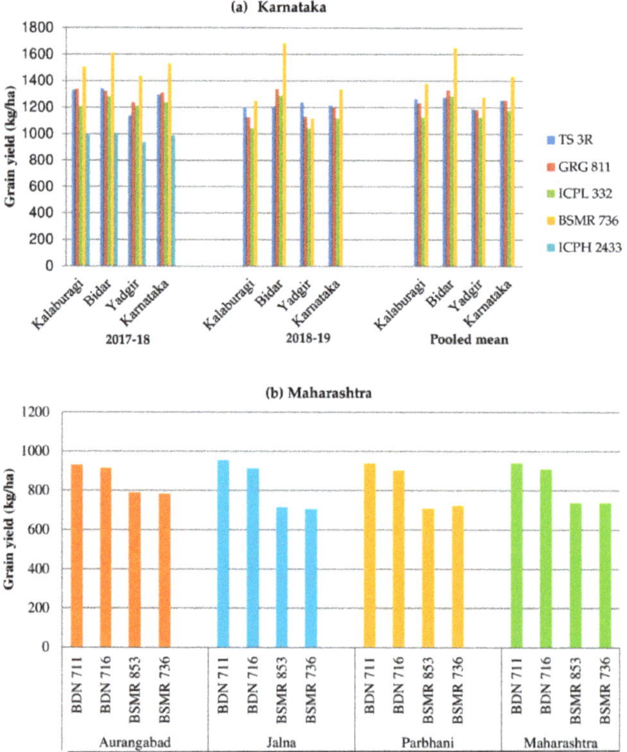

Figure 7. Farmers participatory varietal selection trials of pigeonpea in (**a**) Karnataka: BSMR 736 was identified as the best performing line in FPVs trials in Karnataka state. (**b**) Maharashtra: BDN 711 was identified as the best performing line in FPVs trials in Maharashtra state.

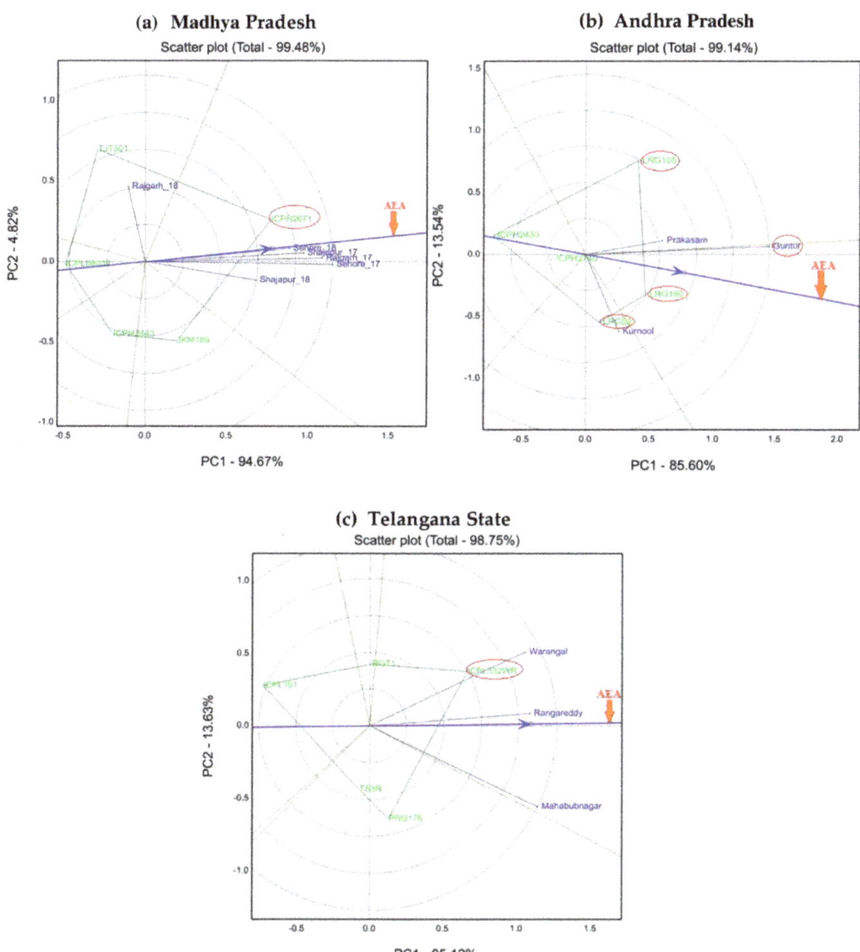

Figure 8. GGE biplot showing ranking of genotypes and environments in FPVS trial for mean and stability for grain yield in (**a**) Madhya Pradesh (**b**) Andhra Pradesh and (**c**) Telangana State.

6. Summary and Outlook

By using the TPGC as the multi-institutional team, we established a data-driven crop improvement program in pigeonpea. The TPGC has been engaged on different fronts starting from upstream research such as development of traits associated markers to very downstream work such as FPVS trials with farmers. For instance, significant efforts have been made to establish relationships between observed phenotype and genomic constitution in pigeonpea. This will enhance the precision and efficiency of the prediction of phenotypes from genotypes and subsequently in developing superior genotypes and varieties. The TPGC has been successful in developing new genetic stock, trait-associated molecular markers, and is currently working on developing new, promising lines through the MABC approach. Similarly, new genetic stock in the form of MAGIC and NAM populations have been developed. These multi-parent populations not only overcome the limitations of traditional trait mapping but also offer new potential to accurately define the genetic basis of complex crop traits [55]. NAM and MAGIC populations allow intensive genome reshuffling, making them suitable for high-resolution mapping due to broad genetic diversity created through high numbers of recombination events. Genotypes in the

NAM population exhibited a normal distribution for the majority of the traits, indicating quantitative genetic control. The significant variations observed in NAM and MAGIC populations will be harnessed in coming years for identification of QTLs and candidate genes for important traits, like pod numbers, growth habit, protein content, flowering, maturity, seed size, seed colour, etc. As success stories of MABC are available in other legume crops, like chickpea [21] and peanut [56,57], the TPGC has also initiated the introgression QTLs for diseases (FW and SMD) resistance in 11 mega varieties of pigeonpea. We anticipate some improved lines through MABC for commercial release in the near future.

Multi-location trials and FPVS trials were used for identification of high-performing varieties in station plots and farmers' fields, respectively, in different states. In early maturity group multi-location trials, IPA 15-05 revealed the highest grain yield at Tandur, Lam and across the locations while, from medium maturity trials, TDRG 60 followed by TDRG 4 and LRG 105 at Tandur, LRG 105 followed by LRG 52 and JKM 189 at Lam, ICPL 99050 at Kalaburagi and Sehore, and GRG 152 across the locations revealed the highest grain yield over check. Likewise, in super-early, ICPL 20325 followed by ICPL 11245 and ICPL 11292 at Tandur, ICPL 11300 followed by ICPL 20325 and ICPL 20327 at Lam, ICPL 11301 followed by ICPL 11244 and ICPL 20325 at Patancheru and ICPL 20325 across the locations revealed the higher grain yield over check with higher grain yield and high stability. The test environments that are highly discriminating are good for selecting adapted genotypes [58].

FPVS trials offer farmers the ability to adopt high-yielding, improved cultivars within a short time on a larger scale. Based on these FPVS trials, the cultivars BSMR 736 and BDN 711 showed constant better performance across the locations in Karnataka and Maharashtra states, respectively, due to the non-significant genotype × environment interaction (GEI) effect, so these high-yielding and stable cultivars would be mass multiplied and adopted in the respective regions. FPVS trials from other states and multi-location trials revealed significant GEI effects, indicating a differential response of genotypes in different environments. Significant genotype, environment, and GEI in pigeonpea for grain yield were reported earlier by Muniswamy et al. [59] and Arunkumar et al. [60]. For instance, FPVS trials in Andhra Pradesh, at Guntur and Prakasam, LRG 105 followed by LRG 160 while at Kurnool, LRG 52 followed by LRG 160 but across the locations, LRG 160 followed by LRG 105 revealed a significantly higher grain yield. The cultivar ICPH 2671 and ICPL 332 WR exhibited the highest grain yield at test sites and across the locations of Madhya Pradesh and Telangana, respectively.

In conclusion, advances in genetics and genomics made through TPGC are being utilized for developing new cultivars with desirable combinations of traits. The advanced backcross lines resistance to FW and SMD will be evaluated for grain yield for varietal release. The high-yielding and stably-performing genotypes in multi-location trials may be recommended for varietal release following AICRP-Pigeonpea guidelines. Similarly, high-performing and farmer-preferred varieties may be adopted in the respective districts and states by integrated efforts of different agriculture authorities, including state agricultural universities, and Kisan Vigyan Kendras.

Supplementary Materials: The following are available online at http://www.mdpi.com/2073-4395/10/9/1289/s1. Table S1: Details on QTLs/genomic segments identified for target traits in pigeonpea. Table S2: List of 100 elite lines used for evaluation for grain yield in pigeonpea. Table S3: Mean, range and ANOVA for grain yield of multi-location trial during cropping season 2017–2018 and 2018–2019. Table S4: Mean grain yield data of super-early duration trial during cropping season 2017–2018 and 2018–2019. Table S5: Mean grain yield data of early duration trial during cropping season 2017–2018 and 2018–2019. Table S6: Mean grain yield data of medium duration trial during cropping season 2017–2018 and 2018–2019. Table S7: Mean performance of selected cultivars of FPVS trial conducted in Karnataka for grain yield during cropping season 2017–2018 and 2018–2019. Table S8: Mean performance of selected cultivars of FPVS trials conducted in Maharashtra for grain yield during cropping season 2018–2019. Table S9: Mean performance of selected cultivars of FPVS trials conducted in Madhya Pradesh for grain yield during cropping season 2017–2018 and 2018–2019. Table S10: Mean performance of selected cultivars of FPVS trials conducted in Andhra Pradesh for grain yield during cropping season 2018–2019. Table S11: Mean performance of selected cultivars of FPVS trials conducted in Telangana State for grain yield during cropping season 2018–2019.

Author Contributions: Conceptualization: R.K.V. and R.K.S. along with N.S.; Methodology: A.H., C.V.S.K., A.B., A.J.H., M.S., A.R., A.M., A.N.T., C.S., S.R., D.K.P., I.P.S.; formal analysis: A.H., A.R., A.V.K.; data curation: A.H., A.R., A.V.K.; writing—original draft preparation: R.K.S., A.H., R.K.V., A.B.; writing—review and editing: R.K.S., A.H., R.K.V.; supervision: R.K.V., R.K.S., N.P.S., A.B., I.P.S.; project administration: R.K.V., N.P.S., R.K.S.; funding acquisition: R.K.V., N.P.S. All authors have read and agreed to the published version of the manuscript.

Funding: This research was funded by the Department of Agriculture Cooperation and Farmers Welfare, Ministry of Agriculture and Farmers Welfare, the Government of India and partly funded by Bill and Melinda Gates Foundation. This work has been undertaken as part of the CGIAR Research Program on Grain Legumes and Dryland Cereals (GLDC). ICRISAT is a member of CGIAR Consortium.

Acknowledgments: Authors are thankful to KB Saxena for useful discussions and suggestions on project activities and to V Suryanarayana and M Sudhakar for technical assistance in various activities.

Conflicts of Interest: The authors declare no conflict of interest.

References

1. Food and Agriculture Organization of the United Nations Database (FAOSTAT). 2020. Available online: http://faostat.fao.org/database (accessed on 16 January 2020).

2. Bhatia, V.S.; Singh, P.; Wani, S.P.; Rao, A.K.; Srinivas, K. *Yield Gap Analysis of Soybean, Groundnut, Pigeonpea and Chickpea in India Using Simulation Modeling: Global Theme on Agroecosystems*; Report no. 31. Patancheru 502 324; International Crops Research Institute for the Semi-Arid Tropics (ICRISAT): Andhra Pradesh, India, 2006; p. 156.

3. Yang, D.; Tang, J.; Yang, D.; Chen, Y.; Ali, J.; Mou, T. Improving rice blast resistance of Feng39S through molecular marker-assisted backcrossing. *Rice* **2019**, *12*, 70. [CrossRef] [PubMed]

4. Yugander, A.; Sundaram, R.M.; Singh, K.; Ladhalakshmi, D.; Rao, L.V.S.; Madhav, M.S.; Badri, J.; Prasad, M.S.; Laha, G.S. Incorporation of the novel bacterial blight resistance gene *Xa38* into the genetic background of elite rice variety Improved Samba Mahsuri. *PLoS ONE* **2018**, *13*, e0198260. [CrossRef] [PubMed]

5. Miah, G.; Rafii, M.Y.; Ismail, M.R.; Puteh, A.B.; Rahim, H.A.; Latif, M.A. Marker-assisted introgression of broad-spectrum blast resistance genes into the cultivated MR219 rice variety. *J. Sci. Food Agric.* **2017**, *97*, 2810–2818. [CrossRef] [PubMed]

6. Linh, L.H.; Linh, T.H.; Xuan, T.D.; Ham, L.H.; Ismail, A.M.; Khanh, T.D. Molecular breeding to improve salt tolerance of rice (*Oryza sativa* L.) in the Red River Delta of Vietnam. *Int. J. Plant Genom.* **2012**, *2012*. [CrossRef]

7. Randhawa, M.S.; Bains, N.S.; Sohu, V.S.; Chhuneja, P.; Trethowan, R.M.; Bariana, H.S.; Bansal, U. Marker assisted transfer of stripe rust and stem rust resistance genes into four wheat cultivars. *Agronomy* **2019**, *9*, 497. [CrossRef]

8. Tyagi, S.; Mir, R.R.; Kaur, H.; Chhuneja, P.; Ramesh, B.; Balyan, H.S.; Gupta, P.K. Marker-assisted pyramiding of eight QTLs/genes for seven different traits in common wheat (*Triticum aestivum* L.). *Mol. Breed.* **2014**, *34*, 167–175. [CrossRef]

9. Charpe, A.; Koul, S.; Gupta, S.K.; Singh, A.; Pallavi, J.K.; Prabhu, K.V. Marker assisted gene pyramiding of leaf rust resistance genes *Lr 9*, *Lr 24* and *Lr 28* in a bread wheat cultivar HD 2329. *J. Wheat Res.* **2012**, *4*, 20–28.

10. Kumar, J.; Jaiswal, V.; Kumar, A.; Kumar, N.; Mir, R.R.; Kumar, S.; Dhariwal, R.; Tyagi, S.; Khandelwal, M.; Prabhu, K.V.; et al. Introgression of a major gene for high grain protein content in some Indian bread wheat cultivars. *Field Crops Res.* **2011**, *123*, 226–233. [CrossRef]

11. Barloy, D.; Lemoine, J.; Abelard, P.; Tanguy, A.M.; Rivoal, R.; Jahier, J. Marker-assisted pyramiding of two cereal cyst nematode resistance genes from *Aegilops variabilis* in wheat. *Mol. Breed.* **2007**, *20*, 31–40. [CrossRef]

12. Afolayan, G.; Aladele, S.E.; Deshpande, S.P.; Oduoye, O.T.; Nwosu, D.J.; Michael, C.; Blay, E.T.; Danquah, E.Y. Marker assisted foreground selection for identification of striga resistant backcross lines in *Sorghum Bicolor*. *Covenant J. Phys. Life Sci.* **2019**, *7*, 29–36.

13. Kadam, S.R.; Fakrudin, B. Marker assisted pyramiding of root volume QTLs to improve drought tolerance in rabi sorghum. *Res. Crops* **2017**, *18*, 683–692. [CrossRef]

14. Ngugi, K.; Kimani, W.; Kiambi, D. Introgression of stay-green trait into a Kenyan farmer prefered sorghum variety. *Afr. Crop Sci. J.* **2010**, *18*, 141–146. [CrossRef]

15. Krishna, M.S.R.; Reddy, S.S.; Satyanarayana, S.D. Marker-assisted breeding for introgression of opaque-2 allele into elite maize inbred line BML-7. *3 Biotech* **2017**, *7*, 165. [CrossRef] [PubMed]

16. Muthusamy, V.; Hossain, F.; Thirunavukkarasu, N.; Choudhary, M.; Saha, S.; Bhat, J.S.; Prasanna, B.M.; Gupta, H.S. Development of β-carotene rich maize hybrids through marker-assisted introgression of β-carotene hydroxylase allele. *PLoS ONE* **2014**, *9*, e113583. [CrossRef]

17. Babu, R.; Nair, S.K.; Kumar, A.; Venkatesh, S.; Sekhar, J.C.; Singh, N.N.; Srinivasan, G.; Gupta, H.S. Two-generation marker-aided backcrossing for rapid conversion of normal maize lines to quality protein maize (QPM). *Theor. Appl. Genet.* **2005**, *111*, 888–897. [CrossRef] [PubMed]

18. Taunk, J.; Rani, A.; Yadav, N.R.; Yadav, D.V.; Yadav, R.C.; Raj, K.; Kumar, R.; Yadav, H.P. Molecular breeding of ameliorating commercial pearl millet hybrid for downy mildew resistance. *J. Genet.* **2018**, *97*, 1241–1251. [CrossRef] [PubMed]

19. Hash, C.T.; Sharma, A.; Kolesnikova-Allen, M.A.; Singh, S.D.; Thakur, R.P.; Raj, A.G.B.; Rao, M.N.V.R.; Nijhawan, D.C.; Beniwal, C.R.; Sagar, P.; et al. Teamwork delivers biotechnology products to Indian small-holder crop-livestock producers: Pearl millet hybrid "HHB 67 Improved" enters seed delivery pipeline. *SAT Agric. Res.* **2006**, *2*, 1–3.

20. Mannur, D.M.; Babbar, A.; Thudi, M.; Sabbavarapu, M.M.; Roorkiwal, M.; Sharanabasappa, B.Y.; Bansal, V.P.; Jayalakshmi, S.K.; Yadav, S.S.; Rathore, A.; et al. Super Annigeri 1 and improved JG 74: Two Fusarium wilt-resistant introgression lines developed using marker-assisted backcrossing approach in chickpea (*Cicer arietinum* L.). *Mol. Breed.* **2019**, *39*, 2. [CrossRef]

21. Varshney, R.K.; Mohan, S.M.; Gaur, P.M.; Chamarthi, S.K.; Singh, V.K.; Srinivasan, S.; Swapna, N.; Sharma, M.; Pande, S.; Singh, S.; et al. Marker-assisted backcrossing to introgress resistance to Fusarium wilt race 1 and Ascochyta blight in C 214, an elite cultivar of chickpea. *Plant Genome* **2014**, *7*, 1–11. [CrossRef]

22. Varshney, R.K.; Gaur, P.M.; Chamarthi, S.K.; Krishnamurthy, L.; Tripathi, S.; Kashiwagi, J.; Samineni, S.; Singh, V.K.; Thudi, M.; Jaganathan, D. Fast-track introgression of "QTL-hotspot" for root traits and other drought tolerance traits in JG 11, an elite and leading variety of chickpea. *Plant Genome* **2013**, *6*, 1–9. [CrossRef]

23. Taran, B.; Warkentin, T.D.; Vandenberg, A. Fast track genetic improvement of ascochyta blight resistance and double podding in chickpea by marker-assisted backcrossing. *Theor. Appl. Genet.* **2013**, *126*, 1639–1647. [CrossRef] [PubMed]

24. Varshney, R.K. Exciting journey of 10 years from genomes to fields and markets: Some success stories of genomics-assisted breeding in chickpea, pigeonpea and groundnut. *Plant Sci.* **2016**, *242*, 98–107. [CrossRef] [PubMed]

25. Ha, T.; Khang, D.T.; Tuyen, P.T.; Toan, T.B.; Huong, N.N.; Lang, N.T.; Buu, B.C.; Xuan, T.D. Development of new drought tolerant breeding lines for Vietnam using marker-assisted backcrossing. *Int. Lett. Nat. Sci.* **2016**, *59*, 1–13.

26. Kumar, V.; Rani, A.; Rawal, R.; Mourya, V. Marker assisted accelerated introgression of null allele of kunitz trypsin inhibitor in soybean. *Breed Sci.* **2015**, *65*, 447–452. [CrossRef] [PubMed]

27. Landau-Ellis, D.; Pantalone, V.R. Marker-assisted backcrossing to incorporate two low phytate alleles into the Tennessee soybean cultivar 5601T. In *Induced Plant Mutations in the Genomics Era, Proceedings of an International Joint FAO/IAEA Symposium*; Food and Agriculture Organization of the United Nations: Rome, Italy, 2009; pp. 316–318.

28. Varshney, R.K.; Chen, W.; Li, Y.; Bharti, A.K.; Saxena, R.K.; Schlueter, J.A.; Donoghue, M.T.; Azam, S.; Fan, G.; Whaley, A.M.; et al. Draft genome sequence of pigeonpea (*Cajanus cajan*), an orphan legume crop of resource-poor farmers. *Nat. Biotechnol.* **2012**, *30*, 83. [CrossRef]

29. Saxena, R.K.; Prathima, C.; Saxena, K.B.; Hoisington, D.A.; Singh, N.K.; Varshney, R.K. Novel SSR markers for polymorphism detection in pigeonpea (*Cajanus* spp.). *Plant Breed.* **2010**, *129*, 142–148. [CrossRef]

30. Bohra, A.; Saxena, K.B.; Varshney, R.K.; Saxena, R.K. Genomics-assisted breeding for pigeonpea improvement. *Theor. Appl. Genet.* **2020**, *133*, 1–17. [CrossRef]

31. Saxena, R.K.; Kale, S.M.; Kumar, V.; Parupali, S.; Joshi, S.; Singh, V.; Garg, V.; Das, R.R.; Sharma, M.; Yamini, K.N.; et al. Genotyping-by-sequencing of three mapping populations for identification of candidate genomic regions for resistance to sterility mosaic disease in pigeonpea. *Sci. Rep.* **2017**, *7*, 1–10. [CrossRef]

32. Saxena, R.K.; Rathore, A.; Bohra, A.; Yadav, P.; Das, R.R.; Khan, A.W.; Singh, V.K.; Chitikineni, A.; Singh, I.P.; Kumar, C.V.; et al. Development and application of high-density axiom *Cajanus* SNP array with 56k SNPs to understand the genome architecture of released cultivars and founder genotypes. *Plant Genome* **2018**, *11*, 1–10. [CrossRef]

33. Parupalli, S.; Saxena, R.K.; Kumar, C.V.; Sharma, M.; Singh, V.K.; Vechalapu, S.; Kavikishor, P.B.; Saxena, K.B.; Varshney, R.K. Genetics of fusarium wilt resistance in pigeonpea as revealed by phenotyping of RILs. *J. Food Legumes* **2017**, *30*, 241–244.

34. Daspute, A.; Fakrudin, B.; Bhairappanavar, S.B.; Kavil, S.P.; Narayana, Y.D.; Muniswamy Kaumar, A.; Krishnaraj, P.U.; Yerimani, A.; Khadi, B.M. Inheritance of pigeonpea sterility mosaic disease resistance in pigeonpea. *Plant Pathol. J.* **2014**, *30*, 188–194. [CrossRef] [PubMed]

35. Varshney, R.K.; Penmetsa, R.V.; Dutta, S.; Kulwal, P.L.; Saxena, R.K.; Datta, S.; Sharma, T.R.; Rosen, B.; Carrasquilla-Garcia, N.; Farmer, A.D.; et al. Pigeonpea genomics initiative (PGI): An international effort to improve crop productivity of pigeonpea (*Cajanus cajan* L.). *Mol. Breed.* **2010**, *26*, 393–408. [CrossRef] [PubMed]

36. Upadhyaya, H.D.; Reddy, K.N.; Sharma, S.; Varshney, R.K.; Bhattacharjee, R.; Singh, S.; Gowda, C.L.L. Pigeonpea composite collection and identification of germplasm for use in crop improvement programmes. *Plant Genet. Resour.* **2011**, *9*, 97–108. [CrossRef]

37. Reddy, L.J.; Upadhyaya, H.D.; Gowda, C.L.L.; Singh, S. Development of core collection in pigeonpea [*Cajanus cajan* (L.) Millspaugh] using geographic and qualitative morphological descriptors. *Genet. Resour. Crop Evol.* **2005**, *52*, 1049–1056. [CrossRef]

38. Upadhyaya, H.D.; Reddy, L.J.; Gowda, C.L.L.; Reddy, K.N.; Singh, S. Development of a mini core subset for enhanced and diversified utilization of pigeonpea germplasm resources. *Crop Sci.* **2006**, *46*, 2127–2132. [CrossRef]

39. Saxena, R.K.; Kale, S.; Mir, R.R.; Mallikarjuna, N.; Yadav, P.; Das, R.R.; Molla, J.; Sonnappa, M.; Ghanta, A.; Narasimhan, Y.; et al. Genotyping-by-sequencing and multilocation evaluation of two interspecific backcross populations identify QTLs for yield-related traits in pigeonpea. *Theor. Appl. Genet.* **2020**, *133*, 737–749. [CrossRef]

40. IBPGR and ICRISAT. *Descriptors for Pigeonpea [Cajanus cajan (L.) Millsp.]*; IBPGR: Rome, Italy; ICRISAT: Patancheru, India, 1993.

41. Cavanagh, C.; Morell, M.; Mackay, I.; Powell, W. From mutations to MAGIC: Resources for gene discovery, validation and delivery in crop plants. *Curr. Opin. Plant Biol.* **2008**, *11*, 215–221. [CrossRef]

42. Prasad, P.S.; Muhammad, S.; Mahesh, M.; Kumar, G.N.V. Management of pigeonpea wilt caused by *Fusarium udum* Butler through integrated approaches. *J. Biol. Control.* **2012**, *26*, 361–367.

43. Reddy, M.V.; Raju, T.N.; Lenne, J.M. Diseases of pigeonpea. In *The Pathology of Food and Pasture Legumes*; Allen, D.J., Lenne, J.M., Eds.; CAB International: Wallingford, UK, 1998; pp. 517–558.

44. Reddy, M.V.; Sharma, S.B.; Nene, Y.L. Pigeonpea: Disease management. In *The Pigeonpea*; Nene, Y.L., Hall, S.D., Sheila, V.K., Eds.; CAB International: Wallingford, UK, 1990; pp. 303–348.

45. Okiror, M.A. Genetics of resistance to *Fusarium udum* in pigeonpea [*Cajanus cajan* (L.) Millsp.]. *Ind. J. Genet. Plant Breed.* **2002**, *62*, 218–220.

46. Gnanesh, B.N.; Ganapathy, K.N.; Ajay, B.C.; Byre Gowda, M. Inheritance of sterility mosaic disease resistance to Bangalore and Patancheru isolates in pigeonpea (*Cajanus cajan* (L.) Millsp.). *Electron. J. Plant Breed.* **2011**, *2*, 218–223.

47. Kulkarni, N.K.; Reddy, A.S.; Kumar, P.L.; Vijayanarasimha, J.; Rangaswamy, K.T.; Muniyappa, V.; Reddy, L.J.; Saxena, K.B.; Jones, A.T.; Reddy, D.V. Broad based resistance to pigeonpea sterility mosaic disease in accessions of *Cajanus scarbaeoides* (L.) Benth. *Ind. J. Plant Pro.* **2003**, *31*, 6–11.

48. Reddy, M.V.; Raju, T.N.; Nene, Y.L.; Ghanekar, A.M.; Amin, K.S.; Arjunan, G.; Astaputre, J.V.; Sinha, B.K.; Reddy, S.V.; Gupta, R.P.; et al. Variability in sterility mosaic pathogen in pigeonpea in India. *Indian Phytopathol.* **1993**, *46*, 206–212.

49. Yadav, P.; Saxena, K.B.; Hingane, A.; Kumar, C.S.; Kandalkar, V.S.; Varshney, R.K.; Saxena, R.K. An "Axiom Cajanus SNP Array" based high density genetic map and QTL mapping for high-selfing flower and seed quality traits in pigeonpea. *BMC Genom.* **2019**, *20*, 1–10. [CrossRef] [PubMed]

50. Obala, J.; Saxena, R.K.; Singh, V.K.; Kumar, C.S.; Saxena, K.B.; Tongoona, P.; Sibiya, J.; Varshney, R.K. Development of sequence-based markers for seed protein content in pigeonpea. *Mol. Genet. Genom.* **2019**, *294*, 57–68. [CrossRef]

51. Obala, J.; Saxena, R.K.; Singh, V.K.; Kale, S.M.; Garg, V.; Kumar, C.S.; Saxena, K.B.; Tongoona, P.; Sibiya, J.; Varshney, R.K. Seed protein content and its relationships with agronomic traits in pigeonpea is controlled by both main and epistatic effects QTLs. *Sci. Rep.* **2020**, *10*, 1–17. [CrossRef] [PubMed]

52. Panse, V.G.; Sukhatme, P.V. *Statistical Methods for Agricultural Workers*, 3rd ed.; ICAR: New Delhi, India, 1954; p. 58.

53. SAS Institute Inc. *SAS/STAT®9.2 User's Guide*; SAS Institute Inc.: Cary, NC, USA, 2008.

54. Yan, W.; Hunt, L.A.; Sheng, Q.; Szlavnics, Z. Cultivar evaluation and mega-environment investigation based on the GGE biplot. *Crop Sci.* **2000**, *40*, 597–605. [CrossRef]

55. Ladejobi, O.; Elderfield, J.; Gardner, K.A.; Gaynor, R.C.; Hickey, J.; Hibberd, J.M.; Mackay, I.J.; Bentley, A.R. Maximizing the potential of multi-parental crop populations. *Appl. Transl. Genom.* **2016**, *11*, 9–17. [CrossRef]

56. Janila, P.; Pandey, M.K.; Shasidhar, Y.; Variath, M.T.; Sriswathi, M.; Khera, P.; Manohar, S.S.; Nagesh, P.; Vishwakarma, M.K.; Mishra, G.P.; et al. Molecular breeding for introgression of fatty acid desaturase mutant alleles (ahFAD2A and ahFAD2B) enhances oil quality in high and low oil containing peanut genotypes. *Plant Sci.* **2016**, *242*, 203–213. [CrossRef]

57. Varshney, R.K.; Pandey, M.K.; Janila, P.; Nigam, S.N.; Sudini, H.; Gowda, M.V.C.; Sriswathi, M.; Radhakrishnan, T.; Manohar, S.S.; Nagesh, P. Marker-assisted introgression of a QTL region to improve rust resistance in three elite and popular varieties of peanut (*Arachis hypogaea* L.). *Theor. Appl. Genet.* **2014**, *127*, 1771–1781. [CrossRef]

58. Yan, W.; Tinker, N.A. Biplot analysis of multi-environment trial data: Principles and applications. *Can. J. Plant Sci.* **2006**, *86*, 623–645. [CrossRef]

59. Muniswamy, S.; Lokesha, R.; Diwan, J.R. Stability for disease, genotype × environment interaction for yield and its components in pigeonpea [*Cajanus cajan* (L.) Millsp.]. *Legume Res.* **2017**, *40*, 624–629.

60. Arunkumar, B.; Muniswamy, S.; Dharmaraj, P.S. Interpretation of genotype × environment interaction and stability analysis for grain yield of pigeon pea (*Cajanus cajan* L.). *J. Appl. Nat. Sci.* **2014**, *6*, 744–747. [CrossRef]

 *agronomy*

Article

Identification of Non-Pleiotropic Loci in Flowering and Maturity Control in Soybean

Eric J. Sedivy [1,2], Abraham Akpertey [3] 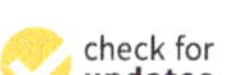, Angela Vela [1], Sandra Abadir [1], Awais Khan [4] and Yoshie Hanzawa [1,*]

[1] Department of Biology, California State University, Northridge, CA 91330, USA; esedivy2@gmail.com (E.J.S.); angela.vela.563@my.csun.edu (A.V.); sandra.abadir.691@my.csun.edu (S.A.)

[2] Department of Crop Sciences, University of Illinois at Urbana-Champaign, Urbana-Champaign, IL 61801, USA

[3] USDA-Agricultural Research Service, Soybean/Maize Germplasm, Pathology, and Genetics Research Unit, Department of Crop Sciences, University of Illinois at Urbana-Champaign, Urbana-Champaign, IL 61801, USA; akperte2@illinois.edu

[4] Plant Pathology and Plant-Microbe Biology Section, Cornell University, Geneva, NY 14456, USA; awais.khan@cornell.edu

* Correspondence: yoshie.hanzawa@csun.edu; Tel.: +1-(818)-677-4478

Received: 30 June 2020; Accepted: 9 August 2020; Published: 17 August 2020

Abstract: Pleiotropy is considered to have a significant impact on multi-trait evolution, but its roles in the evolution of domestication-related traits in crop species have been unclear. In soybean, several known quantitative trait loci (QTL) controlling maturity, called the maturity loci, are known to have major effects on both flowering and maturity in a highly correlated pleiotropic manner. Aiming at the identification of non-pleiotropic QTLs that independently control flowering and maturity and dissecting the effects of pleiotropy in these important agronomic traits, we conducted a QTL mapping experiment by creating a population from a cross between domesticated soybean *G. max* and its wild ancestor *G. soja* that underwent stringent selection for non-pleiotropy in flowering and maturity. Our QTL mapping analyses using the experimental population revealed novel loci that acted in a non-pleiotropic manner: R1-1 controlled primarily flowering and R8-1 and R8-2 controlled maturity, while R1-1 overlapped with QTL, affecting other agronomic traits. Our results suggest that pleiotropy in flowering and maturity can be genetically separated, while artificial selection during soybean domestication and diversification may have favored pleiotropic loci such as *E* loci that control both flowering and maturity. The non-pleiotropic loci identified in this study will help to identify valuable novel genes to optimize soybean's life history traits and to improve soybean's yield potential under diverse environments and cultivation schemes.

Keywords: QTL mapping; *Glycine max*; *Glycine soja*; flowering; Maturation; domestication; pleiotropy

1. Introduction

Genetic pleiotropy, correlation among traits due to genes that possess pleiotropic roles or the linkage of multiple genes, is known to have a major impact on the multi-trait evolution of organisms. Neutral theories and evolutionary models suggest that pleiotropy is a cause of evolutionary constraints, preventing organisms from evolutionary progress [1,2]. In contrast, recent genome-wide observations indicate that pleiotropy promotes the evolution of complexity [3,4]. Crop domestication provides a unique opportunity to dissect the effects of pleiotropy on multi-trait evolution. Intense selection by humans during crop domestication and diversification processes causes rapid and directional changes in a set of traits, typically including non-seed shattering, reduced branching, apical dominance, large fruit or grain size, and synchronized flowering, which is known as the domestication syndrome [5]. However, much remains unknown about the extent of pleiotropy among domestication-related traits and its consequences in the evolution of crop species.

Cultivated soybean (*Glycine max*) is considered to be domesticated from its wild ancestor *Glycine soja* approximately 6000 to 9000 years ago [6,7]. Preceding the domestication event, recent discoveries suggest a prolonged period of low intensity management or semi-domestication of wild soybeans at multiple locations in East Asia [8]. Compared with its ancestor, *G. max* shows reduced pod dehiscence, branching and lodging, larger seeds, and more synchronized flowering and maturation. A handful of genes that are implicated in soybean domestication, diversification, and improvement processes have been identified. Among them, three genes are thought to have played central roles at the early stage of soybean domestication: *SHAT1-5* controlling pod dehiscence [9,10], *GmHs1-1* controlling seed hardness [11], and *GmFT2c* controlling photoperiodic flowering [12].

An emphasis of soybean domestication and diversification was selection for plants adapted to regional photoperiods that confines soybean varieties carrying differing photoperiod sensitivities for flowering and maturity to specific latitudinal boundaries [13,14]. Several QTL controlling diversity in flowering and maturity among cultivated soybean varieties have been identified: the maturity loci *E1–E10* and *J* [15,16]. The *E1* locus encodes a transcription factor carrying a B3 domain [17], *E2* is an ortholog of the Arabidopsis flowering regulator *GIGANTEA* [18], and *E3* and *E4* are homologs of the photoreceptor *PHYTOCHROME A* (*PHYA*) [19,20]. Recently, additional maturity loci *E9* and *J* are identified as the *FLOWERING LOCUS T* homolog *GmFT2a* [21] and the circadian clock gene *EARLY FLOWERING 3* [22], respectively. Major maturity loci are observed to affect both flowering and maturity in a highly correlated manner [23,24]: early flowering is associated with early maturity, while late flowering is associated with late maturity, suggesting that the pleiotropy of these loci controlling both flowering and maturity may have been a limiting factor for the diversification of soybean's life cycle in breeding programs.

In order to dissect the effects of pleiotropy in the evolution of flowering and maturity control in soybean, in this study, we conducted a selection experiment by creating a population from a cross between *G. max* and *G. soja* that underwent stringent selection for non-pleiotropy in flowering and maturity, and carried out QTL mapping analysis to identify novel loci that independently control flowering or maturity in a non-pleiotropic manner.

2. Materials and Methods

2.1. Mapping Population

The initial cross between the *G. max* reference variety Williams 82 (Maturity group III) and the *G. soja* accession PI 549046 (Maturity group III) was made in 1996 (Figure 1). The F1 plants were backcrossed to Williams 82 in 1997, and the BC1F1 plants were grown in 1998. From the 3000 BC1F2 plants, approximately 60 plants were selected based on agronomic appearance similar to Williams 82. From the 60 BC1F3 lines, 87 plants out of 17 BC1F3 lines with the best agronomic appearance similar to Williams 82 were selected in 2000. In the BC1F4 generation, nine lines that showed agronomic appearance similar to Williams 82 were selected and bulk harvested. These lines were grown in the field, and phenotypes were observed in 2001 and 2002. In order to identify non-pleiotropic loci that control either flowering or maturity independently unlike the known major maturity loci that control both flowering and maturity in a highly pleiotropic manner, our selection scheme in this study targeted a type of non-pleiotropy affecting flowering but not maturity. We selected the lines LG01-7909 and LG01-7919 that flowered 10 to 12 days later and matured 7 to 13 days earlier than Williams 82 in 2001, and flowered 9 to 11 days later and matured 11 to 12 days earlier than Williams 82 in 2002. In 2003, these two BC1F5 lines were backcrossed again to Williams 82, and the BC2F1 plants were grown in the following winter in Puerto Rico. These populations were advanced through single-seed descent from the F2 to the F4 generation. In the BC2F4 generation, 109 plants were harvested from the population derived from the LG01-7909 parent, and 79 plants were harvested from the population derived from the LG01-7919 parent, and a single plant was harvested from each row in 2007. From the 166 BC2F5 rows planted in 2008, a total of 115 inbred lines in the maturity date range of 12 days (18–30 September) were selected for the mapping population, consisting of 68 recombinant inbred lines (RILs) from the LG01-7909 pedigree and 47 RILs from the LG01-7919 pedigree.

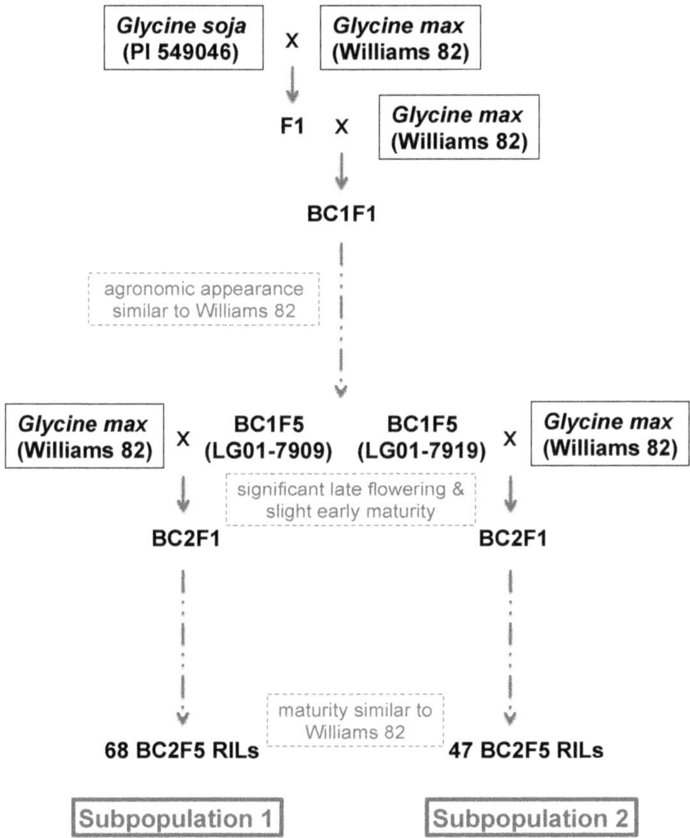

Figure 1. The scheme for mapping population development and selection experiment.

2.2. Field Evaluation

The BC2F6 RILs, the parent line Williams 82, and maturity checks (IA2023 and LD00-3309) were evaluated at Urbana and Stonington, Illinois in 2009 and at Urbana, Villa Grove, Bellflower, and Stonington, Illinois in 2010. All field plots were four rows wide with 0.76 m spacing and 3.1 m long and were planted at a rate of 30 seeds per meter. At all locations and in both years, the lines were planted in a randomized complete block design with two replications. The phenotype data collected include flowering date (R1), which was recorded when approximately 50% of the plants had at least one flower [25], plant maturity date (R8), which was recorded when approximately 95% of the pods had reached mature pod color [25], plant lodging scored at maturity based on a 1 to 10 scale (1 = all plants are erect, 10 = all plants are prostrate), plant height (cm) measured from the soil surface to the top node of the main stem and seed yield (kg ha^{-1}) recorded at 13% moisture, and a stem vininess score between 1 (typical indeterminate cultivar) and 5 (viney stem termination). The 115 BC2F6 RIL population was divided by maturity based on the data collected in 2008. Lines that matured between 18 and 24

September were blocked together (Maturity set 1) and those that matured between 25 and 30 September were blocked together (Maturity set 2) for further analysis.

2.3. Statistical Analysis of Phenotypic Data

Statistical analysis of phenotypic distribution and multiple factor analysis (MFA) were carried out using R scripts. Box plots were created using the Psych package [26] and multiple factor analysis (MFA) was conducted using the FactominR package [27]. Correlation analysis was conducted in R.

2.4. Genotyping

A total of 1536 single nucleotide polymorphisms (SNPs) from the 1536 Universal Soy Linkage Panel (USLP 1.0) [28] was used for genotyping assay. A single leaf from a single plant was collected for each line between 30 and 40 days after sowing. DNA samples from 115 BC2F6 RILs and Williams 82 were obtained using the DNeasy Plant Mini kit (QIAGEN), and genotyping was carried out using the Illumina GoldenGate® Assay [29] by Perry Cregan at USDA-ARS, Maryland.

The *E1* alleles were genotyped as described previously [30], and the SNP marker BARC-056323-14257 was genotyped via the direct sequencing of fragments amplified by polymerase chain reaction using the primers 5′-ACCATCATTGTTGTGAACCCTAC-3′ and 5′-AACGTCATCCACTTGAAACTTGG-3′.

2.5. SNP Selection

SNPs of interest were selected using Illumina's Genome Studio® 1.0. SNPs with low call numbers (below 5%) or no segregation in our mapping population were excluded. In addition, SNPs for which both parents shared the same allele type were removed. Of the 1536 SNPs genotyped, 545 were selected for quantitative trait loci (QTL) mapping (Table S1).

2.6. QTL Mapping

Marker distances and linkages were calculated using the Kosambi function in JoinMap® 4.0 with a minimum logarithm of the odds (LOD) threshold of 3.0. QTL analysis was conducted using composite interval mapping (CIM) in the software Windows QTL Cartographer 2.5 [31]. Stepwise selection was used to identify background co-factors in the model. For CIM, a 1000-iteration permutation test was performed for all traits to generate a genome-wide critical alpha of 0.05. To control the background marker effect, a window size of 1cM was employed.

2.7. Candidate Gene Identification

The physical locations of significant flowering and maturity SNPs were compared with a list of flowering-related genes and previously identified flowering and maturity QTL (http://soybase.org/). Candidate genes were called if they were located within 500 kb of QTL-associated SNPs.

2.8. Plant Growth Condition

The flowering time and maturation time measurement of the BC3F2 population was conducted in the greenhouse during November through February under a moderate short-day photoperiod (12 h light) with supplementary white fluorescence light. Highest temperatures during the day ranged between 24.0 and 26.5 °C, and lowest temperatures in the evening ranged between 20.0 and 22.5 °C.

3. Results

3.1. Selection for Non-Pleiotropy

The *G. max* reference variety Williams 82 and the *G. soja* accession PI 549046 were used as the parents to create the mapping population evaluated in this study (Figure 1). PI 549046 was one of the lines selected from four Chinese provinces that were genetically the most distinct from cultivated soybean [32]. Mimicking the domestication and diversification processes of soybean, two backcrosses to Williams 82 were performed after the initial cross between Williams 82 and PI 549046, and plants that showed agronomic appearance similar to Williams 82 in traits such as lodging, height, and stem vininess were selected through multiple generations. As an attempt to identify non-pleiotropic loci, we implemented a selection scheme for a type of non-pleiotropy that affects flowering but not maturity; in the BC1F5 generation, we selected the lines LG01-7909 and LG01-7919 that flowered significantly later but matured moderately earlier than Williams 82 in both 2001 and 2002. These lines were backcrossed again to Williams 82 (BC2) and advanced to the F5 generation. Then, we further selected lines that showed a maturity time similar to Williams 82, resulting in a total of 115 BC2F5 RILs with a narrow maturity range of 12 days. Strong positive correlation was observed in all comparisons between different locations for each trait, Maturity set, and year (Figure S1), except for height in Urbana and Stonington in Maturity set 2 in 2009, which appeared marginal ($p = 0.088$).

3.2. Phenotypic Distribution

The 115 BC2F6 RIL population was used for phenotyping at multiple field locations: Urbana and Stonington, Illinois in 2009 and Urbana, Villa Grove, Bellflower, and Stonington, Illinois in 2010. All traits displayed a strong yearly and/or location effect in the distribution (Figure 2a,b). The values of flowering time (R1) were distributed in a wider range than that of maturity (R8). Flowering ranged from 39.8 to 68.5 days with a mean of 55.6 days in 2009, and from 18.0 to 41.3 days with a mean of 31.8 days in 2010. Maturity ranged from 118.8 to 133.5 days with the mean of 126.0 days in 2009, and from 102.5 to 120.0 days with the mean of 111.9 days in 2010. The interquartile ranges of R1 and stem vining were above the mean value of Williams 82 in each year, whereas the interquartile ranges of R8 and yield were below Williams 82.

Aiming at identifying loci that affect flowering in a non-pleiotropic manner, the 115 RILs were divided into two sets for further analysis based on their maturity: early maturity (Maturity set 1; $n = 64$) and late maturity (Maturity set 2; $n = 51$). The phenotypic distribution differed significantly between maturity sets 1 and 2 in flowering, maturity, and height (Figure 2c); a difference of 4.6 days in flowering ($p = 2.12 \times 10^{-7}$), 4.8 days in maturity ($p = 8.19 \times 10^{-27}$), and 9.2 cm in height ($p = 1.48 \times 10^{-3}$). In contrast, lodging, stem vining, or yield did not differ between maturity sets 1 and 2. No significant difference in phenotypic distribution was observed between the LG01-7909 progenies (Subpopulation 1; $n = 68$) and the LG01-7919 progenies (Subpopulation 2; $n = 47$) in any traits.

3.3. Multiple Factor Analysis (MFA)

Multiple factor analysis (MFA) further demonstrated the different relationship among the traits (Figure 3). Dimensions 1 and 2 explain 45.3% and 23.1% of the total variation, respectively, and Dimension 3 makes up 16.7%. The traits are observed to cluster into three groups. Height, stem vining, and lodging cluster together and make up major components in Dimension 1 (0.84, 0.80, and 0.86, respectively) (Figure 3; Table 1), while flowering and maturity times are represented together in Dimension 2 (0.82 and 0.56, respectively), consistently with the previous implication regarding the highly correlated behavior of maturity loci in flowering and maturity [23,24,33,34]. Finally, yield is the primary component in Dimension 3 (0.83).

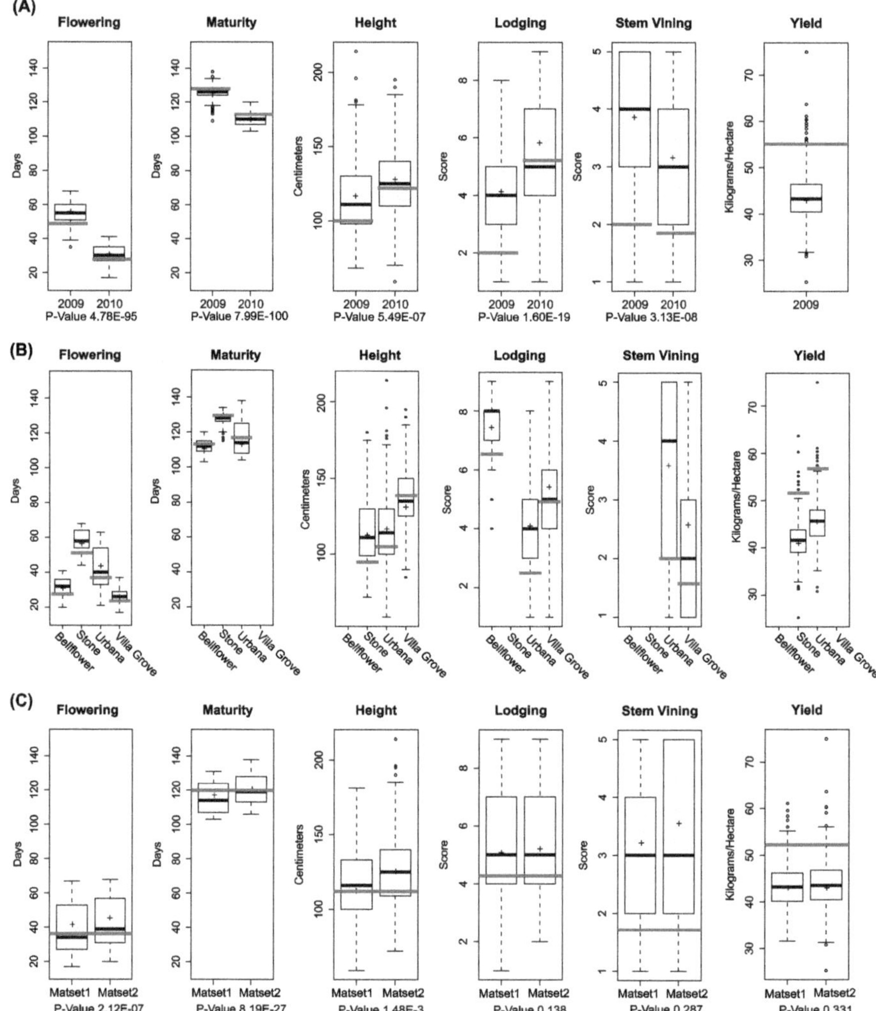

Figure 2. Phenotypic distribution in the 115 BC2F6 RIL population by year and by maturity set. (**A**) Phenotypic distribution in 2009 and 2010 (n = 115). Yield was measured only in 2009. (**B**) Phenotypic distribution at Bellflower, Stone, Urbana, and Villa Grove field locations. (**C**) Phenotypic distribution in the early maturity group (Matset1; $n = 64$) and the late maturity group (Matset2; $n = 51$). The box indicates the interquartile range, and black circles represent outliers. The black horizontal line indicates the distribution median and the cross indicates the distribution mean. The red horizontal line represents the mean of Williams 82. P-values were obtained using a Student's T-test. Flowering shows the time when 50% of the plants have at least one flower on any node (R1), and maturity shows when 95% of the seed pods reach full mature color (R8). Lodging is shown in 1–10 scales where 1 represents Williams 82 and 10 reflects lodging in the *G. soja* parent. Vining is shown in 1–5 scales where 1 is no vining as in Williams 82 and 5 is vining represented by the *G. soja* parent.

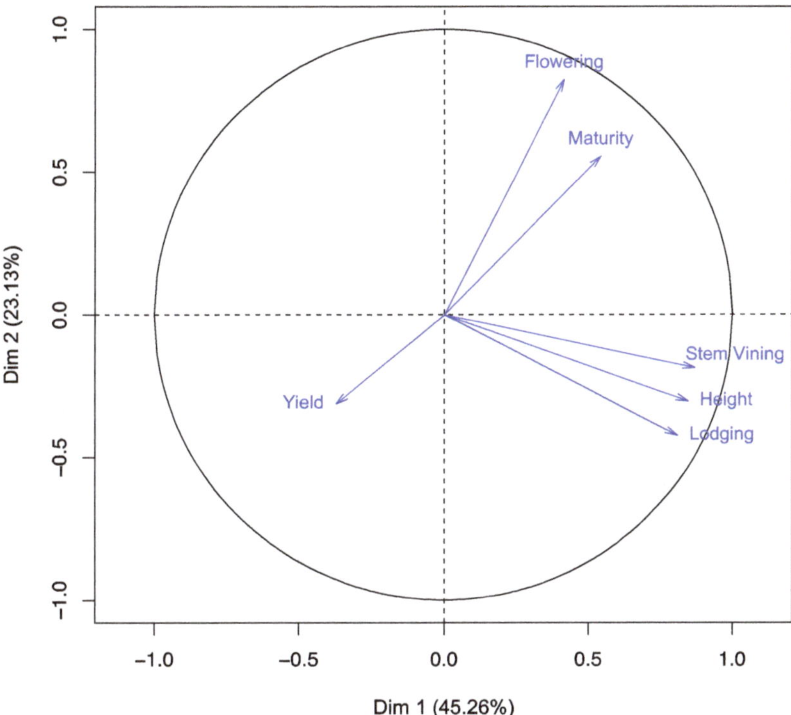

Figure 3. Multiple factor analysis (MFA) of phenotypic variation and correlation. The two-dimensional correlation structure among the 115 BC2F6 RILs is demonstrated by MFA. Dimension 1 is represented by the horizontal axis and Dimension 2 is represented by the vertical axis. The blue arrows represent the strength of trait association within a dimension. Dimension 1 accounts for 45.26% of the total variation, and Dimension 2 accounts for 23.13% of the total variation.

Table 1. Eigen vectors of the first three dimensions of variance for each trait identified in multiple factor analysis (MFA) of phenotypic variation and correlation (Figure 3).

Trait	Dimension 1 (45.3%)	Dimension 2 (23.1%)	Dimension 3 (16.7%)
Flowering	0.414	0.822	0.019
Maturity	0.541	0.555	0.500
Height	0.842	−0.302	0.167
Lodging	0.806	−0.423	0.018
Yield	−0.372	−0.312	0.838
Stem Vining	0.866	−0.186	−0.141

3.4. Construction of Linkage Map

Genetic linkage maps were constructed for this mapping population, resulting in 30 linkage groups with 335 SNP markers spanning 846 cM (Figure S2). Marker locations were generally consistent with the Universal Soy Linkage Panel 1.0 [29]; however, we found that chromosomes 1, 5, 6, 7, 12, 13, 14, and 16 were divided into two or three linkage groups in our maps. This is likely caused by the two backcrosses to Williams 82 and intense selection, limiting the number of SNP markers segregating in the population. On average, each linkage group contained 12 markers with a mean distance of 3.6

cM between the nearest markers. Extensive segregation distortion was observed, resulting in *G. soja* alleles being substantially underrepresented for many markers in our population. Each RIL possessed approximately 3–20% of SNP markers homozygous for *G. soja* alleles (Table S1). Linkage group 7a contained the highest percentage of markers homozygous for *G. soja* alleles at 64%, and linkage group 12a contained the least at 1.3%.

3.5. QTL Mapping

Composite interval mapping (CIM) analysis was carried out on both the data in its entirety and separately for each year (2009 and 2010), location (Urbana 2009, Stonington 2009, Urbana 2010, Bellflower 2010 and Villa Grove 2010), subpopulation, (Subpopulation 1 and Subpopulation 2) and maturity set (Maturity set 1 and Maturity set 2). In total, our analysis has uncovered 19 QTL distributed throughout 6 traits (Figure 4; Table 2). One QTL was found for flowering, two were found for maturity, two were found for height, two were found for yield, seven were found for lodging, and five were found for stem vining. Most of these QTL were found in specific locations or subpopulations.

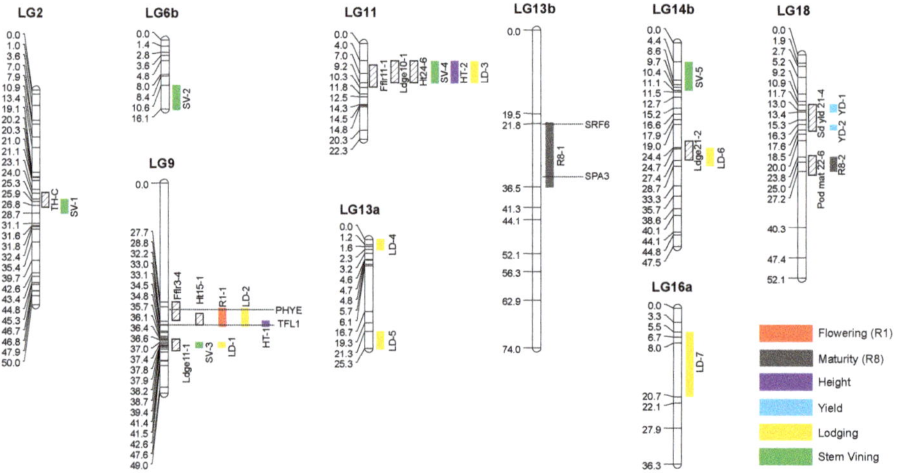

Figure 4. Quantitative trait loci (QTL) identified in this study. Nine linkage groups harbor QTL for flowering (R1), maturity (R8), height (HT), yield (YD), lodging (LD), and stem vining (SV). QTL for R1 are shown in red, while R8 are in gray, HT are in purple, YD are in light blue, LD are in yellow, and SV are in green. Previously reported QTL that overlap with the QTL identified in this study are represented by white boxes with diagonal lines, and the locations of candidate flowering genes are displayed as horizontal hashed lines. The QTL box represents a one-LOD support interval.

The QTL R1-1 on chromosome 9 was significantly associated with flowering (R1). This QTL explains 26% of the flowering variation observed and is centered on the SNP marker BARC-056323-14257 with an LOD score of 4.10 and an additive effect of 6.41. Despite the late flowering parent lines BC1F5 LG01-7909 and LG01-7919 that were selected for the second backcrossing, this result indicates that the *G. soja* allele confers earlier flowering than the *G. max* allele, which is potentially due to epistatic interactions with other loci in the genome.

Table 2. Peak markers that exhibit the highest association for each quantitative trait locus (QTL) identified in this study. Year/location indicate the datasets in which the QTL is identified. The direction of additive effects indicates the effects of *G. max* alleles. Units used for phenotype measurement are flowering and maturity in days, height in cm, seed yield in kg ha^{-1}, lodging on a scale of 1 (all plants are erect) to 10 (all plants are prostrate), and a stem vininess on a scale of 1 (typical indeterminate cultivar) to 5 (viney stem termination).

Trait	QTL	Chromosome	Year/Location	Peak Marker	Position (cM)	Position (Mbp)	LOD	R^2	Additive Effect
Flowering									
	R1-1	9	matset 2	BARC-056323-14257	32.24	20.99	4.10	0.26	6.41
Maturity									
	R8-1	13	pop2(7919)	BARC-041649-08056	21.76	33.28	5.88	0.32	−1.31
	R8-2	18	matset2	BARC-015067-02556	25.04	2.15	3.14	0.17	−1.18
Height									
	HT-1	9	All years	BARC-056323-14257	32.24	20.99	3.15	0.10	8.29
		9	2009	BARC-056323-14257	32.24	20.99	3.15	0.10	8.29
	HT-2	11	pop1(7909)	BARC-040851-07854	7.00	7.61	5.48	0.26	13.87
		11	Urbana09	BARC-040851-07854	7.00	7.61	4.92	0.15	15.13
Yield									
	YD-1	18	Stone09	BARC-014395-01348	13.02	3.44	4.71	0.15	3.83
	YD-2	18	Stone09	BARC-060195-16470	16.32	0.47	3.77	0.13	3.60
Lodging									
	LD-1	9	pop1(7909)	BARC-028249-05804	38.15	7.47	3.18	0.16	1.08
		9	Urbana09	BARC-028249-05804	38.15	7.47	3.60	0.13	1.39
	LD-2	9	Urbana10	BARC-056323-14257	32.24	20.99	3.59	0.11	0.66
	LD-3	11	2010	BARC-040851-07854	6.98	7.61	5.36	0.17	0.94
	LD-4	13	matset2	BARC-062009-17616	0.00	19.33	3.64	0.16	−0.52
	LD-5	13	matset2	BARC-065851-19789	25.32	2.44	3.72	0.18	0.51
	LD-6	14	matset2	BARC-017933-02457	24.71	46.13	3.39	0.14	−0.43
	LD-7	16	pop2(7919)	BARC-054099-12340	6.74	3.56	3.18	0.20	0.43
Stem vining									
	SV-1	2	Matset2	BARC-021647-04164	25.85	46.04	4.54	0.23	1.36
	SV-2	6	VG10	BARC-025179-06455	10.62	50.32	4.37	0.16	0.97
	SV-3	9	All years	BARC-028249-05804	38.15	7.47	2.90	0.09	1.02
		9	2009	BARC-028249-05804	38.15	7.47	2.90	0.09	1.02
	SV-4	11	pop1(7909)	BARC-040851-07854	6.98	7.61	4.96	0.24	1.09
		11	Urbana09	BARC-040851-07854	6.98	7.61	4.38	0.14	0.99
		11	Urbana10	BARC-040851-07854	6.98	7.61	4.21	0.13	0.92
		11	2010	BARC-040851-07854	6.98	7.61	3.34	0.10	0.85
	SV-5	14	Matset2	BARC-052759-11611	4.40	10.01	3.05	0.14	−0.42

Two QTL, R8-1 and R8-2, were associated with maturity (R8) and located near the markers BARC-041649-08056 and BARC-015067-02556 on chromosomes 13 and 18, respectively. R8-1 explains 32% of the variation observed with an LOD score of 5.88. R8-2 explains 17% of the variation observed with an LOD score of 3.14. The additive effects of R8-1 and R8-2 are −1.31 and −1.18, respectively, with the *G. soja* allele conferring faintly later maturation than the *G. max* allele.

Two QTL, HT-1 and HT-2, were associated with height and located near the markers BARC-056323-14257 and BARC-040851-07854 on chromosomes 9 and 11, respectively. HT-1 explains 10% of the variation observed with an LOD score of 3.15. HT-2 explains between 15% and 26% of the variation observed with a LOD score between 4.92 and 5.48. The additive effects of HT-1 and HT-2 are above 8.29 with both *G. soja* alleles conferring a large reduction in height.

Two QTL, YD-1 and YD-2, were associated with yield and located near BARC-014395-01348 and BARC-060195-16470, respectively, both on chromosome 18. YD-1 explains 15% of the variation observed with an LOD score of 4.71. YD-2 explains 13% of the variation observed with an LOD score of 3.77. The additive effects for YD-1 and YD-2 are 3.83 and 3.60, respectively, with the *G. soja* allele conferring a reduction in yield.

Seven QTL were associated with lodging, locating on chromosomes 9, 11, 13, 14, and 16. R^2 values ranged from 13% to 20%, with LD-7 explaining the highest percentage of the variation at 20%. LOD scores spanned between 3.18 and 5.36, with LD-3 showing the strongest peak. The additive effects of QTL associated with lodging range between −0.52 and 1.39, with the majority of *G. soja* allele conferring a decline in lodging. LD-4 and LD-6 are the exceptions to this trend.

Five QTL were associated with stem vining, locating on chromosomes 2, 6, 9, 11, and 14. R^2 values ranged from 9% to 24% variation explained, with SV-1 showing the highest value. LOD scores were spread between 2.90 and 4.96, with SV-5 showing the strongest peak. The additive effects of QTL affecting stem vining ranged from −0.42 to 1.36. With the exception of SV-5, all *G. soja* alleles confer a reduction in stem vining.

3.6. Pleiotropic QTL and Verification

Unlike the pleiotropic *E* loci, the flowering QTL R1-1 identified in this study did not overlap with the maturity QTL R8-1 and R8-2, nor with previously reported maturity QTL (Figure 4; Table 2). However, the R1-1 locus overlapped with the QTL controlling other agronomic traits: HT-1 and LD-2, co-localizing on chromosome 9. We found another cluster of QTL controlling multiple traits: HT-2, LD-3, and SV-4 on chromosome 11.

To verify the QTL identified in this study, previously reported QTL listed in SoyBase (http://soybase. org/) were compared with the locations of our QTL. QTL for flowering (R1-1), height (HT-1 and HT-2), and yield (YD-1 and YD-2) identified in this study, as well as one of maturity QTL (R8-2) and 3 of lodging QTL (LD-1, LD-3 and LD-6), overlapped with previously reported QTL (Figure 4; Table 3). The stem vining QTL SV-1 overlaps with the QTL TH-C for twinning habit [35], but none of the other stem vining QTL has been previously reported. To verify our QTL controlling flowering or maturity, we compared their locations with those of known flowering gene homologs [8]. Two genes homologous to *Phytochrome E* (*PHYE*) and *TFL1/FT* that are known to play a role in flowering control in Arabidopsis are located within the close range of the flowering QTL R1-1 (Figure 4; Table 4). In addition, one of the maturity QTL, R8-1, appears to be in the vicinity of homologs of the flowering-related genes *SRF6* and *SPA3*.

To provide further verification for the flowering QTL R1-1, the RIL LG08-7379 carrying the *G. soja* allele of the peak marker BARC-056323-14257 was backcrossed with Williams 82. In the F2 population, the plants carrying the homozygous *G. soja* allele at the marker flowered earlier than the plants carrying the homozygous *G. max* allele by 5.8 days ($p < 0.01$), while the maturation time of these plants was statistically indifferent (Figure 5).

Table 3. Previously reported QTL (http://soybase.org/) that overlap with the QTL identified in this study.

Trait	QTL	Chromosome	Peak Marker	Previously Identified QTL
R1	R1-1	9	BARC-056323-14257	Fflr 3-4
R8	R8-2	18	BARC-015067-02556	Pod mat 22-6
Height	HT-1	9	BARC-056323-14257	Pl ht 15-1
	HT-2	11	BARC-040851-07854	Pl ht 24-6
Yield	YD-1	18	BARC-014395-01348	Sd yld 21-4
	YD-2	18	BARC-060195-16470	Sd yld 21-4
Lodging	LD-1	9	BARC-028249-05804	Ldge 11-1
	LD-3	11	BARC-040851-07854	Ldge 10-1
	LD-6	14	BARC-017933-02457	Ldge 21-2
Stem vining	SV-1	2	BARC-021647-04164	TH-C

Table 4. Flowering-related genes in the vicinity of the QTL identified in this study.

QTL	Chromosome	Year/Location	Peak Marker	Gene	ID Number	Start Position
R8-1	13	pop2(7919)	BARC-041649-08056	SPA3	Glyma.13g276700	37212524
				SRF6	Glyma.13g241100	34503402
R1-1	9	matset 2	BARC-056323-14257	PHYE	Glyma.09g088500	11657930
				TFL1	Glyma.09g143500	36559409

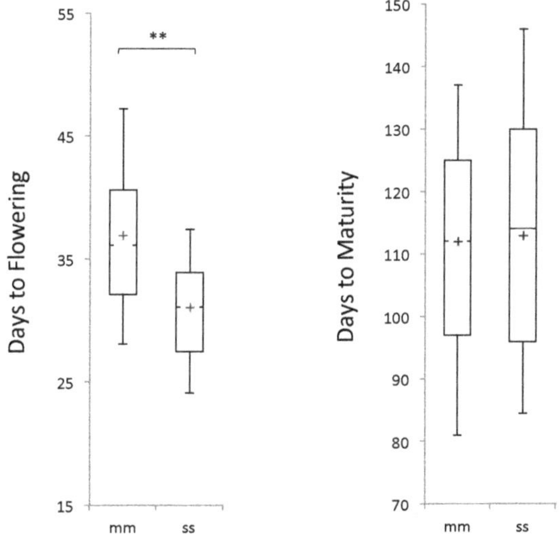

Figure 5. Flowering (R1) and maturation (R8) of the QTL R1-1 in the BC3F2 population. The BC2F6 RIL LG08-7379 carrying the *G. soja* allele of the QTL R1-1 peak marker BARC-056323-14257 was backcrossed with Williams 82. Twenty-eight F2 plants carrying homozygous *G. max* alleles (mm) and 30 F2 plants carrying homozygous *G. soja* alleles (ss) were grown in a greenhouse under 12 h light at the average highest day-time temperature 25 °C. The box indicates the interquartile range, the black horizontal line indicates the distribution median, and the cross indicates the distribution mean. ** $p < 0.01$.

4. Discussion

Pleiotropy is considered to have significant implications for multi-trait evolution. In this study, we focused on the key domestication-related traits flowering and maturity that were known to behave in a highly correlated manner in domesticated soybean. Our selection experiment for non-pleiotropy resulted in a different evolutionary trajectory from that of soybean domestication. Our QTL mapping analysis using the experimental populations did not identify the *E* loci that were selected during soybean domestication and improvement, but it did identify novel loci affecting either flowering or maturity in a non-pleiotropic manner. This result suggests that artificial selection during the soybean domestication and diversification processes favored pleiotropic loci over non-pleiotropic loci in the evolution of flowering and maturity control, despite the fact that genetic variations in the wild progenitor that allow diverse patterns of life history are available.

In addition to flowering and maturity, pleiotropy likely played an important role in the evolution of multiple agronomic traits. A previous study reported a QTL on chromosome 11 in a mapping population derived from *G. max* parents that affected flowering, maturity, height, and lodging, and as well as another QTL on chromosome 6 that affected flowering, height, and lodging [36]. The latter locus maps near the known major maturity locus *E1*. The *E1* locus has also been identified in multiple mapping populations derived from a cross between *G. max* and *G. soja* parents, and it affects several agronomic traits [35,37]. Our QTL mapping did not identify this locus, which was likely due to the selection for non-pleiotropy between flowering and maturity. Indeed, we found zero lines in Subpopulation 2 and only 2 homozygous and 1 heterozygous lines in Subpopulation 1 carrying the *G. soja E1* allele (Table S2), suggesting that our selection scheme effectively purged pleiotropic loci such as *E1* from the population. In addition, segregation distortion may have played a role in removing certain *G. soja* alleles, including *E1*, from the population. Multiple studies have reported segregation distortion in mapping populations derived from a cross between *G. max* and *G. soja* that results in disproportionally low percentages of *G. soja* alleles at heterologous segregating loci [35]. The causes of segregation distortion are yet to be clarified, but in our selection scheme, it is possible that selection for agronomic appearance while advancing generations through single seed descent, as well as selection for a narrow maturity range at the BC2F5 stage, may have skewed the proportion of *G. soja* alleles. Despite the strict selection scheme for agronomic appearance and segregation distortion, our QTL mapping found two genomic locations controlling multiple agronomic traits (Figure 4), suggesting the prevalence of pleiotropy among loci controlling agronomic traits. It has been reported that artificial selection during domestication causes rapid evolutionary changes [38]. Selection for prevalent pleiotropic loci may have accelerated the evolution of cultivated soybean during the domestication and diversification processes.

While our selection experiment successfully identified non-pleiotropic novel loci, it allowed only a small fraction of genetic variation subjected to the QTL mapping analysis due to our selection scheme and the biparental nature of our mapping population. In addition, the relatively small size of our mapping population may have limited the power of QTL mapping. Future experiments to follow-up this work include parallel evolutionary experiments under different selection regimes using expanded mapping populations created from diverse parental varieties in *G. max* and *G. soja*. The use of a larger mapping population, as well as a gentler selection scheme for agronomic appearance would also help maintain the level of *G. soja* alleles in a mapping population.

Another limitation of our experiment is that it does not assess the magnitude of factors other than pleiotropy that account for phenotypic evolution, nor their interaction with pleiotropy. For example, artificial selection toward pleiotropic loci may be in part a result of widespread epistatic interactions in the genome, which are shown to have a prominent effect on narrowing the evolutionary path [39]. Moreover, the possibility in which pleiotropy was favored as an indirect result of selection for gene-by-environment effects that eliminates alleles sensitive to environmental variants cannot be ruled out, as genetic loci that exhibit phenotypic robustness to diverse environments are shown to have a tendency to affect multiple traits [38]. Indeed, several QTL identified in this study appeared under specific years or locations.

5. Conclusions

This study represents one of the first attempts to experimentally dissect the roles of pleiotropy in the multi-trait evolution of crops. Our work demonstrates that pleiotropy in flowering and maturity can be genetically separated. Non-pleiotropic loci identified in this work provide novel genetic resources for soybean breeding programs and ultimately allow the diversification of soybean's growth habit and life history optimized for regional environments and different cultivation strategies.

Supplementary Materials: The following are available online at http://www.mdpi.com/2073-4395/10/8/1204/s1. Figure S1: Phenotype correlation between different field sites; Figure S2: Complete linkage map; Table S1: SNP alleles in the mapping population; Table S2: E1 alleles in the mapping population.

Author Contributions: Y.H. designed research. E.J.S., A.A., A.V., S.A. and A.K. carried out experiments and analyzed the data. E.J.S. and Y.H. wrote the manuscript. All authors have read and agreed to the published version of the manuscript.

Funding: This work was supported by the Plant Genome Research Program from the National Science Foundation (NSF-IOS-PGRP-1823145) to Y.H.

Acknowledgments: We thank Randall L. Nelson (USDA-ARS, Urbana) for providing with the mapping population.

Conflicts of Interest: The authors declare no conflict of interest. The funders had no role in the design of the study; in the collection, analyses, or interpretation of data; in the writing of the manuscript, or in the decision to publish the results.

References

1. Kimura, M.; Ohta, T. On some principles governing molecular evolution. *Proc. Natl. Acad. Sci. USA* **1974**, *71*, 2848–2852. [CrossRef]

2. Ohta, T.; Gillespie, J.H. Development of neutral and nearly neutral theories. *Theor. Popul. Biol.* **1996**, *49*, 128–142. [CrossRef]

3. Wagner, G.P.; Kenney-Hunt, J.P.; Pavlicev, M.; Peck, J.R.; Waxman, D.; Cheverud, J.M. Pleiotropic scaling of gene effects and the 'cost of complexity. *Nature* **2008**, *27*, 470–472. [CrossRef]

4. Wang, Z.; Liao, B.-Y.; Zhang, J. Genomic patterns of pleiotropy and the evolution of complexity. *Proc. Natl. Acad. Sci. USA* **2010**, *107*, 18034–18039. [CrossRef] [PubMed]

5. Hammer, K. Das domestikationssyndrom. *Die Kult.* **1984**, *32*, 11–34. (In German) [CrossRef]

6. Carter, T., Jr.; Hymowitz, T.; Nelson, R. Biogeography, local adaptation, Vavilov, and genetic diversity in soybean. In *Biological Resources and Migration*; Springer: Berlin/Heidelberg, Germany, 2004; pp. 47–59.

7. Kim, M.Y.; Van, K.; Kang, Y.J.; Kim, K.H.; Lee, S.-H. Tracing soybean domestication history: From nucleotide to genome. *Breed. Sci.* **2012**, *61*, 445–452. [CrossRef] [PubMed]

8. Sedivy, E.J.; Wu, F.; Hanzawa, Y. Soybean domestication, the origin, genetic architecture and molecular bases. *New Phytol.* **2017**, *214*, 539–553. [CrossRef] [PubMed]

9. Dong, Y.; Yang, X.; Liu, J.; Wang, B.-H.; Liu, B.-L.; Wang, Y.-Z. Pod shattering resistance associated with domestication is mediated by a NAC gene in soybean. *Nat. Commun.* **2014**, *5*, 1–11. [CrossRef] [PubMed]

10. Funatsuki, H.; Suzuki, M.; Hirose, A.; Inaba, H.; Yamada, T.; Hajika, M.; Komatsu, K.; Katayama, T.; Sayama, T.; Ishimoto, M.; et al. Molecular basis of a shattering resistance boosting global dissemination of soybean. *Proc. Natl. Acad. Sci. USA* **2014**, *111*, 17797–17802. [CrossRef] [PubMed]

11. Sun, L.; Miao, Z.; Cai, C.; Zhang, D.; Zhao, M.; Wu, Y.; Zhang, X.; Swarm, S.A.; Zhou, L.; Zhang, Z.J.; et al. GmHs1-1, encoding a calcineurin-like protein, controls hard-seededness in soybean. *Nat. Genet.* **2015**, *47*, 939–943. [CrossRef] [PubMed]

12. Wu, F.; Sedivy, E.J.; Price, W.B.; Haider, W.; Hanzawa, Y. Evolutionary trajectories of duplicated FT homologues and their roles in soybean domestication. *Plant J.* **2017**, *90*, 941–953. [CrossRef] [PubMed]

13. Cober, E.R.; Morrison, M.J. Regulation of seed yield and agronomic characters by photoperiod sensitivity and growth habit genes in soybean. *Theor. Appl. Genet.* **2010**, *120*, 1005–1012. [CrossRef] [PubMed]

14. Kim, M.Y.; Shin, J.H.; Kang, Y.J.; Shim, S.R.; Lee, S.-H. Divergence of flowering genes in soybean. *J. Biosci.* **2012**, *37*, 857–870. [CrossRef] [PubMed]

15. Cao, D.; Takeshima, R.; Zhao, C.; Liu, B.; Jun, A.; Kong, F. Molecular mechanisms of flowering under long days and stem growth habit in soybean. *J. Exp. Bot.* **2017**, *68*, 1873–1884. [CrossRef] [PubMed]

16. Samanfar, B.; Molnar, S.J.; Charette, M.; Schoenrock, A.; Dehne, F.; Golshani, A. Mapping and identification of a potential candidate gene for a novel maturity locus, E10, in soybean. *Theor. Appl. Genet.* **2017**, *130*, 377–390. [CrossRef] [PubMed]

17. Xia, Z.; Watanabe, S.; Yamada, T.; Tsubokura, Y.; Nakashima, H.; Zhai, H.; Anai, T.; Sato, S.; Yamazaki, T.; Lu, S.; et al. Positional cloning and characterization reveal the molecular basis for soybean maturity locus E1 that regulates photoperiodic flowering. *Proc. Natl. Acad. Sci. USA* **2012**, *109*, E2155–E2164. [CrossRef]

18. Watanabe, S.; Xia, Z.; Hideshima, R.; Tsubokura, Y.; Sato, S.; Yamanaka, N.; Takahashi, R.; Anai, T.; Tabata, S.; Kitamura, K. A map-based cloning strategy employing a residual heterozygous line reveals that the GIGANTEA gene is involved in soybean maturity and flowering. *Genetics* **2011**, *188*, 395–407. [CrossRef]

19. Liu, B.; Kanazawa, A.; Matsumura, H.; Takahashi, R.; Harada, K.; Abe, J. Genetic redundancy in soybean photoresponses associated with duplication of the phytochrome A gene. *Genetics* **2008**, *180*, 995–1007. [CrossRef]

20. Watanabe, S.; Hideshima, R.; Xia, Z.; Tsubokura, Y.; Sato, S.; Nakamoto, Y.; Yamanaka, N.; Takahashi, R.; Ishimoto, M.; Anai, T. Map-based cloning of the gene associated with the soybean maturity locus E3. *Genetics* **2009**, *182*, 1251–1262. [CrossRef]

21. Zhao, C.; Takeshima, R.; Zhu, J.; Xu, M.; Sato, M.; Watanabe, S.; Kanazawa, A.; Liu, B.; Kong, F.; Yamada, T.; et al. A recessive allele for delayed flowering at the soybean maturity locus E9 is a leaky allele of FT2a, a FLOWERING LOCUS T ortholog. *BMC Plant Biol.* **2016**, *16*, 20. [CrossRef]

22. Lu, S.; Zhao, X.; Hu, Y.; Liu, S.; Nan, H.; Li, X.; Fang, C.; Cao, D.; Shi, X.; Kong, L.; et al. Natural variation at the soybean J locus improves adaptation to the tropics and enhances yield. *Nat. Genet.* **2017**, *49*, 773–779. [CrossRef] [PubMed]

23. Nelson, R.L. Response to Selection for Time of Flowering in Soybean. *Crop Sci.* **1988**, 623–626. [CrossRef]

24. Smith, J.R.; Nelson, R.L. Selection for Seed-Filling Period in Soybean. *Crop Sci.* **1986**, *26*, 466–469. [CrossRef]

25. Fehr, W.R.; Caviness, C.E.; Burmood, D.T.; Pennington, J.S. Stage of development descriptions for soybeans, Glycine max (L.) Merrill. *Crop Sci.* **1971**, *11*, 929–931. [CrossRef]

26. Revelle, W. An Overview of the Psych Package. Department of Psychology, Northwestern University. 2017. Available online: http://personality-project.org/r/overview.pdf (accessed on 1 July 2017).

27. Pagès, J. Multiple Factor Analysis, Main Features and Application to Sensory Data. Agrocampus Ouest, Rennes, France. Available online: http://factominer.free.fr/more/PagesAFM.pdf (accessed on 1 April 2016).

28. Hyten, D.L.; Cannon, S.B.; Song, Q.; Weeks, N.; Fickus, E.W.; Shoemaker, R.C.; Specht, J.E.; Farmer, A.D.; May, G.D.; Cregan, P.B. High-throughput SNP discovery through deep resequencing of a reduced representation library to anchor and orient scaffolds in the soybean whole genome sequence. *BMC Genom.* **2010**, *11*, 38. [CrossRef]

29. Hyten, D.L.; Song, Q.; Choi, I.Y.; Yoon, M.S.; Specht, J.E.; Matukumalli, L.K.; Nelson, R.L.; Shoemaker, R.C.; Young, N.D.; Cregan, P.B. High-throughput genotyping with the GoldenGate assay in the complex genome of soybean. *Theor. Appl. Genet.* **2008**, *116*, 945–952. [CrossRef]

30. Wu, F.; Kang, X.; Wang, M.; Haider, W.; Price, W.B.; Hajek, B.; Hanzawa, Y. Transcriptome-enabled network inference revealed the GmCOL1 feed-forward look and its roles in photoperiodic flowering of soybean. *Front. Plant Sci.* **2019**. [CrossRef]

31. Wang, S.C.; Basten, C.J.; Zeng, Z.-B. Windows QTL Cartographer 2.5. Department of Statistics, North Carolina State University, Raleigh, NC. 2012. Available online: http://statgen.ncsu.edu/qtlcart/WQTLCart.htm (accessed on 1 April 2016).

32. Li, Z.; Nelson, R.L. RAPD Marker Diversity among Cultivated and Wild Soybean Accessions from Four Chinese Provinces. *Crop Sci.* **2002**, *42*, 1737–1744. [CrossRef]

33. McBlain, B.A.; Bernard, R.L. A new gene affecting the time of flowering and maturity in soybeans. *J. Hered.* **1987**, *78*, 160–162. [CrossRef]

34. Buzzell, R.V.H. Inheritance of insensitivity to long day length. *Soybean Genet. Newsl.* **1980**, *7*, 26–29.

35. Liu, B.; Fujita, T.; Yan, Z.H.; Sakamoto, S.; Xu, D.; Abe, J. QTL mapping of domestication-related traits in soybean (Glycine max). *Ann. Bot.* **2007**, *100*, 1027–1038. [CrossRef] [PubMed]

36. Zhang, W.K.; Wang, Y.J.; Luo, G.Z.; Zhang, J.S.; He, C.Y.; Wu, X.L.; Gai, J.Y.; Chen, S.Y. QTL mapping of ten agronomic traits on the soybean (Glycine max, L. Merr.) genetic map and their association with EST markers. *Theor. Appl. Genet.* **2004**, *108*, 1131–1139. [CrossRef] [PubMed]

37. Guzman, P.S.; Diers, B.W.; Neece, D.J.; St Martin, S.K.; LeRoy, A.R.; Grau, C.R.; Hughes, T.J.; Nelson, R.L. QTL associated with yield in three backcross-derived populations of soybean. *Crop Sci.* **2007**, *12*, 111–122. [CrossRef]

38. Doust, A.N.; Lukens, L.; Olsen, K.M.; Mauro-Herrera, M.; Meyer, A.; Rogers, K. Beyond the single gene, How epistasis and gene-by-environment effects influence crop domestication. *Proc. Natl. Acad. Sci. USA* **2014**, *111*, 6178–6183. [CrossRef] [PubMed]

39. Kryazhimskiy, S.; Rice, D.P.; Jerison, E.R.; Desai, M.M. Global epistasis makes adaptation predictable despite sequence-level stochasticity. *Science* **2014**, *344*, 1519–1522. [CrossRef]

 *agronomy*

Article

Integration of Extra-Large-Seeded and Double-Podded Traits in Chickpea (*Cicer arietinum* L.)

Kamile Gul Kivrak [1], Tuba Eker [1], Hatice Sari [1], Duygu Sari [1], Kadir Akan [2], Bilal Aydinoglu [1], Mursel Catal [3] and Cengiz Toker [1,*]

[1] Department of Field Crops, Faculty of Agriculture, Akdeniz University, 07070 Antalya, Turkey; kamilegul92@gmail.com (K.G.K.); ekertuba07@gmail.com (T.E.); haticesari@akdeniz.edu.tr (H.S.); duygusari@akdeniz.edu.tr (D.S.); aydinoglu@akdeniz.edu.tr (B.A.)

[2] Department of Plant Protection, Faculty of Agriculture, Kirsehir Ahi Evran University, 40200 Kirsehir, Turkey; kadir_akan@hotmail.com

[3] Department of Plant Protection, Faculty of Agriculture, Akdeniz University, 07070 Antalya, Turkey; mcatal@akdeniz.edu.tr

* Correspondence: toker@akdeniz.edu.tr; Tel.: +90-242-310-24-21 or +90-537-543-10-37

Received: 1 June 2020; Accepted: 22 June 2020; Published: 24 June 2020

Abstract: A large seed size in the kabuli chickpea (*Cicer arietinum* L.) is important in the market not only due to its high price but also for its superior seedling vigor. The double-podded chickpea has a considerable yield and stability advantage over the single-podded chickpea. The study aimed at (i) integrating extra-large-seeded and double-podded traits in the kabuli chickpea, (ii) increasing variation by transgressive segregations and (iii) estimating the heritability of the 100-seed weight along with important agro-morphological traits in F_2 and F_3 populations. For these objectives, the large-seeded chickpea, Sierra, having a single pod and unifoliolate leaves, was crossed with the small-seeded CA 2969, having double pods and imparipinnate leaves. The inheritance pattern of the extra-large-seeded trait was polygenically controlled by partial dominant alleles. Transgressive segregations were found for all agro-morphological traits. Some progeny with 100-seed weights of ≥55 g and two pods had larger seed sizes than those of the best parents. As outputs of the epistatic effect of the double-podded gene in certain genetic backgrounds, three or more flowers or pods were found in some progeny. Progeny having imparipinnate leaves or two or more pods should be considered in breeding, since they had higher numbers of pods and seeds per plant and seed yields than their counterparts.

Keywords: *Cicer arietinum*; intraspecific crosses; transgressive segregations; large seed; double pod

---

1. Introduction

The domesticated chickpea, *Cicer arietinum* L., is not only an important food legume but is also one of the most important crops on the basis of drought resistance [1–3]. It is used as a cash crop in trade, having a high level of protein in its seeds, a rotation crop due to its ability to fix atmospheric nitrogen to soil, and a cover crop in sustainable agriculture [4–6]. Globally, it has the widest sowing area among cool season food legumes, with a 17.8 million ha cultivation area in 2018 [7].

The domesticated chickpea is well-defined in two classes as "*macrosperma*" or "kabuli" and "*microsperma*" or "desi" according to the pigmentation of the plant, the flower and the seed size, shape and color. The plants in the former class do not possess pigment on the vegetative green parts and flowers, and these plants generally produce larger cream and whitish-cream seeds. On the other hand, the plants of the latter class possess pigments with purple-pink and pinkish-blue flowers, and these generally produce smaller seeds with an angular and rough appearance with different colors including brown, black and green [8,9]. Seed size and color are important criteria in the market [10]. The seed size in chickpeas is referred to as the single seed weight [11,12], 100-seed weight [13] and scale of the sieve size in mm [14,15]. In "kabuli" chickpeas, three distinct seed shapes have been recognized—ram-headed, owl-headed and pea-shaped [16]—and chickpeas with ram-head-shaped seeds generally have the largest seed size, as high as 60 g $\geq$ per 100-seed weight [17,18]. Seed size in chickpeas is not only governed by genetic factors but also affected by environment [9,12,19], generally having a high heritability [9,20]. Farmers (producers) prefer to produce large-seeded chickpeas due to consumer preference, since a larger seed size commands a higher price in regional and international markets [21–23]. In addition to a higher price, a large seed size confers an advantage during germination, higher seedling vigor, allowing deeper sowing than that of small seeds in order to escape drought [24]. Large seed size has therefore been considered to be a noteworthy trait in breeding programs [12]. To investigate the genetics of large seed size, several studies have been carried out from the 1950s to date [13,25–29]. The inheritance of seed size was determined as monogenic, digenic and polygenic [12,13,29–33]. Seed size in chickpeas was mapped using recombinant inbred lines (RILs) and some quantitative trait loci (QTL), and candidate genes were located in LG1, LG2, LG4, LG5, LG6, LG7 and LG8 [32,34–37].

Like large seed size, the double-podded trait in chickpea is a significant trait for increasing yield [20,38–41] and seed yield stability [42,43]. It was first described in a mutant desi chickpea in the 1930s and was determined to be governed by a single recessive gene "*s*" or "*sfl*" [9,44]. QTLs were identified on LG 4 and LG 6 [45–49]. For the integration of the gene conferring the double-podded trait, interspecific and intraspecific crosses in chickpeas were made [40,42,43,48,49] and transgressive segregations were reported for quantitative traits [20,41,50]. Progeny in segregated populations having higher or lower values than their parents are transgressive segregations [51]. Some example studies on the "kabuli" chickpea have not only been conducted to increase seed size [52,53] but also integrated with resistance to ascochyta blight, caused by *Ascochyta rabiei* (Pass.) Labr. [54]. Thus, some improved cultivars have been described [55–62]. None of these studies have reported the integration of a large seed size and the double-podded trait. Therefore, the present study aims (i) to integrate extra-large-seeded and double-podded traits in the "kabuli" chickpea, (ii) to increase variation by transgressive segregations in intraspecific crosses and (iii) to estimate the heritability of 100-seed weight along with important agro-morphological traits.

2. Materials and Methods

2.1. Parents in Intraspecific Crosses

The female parent Sierra (PI 631078) is an extra-large-seeded "kabuli" chickpea improved by USDA-ARS in cooperation with the Washington Agricultural Research Center, Pullman, WA based on a large seed size and resistance to ascochyta blight. Sierra was derived from F_8 of a three-way cross, "Dwelley"//FLIP 85-58C/Spanish or Mexican White, and released according to pedigree breeding in 2004. Its prominent plant traits were reported to be a plant height of 53 cm, branching at the base, a simple or unifoliolate leaf, a single flower per axil and a 100-seed weight of 61.4 g [55]. The pollen donor parent CA 2969 (PI632396) is also a "kabuli" chickpea developed by CIFA, Cordoba, Spain since it has a good resistance to ascochyta blight. It was selected from [CA 2156/JG 62 (PI 439821)]//ILC 3279 (PI 471915). CA 2156, JG 62 (ICC 5149) and ILC 3279 are large-seeded, double pods and resistant to ascochyta blight, respectively. CA 2969 possesses imparipinnate or fern-like (normal) leaves, double pods per axil and a seed weight of 30.1 g per 100 seeds [63]. The leaves and flowers of Sierra and CA 2969 are shown in Figure 1.

Figure 1. Morphological traits of parents, Sierra (single-podded and unifoliolate leaf, left side) and CA 2969 (double-podded and imparipinnate leaf, right side).

As reported by Auckland and van der Maesen [64], flowers of the female Sierra were emasculated at early morning and then pollinations were done using flowers of the pollen donor CA 2969 within one hour at the campus of Akdeniz University, Antalya, Turkey (30°38′ E, 36°53′ N, 51 m above sea level). F_1, F_2 and F_3 plants were grown as single plant progeny and individually harvested in 2017, 2018 and 2019, respectively. The present study consists of F_1 progeny and F_2 and F_3 populations.

2.2. Agronomic Applications

Parents and progeny were planted in rows spaced 50 cm apart with a within-row plant spacing of 10 cm. The parent plants were grown as four replicates (about 40 plants), while F_1, F_2 and F_3 progeny were grown as progeny rows in the same field. F_1 progeny and the F_2 population were grown under rainfed conditions, while drip irrigation was used for the F_3 population. Weeds were controlled by hand during the seedling and before the flowering stages. No input such as fertilization was used.

2.3. Soil Properties

Soil in the experimental area was sampled between 0 and 30 cm and then analyzed to determine the experimental soil traits. Some plant nutrition elements were found to be at a sufficient level, while organic matter and nitrogen were determined to be at low levels. Like organic matter and nitrogen, iron and zinc levels were considered to be possibly deficient due to high pH. The soil texture was loam, with a $CaCO_3$ content of 26.5%, whereas the pH was high, at 7.69.

2.4. Weather Conditions

When the growing periods of F_1, F_2 and F_3 were considered, from February to July, the total precipitation in 2017, 2018 and 2019 was recorded as 405.7 mm, 545.1 mm and 653.1 mm, respectively. The extreme maximum temperatures during the flowering stage, when the F_1, F_2 and F_3 progeny were grown, were 33.9, 32.4 and 31.9 °C, respectively, whereas during the pod formation stage, they were 44.8, 38.9 and 39.7 °C, respectively (Table S1). Due to the sudden increase in extreme temperature (38.9 °C) during the pod setting (Table S1), a considerable number of progeny produced empty pods.

2.5. Data Collection

The days to first flowering and days to 50% flowering were recorded as phenological traits, whereas the plant height, first pod height, number of main stems, pods and seeds per plant, seed yield per plant, seed weight or seed size, number of pods per axil (as single or double pods) and leaf shape (as fern-like or unifoliolate leaves) were recorded as agro-morphological traits of each parent and progeny. The seed size in the present study as hereafter referred to is the 100-seed weight, determined by using following formula [13,33]:

100-seed weight (g) = [Total seed weight per plant (g)/Total number of seeds per plant] × 100

2.6. Data Analyses

All the agro-morphological data were analyzed to determine descriptive statistics using the MINITAB 17 software [65]. Transgressive segregations, the progeny with higher or lower values than those of their parents in the F_2 and F_3 populations, were determined using minimum and maximum values of the F_2 and F_3 populations. Besides, progeny in the F_2 and F_3 populations were divided into four classes to compare agro-morphological traits, as (i) imparipinnate leaf and single-podded, (ii) imparipinnate leaf and double-podded, (iii) unifoliolate leaf and sing-podded and (iv) unifoliolate leaf and double-podded.

Narrow-sense heritability (h^2) in the F_2 population was estimated using progeny–parent regression according to Poehlman and Sleper [66]:

$$h^2 = b,$$

$$b = \sum(X - \overline{X})(Y - \overline{Y})/\sum(X - \overline{X})^2,$$

where b is the regression coefficient, and X, Y, $\overline{X}$ and $\overline{Y}$ are the values and means for the progeny and parents, respectively.

The chi-square test (χ^2) [67] was used to test the goodness of fit to the expected 9:3:3:1 ratio of segregation in the F_2 populations:

$$\chi^2 = \frac{(O - E)^2}{E},$$

where O and E are the observed and expected values, respectively.

3. Results

3.1. F_1 Progeny and F_2 and F_3 Populations

In F_1, negative selection was applied for untrue hybrids according to dominant morphological traits such as a single pod per axil and imparipinnate leaf shape. Hence, progeny with unifoliolate leaves due to selfing such as the female Sierra were discarded from rows in 2017. The true F_1 had imparipinnate leaves and produced single flowers and pods per axil. Fifty-one F_1 progeny were produced from intraspecific crosses between Sierra and CA 2969 in 2017. Each F_1 progeny had about 18 seeds per plant. Thus, a total 915 progeny were sown in F_2 in 2018. F_1 plants were single-podded with imparipinnate leaves, while F_2 plants segregated as imparipinnate leaves and a single-podded, imparipinnate leaves and double-podded, unifoliolate leaves and single-podded, and unifoliolate leaves and double-podded. The segregation was found to be a good fit to a 9:3:3:1 ratio (Table 1).

Table 1. Inheritance of the leaf shape (imparipinnate vs. unifoliolate) and number of pods per axil (single vs. double) in F_2 population derived from intraspecific crosses between Sierra and CA 2969.

Phenotype of F_{1s}	F_2			χ^2	p
	Phenotype of F_2 Population	Observed	Expected		
	Imparipinnate leaf and single pod	475			
Imparipinnate leaf and single pod	Imparipinnate leaf and double pods	181	9:3:3:1	8.25	0.50–0.10
	Unifoliolate leaf and single pod	189			
	Unifoliolate leaf and double pods	70			

3.2. Transgressive Segregations

According to data analyses to determine descriptive statistics in the F_2 and F_3 populations, transgressive segregations were found for all the agro-morphological traits, including 100-seed weight (Tables 2 and 3). The minimum and maximum values of the days to the first flowering of the F_2 population were found to be 45 days and 66 days, respectively, whereas the days to 50% flowering in the F_2 population ranged from 45 days to 75 days, respectively (Table 2). The days to the first flowering and days to 50% flowering of Sierra and CA 2969 were 48 and 50 days and 50 and 52 days, respectively. The plant height of the F_2 population varied from 17 cm to 59 cm, whereas the plant heights of Sierra and CA 2969 were 41 cm and 34.3 cm, respectively. The mean first pod height in the F_2 population was 31.7 cm, whereas it was 21 cm for Sierra and 19.7 cm for CA 2969. The mean number of main stems per plant in the F_2 population was 2.3, while it was 2.7 in Sierra and 2.3 in CA 2969. The number of pods per plant in the F_2 population ranged from 1 to 25, whereas the number of pods per plant was 5.2 in Sierra and 3 in CA 2969. The number of seeds per plant in the F_2 population ranged from 1 to 24, whereas the means of this trait were 4.2 in Sierra and 11.7 in CA 2969. The seed yield per plant was recorded as 0.1–7 g in the F_2 population, but it was 1.8 in Sierra and 3.7 in CA 2969. As for the 100-seed weight, this ranged from 9.5 g to 69 g in the F_2 population, while the 100-seed weights of Sierra and CA 2969 were 49.9 g and 31 g, respectively (Table 2).

The minimum and maximum values of the days to the first flowering of the F_3 population were between 36 and 75 days, whereas the days to 50% flowering ranged from 38 days to 82 days (Table 3). The plant height varied from 19 cm to 68 cm, whereas the plant heights of Sierra and CA 2969 were 52.3 cm and 42.7 cm, respectively. The first pod height ranged from 13 cm to 46 cm, whereas it was 31.3 cm for Sierra and 33.3 cm for CA 2969. The number of main stems ranged from 1 to 6, whereas it was found to be 2.3 in both Sierra and CA 2969. The number of pods per plant varied from 1 to 254, whereas it was 36.7 in Sierra and 42.7 in CA 2969. The number of seeds per plant ranged from 1 to 267, whereas the means of this trait were 32.7 in Sierra and 50.7 in CA 2969. The seed yield per plant was between 0.1 and 79 g, but it was 15.2 g in Sierra and 13.9 g in CA 2969. As the seed size in the F_3 population, the 100-seed weight ranged from 7 g to 64 g, while it was 46.9 g in Sierra and 27.4 g in CA 2969 (Table 3).

Table 2. Means ± standard errors, ranges and narrow sense heritability for agro-morphological traits in F_2 derived from intraspecific crosses between Sierra (single-podded and unifoliolate leaf) and CA 2969 (double-podded and imparipinnate leaf).

| Traits | Sierra | CA 2969 | Imparipinnate Leaf | | Unifoliolate Leaf | | F_2 Population | | h^2 |
| | | | Single Pod | Double Pods | Single Pod | Double Pods | | | |
	$\overline{X}\pm S_{\overline{x}}$	$\overline{X}\pm S_{\overline{x}}$	$\overline{X}\pm S_{\overline{x}}$	$\overline{X}\pm S_{\overline{x}}$	$\overline{X}\pm S_{\overline{x}}$	$\overline{X}\pm S_{\overline{x}}$	$\overline{X}\pm S_{\overline{x}}$	Range	
Days to first flowering (days)	48.0 ± 0.97	50.0 ± 0.90	41.0 ± 0.29	41.5 ± 0.50	41.3 ± 0.47	40.8 ± 0.70	41.2 ± 0.21	45–66	0.80
Days to 50% flowering (days)	50.0 ± 0.97	52.0 ± 0.90	47.6 ± 0.29	47.8 ± 0.50	48.1 ± 0.46	47.5 ± 0.77	47.8 ± 0.24	45–75	0.80
Plant height (cm)	41.0 ± 0.97	34.3 ± 1.17	37.7 ± 0.31	37.6 ± 0.50	38.9 ± 0.43	41.7 ± 0.70	38.9 ± 0.21	17–59	0.43
First pod height (cm)	21.0 ± 0.73	19.7 ± 2.35	30.5 ± 0.32	30.2 ± 0.44	32.2 ± 0.49	34.0 ± 0.85	31.7 ± 0.23	11–58	0.65
Main stems per plant (No.)	2.7 ± 0.49	2.3 ± 0.21	2.0 ± 0.04	2.1 ± 0.06	2.4 ± 0.07	2.7 ± 0.15	2.3 ± 0.03	1–7	0.38
Pods per plant (No.)	5.2 ± 1.30	3.0 ± 2.54	3.6 ± 0.16	4.5 ± 0.30	3.6 ± 0.28	4.5 ± 0.55	4.1 ± 0.11	1–25	0.49
Seeds per plant (No.)	4.2 ± 1.08	11.7 ± 1.96	3.5 ± 0.14	4.2 ± 0.28	3.3 ± 0.22	4.2 ± 0.44	3.8 ± 0.11	1–24	0.66
Seed yield (g)	1.8 ± 0.56	3.7 ± 0.72	1.3 ± 0.05	1.5 ± 0.10	1.3 ± 0.10	1.5 ± 0.16	1.4 ± 0.04	0.1–6.9	0.60
100-seed weight (g)	49.9 ± 1.61	31.0 ± 1.26	37.2 ± 0.54	35.2 ± 0.76	37.3 ± 1.70	36.0 ± 0.9	36.4 ± 0.36	9.5–69.0	0.45

Table 3. Means ± standard errors and ranges for agro-morphological traits in F_3 derived from intraspecific crosses between Sierra (single-podded and unifoliolate leaf) and CA 2969 (double-podded and imparipinnate leaf).

| Traits | Sierra | CA 2969 | Imparipinnate Leaf | | Unifoliolate Leaf | | F_3 Population | |
| | | | Single Pod | Double Pods | Single Pod | Double Pods | | |
	$\overline{X}\pm S_{\overline{x}}$	$\overline{X}\pm S_{\overline{x}}$	$\overline{X}\pm S_{\overline{x}}$	$\overline{X}\pm S_{\overline{x}}$	$\overline{X}\pm S_{\overline{x}}$	$\overline{X}\pm S_{\overline{x}}$	$\overline{X}\pm S_{\overline{x}}$	Range
Days to first flowering (days)	48.3 ± 0.08	50.0 ± 0.50	46.1 ± 0.23	46.2 ± 0.19	45.0 ± 0.29	46.4 ± 0.27	46.0 ± 0.12	36–75
Days to 50% flowering (days)	50.3 ± 0.08	52.3 ± 0.42	48.9 ± 0.20	49.2 ± 0.17	47.9 ± 0.23	49.2 ± 0.21	48.9 ± 0.10	38–82
Plant height (cm)	52.3 ± 0.61	42.7 ± 0.08	43.8 ± 0.27	44.5 ± 0.25	48.9 ± 0.42	50.8 ± 0.40	46.3 ± 0.18	19–68
First pod height (cm)	31.3 ± 0.38	33.3 ± 0.79	29.9 ± 0.26	30.7 ± 0.25	31.6 ± 0.41	32.4 ± 0.37	30.9 ± 0.15	13–46
Main stems per plant (No.)	2.3 ± 0.09	2.3 ± 0.08	2.7 ± 0.04	2.8 ± 0.04	2.8 ± 0.05	2.9 ± 0.04	2.8 ± 0.02	1–6
Pods per plant (No.)	36.7 ± 2.05	42.7 ± 2.7	62.7 ± 1.87	69.9 ± 1.85	48.5 ± 1.75	48.8 ± 1.35	59.5 ± 0.96	1–254
Seeds per plant (No.)	32.7 ± 2.02	50.7 ± 2.46	68.0 ± 2.06	77.9 ± 2.09	51.4 ± 1.85	52.3 ± 1.46	64.9 ± 1.07	1–267
Seed yield (g)	15.2 ± 0.86	13.9 ± 0.72	24.3 ± 0.74	26.4 ± 0.69	16.3 ± 0.58	16.3 ± 0.46	21.9 ± 0.37	0.1–79.0
100-seed weight (g)	46.9 ± 0.31	27.4 ± 0.34	42.1 ± 0.38	41.4 ± 0.38	38.9 ± 0.45	38.5 ± 0.41	40.7 ± 0.21	7.0–64.0

3.3. Comparisons of Number of Pods per Node and Leaf Shapes

The numbers of pods, seeds per plant and seed yields were higher in the double-podded progeny than those in the single-podded ones, while the 100-seed weight was higher in the single-podded progeny than that in the double-podded progeny (Tables 2 and 3). Progeny having unifoliolate leaves had higher plant heights, first pod heights and numbers of main stems per plant than those of progeny having imparipinnate leaves (Tables 2 and 3).

3.4. Inheritance of Seed Size and Agro-Morphological Traits

The narrow-sense heritability for the days to first flowering; days to 50% flowering; plant height; first pod height; numbers of main stems, pods and seeds per plant; and seed yield per plant were estimated to be 0.80, 0.80, 0.43, 0.65, 0.38, 0.49, 0.66, and 0.60, respectively (Table 2). The narrow-sense heritability for the 100-seed weight was found to be $h^2 = 0.45$. The inheritance pattern of the 100-seed weight was found to be polygenic and governed by partial dominant genes in the present population (Figure 2).

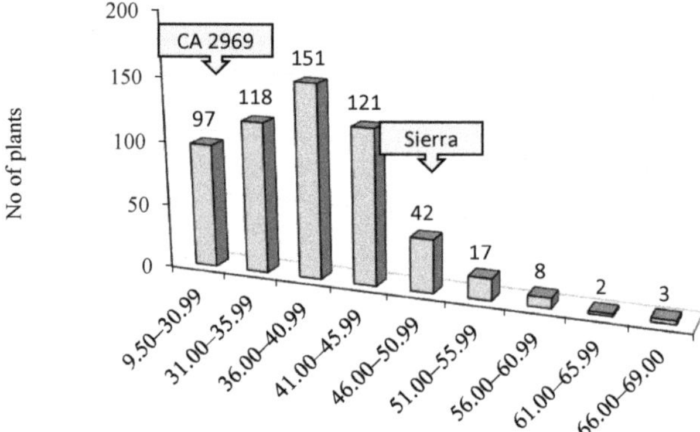

Figure 2. Distribution of F_2 progeny derived from intraspecific crosses between Sierra (single-podded and unifoliolate leaf) and CA 2969 (double-podded and imparipinnate leaf) according to 100-seed weight.

4. Discussion

Agro-morphological traits were negatively affected by drought and heat stress (Table 1), since the progeny from F_1 to F_2 were not only grown under rainfed conditions without using inputs but they were also grown in low-quality organic matter and nitrogen. However, the F_3 progeny had higher yields and more pods and seeds per plant than those of the F_1 and F_2 progeny, due to irrigation (Tables 2 and 3). The domesticated chickpea prefers to grow at temperatures of less than 30 °C [2,8,24], because heat stress may trigger drought stress, which is the major reason for the shedding of flowers.

All the progeny in F_1 had imparipinnate leaves and a single flower or pod per axil as in the male parent CA 2969, indicating that imparipinnate leaves and a single flower or pod per axil were dominant over unifoliolate leaves and double flowers or pods per axil. The F_2 progeny segregated by leaf shape and the number of pods per axil. With a non-significant χ^2 value ($\chi^2 = 8.25 < p$ 0.05), the segregation ratio was found to be a good fit to a ratio of 9 (imparipinnate leaf and single-podded):3 (imparipinnate leaf and double-podded):3 (unifoliolate leaf and single-podded):1 (unifoliolate leaf and double-pooded). The inheritance of leaf shape in domesticated chickpeas in the present study was in agreement with the results of studies [9,68–71]. As in the present study, the presence of double pods per axil was governed by a recessive single gene [9,39,40,42]. Once progeny in the F_2 and F_3 populations having two pods per axil are selected for this unique trait, this trait, controlled by a single recessive gene, will not segregate in later generations [9].

Transgressive segregations were found for all the agro-morphological traits (Tables 2 and 3) and the 100-seed weight in F_2 (Figure 2) and F_3 (Figure 3). Among the progeny, 30 progeny in the F_2 population had a higher 100-seed weight than that of the best parent Sierra, which had a 49.9 g one (Figure 2). A total of 131 progeny in the F_3 population had a 100-seed weight higher than 50 g (Table 3). Some of them had produced two pods and extra-large seed sizes as large as 63 g (Figure 3). Transgressive segregations were considered to be derived from the complementary action of genes and the expression of suppressed recessive genes in the parents [20,41,50,51]. As promising progeny, extra-large-seeded progeny with two or three pods per axil were isolated in F_3 (Figures 3 and 4). As outputs of the epistatic effect of the double-podded trait in a different background, three or more flowers or pods per axil were discovered in some progeny (Figure 4).

Figure 3. Seeds of a progeny (with two pods per axil and 63 g per 100 seeds, left side) in F_3 derived from intraspecific crosses between Sierra (single-podded, unifoliolate leaf and 46.9 g per 100 seeds, right side) and CA 2969 (double-podded, imparipinnate leaf and 27.4 g per 100 seeds, middle).

Figure 4. A progeny in F_3 (with three flowers per axil, unifoliolate leaves and large seeds) derived from intraspecific crosses between Sierra (single-podded and unifoliolate leaf) and CA 2969 (double-podded and imparipinnate leaf). Red circles indicate three flowers/pods.

Single-podded progeny had larger seed sizes than double-podded progeny, while double-podded progeny had higher seed yields, numbers of pods and numbers of seeds per plant than single-podded progeny (Tables 2 and 3). The triple-flowered trait (sfl^t) in the domestic chickpea is controlled by a single recessive gene. The double-flowered trait (sfl^d) is dominant over the triple-flowered trait, and the dominance relationship of these alleles at the *Sfl* locus was reported to be *Sfl* (single flower) > sfl^d (double flower) > sfl^t (triple flower) [44].

Progeny having imparipinnate leaves had higher seed yields than those of progeny having unifoliolate leaves (Tables 2 and 3). This result was in agreement with the findings of Abbo et al. [72] in chickpeas having compound leaves. Additionally, progeny having imparipinnate leaves attained more pods and seeds per plant and higher 100-seed weights than those of unifoliolate progeny (Tables 2 and 3). The higher seed yields and larger seed sizes of the progeny having imparipinnate leaves may stem from the wider photosynthetic area of the progeny having imparipinnate leaves than their counterparts. Abbo et al. [72] indicated that compound leaf lines of chickpeas attained higher leaf area indices at both low and high sowing densities.

Regarding seed size, the 100-seed weight had one of the smallest values of narrow-sense heritability, at $h^2 = 0.45$, showing that the 100-seed weight was one of the most affected agro-morphological traits by genotype-by-environment or genotype-by-year interactions. When considering the environment, heat and drought stress (Table 1) had the greatest effects on seed size (Tables 2 and 3). In the present study, the inheritance pattern for the 100-seed weight was revealed to be polygenic due to continuous distribution (Figure 2). Most of the progeny in F_2 had higher 100-seed weights than that 100-seed weight of the parent CA 2969 (Figure 2), indicating that partial dominant genes played a crucial role in the inheritance of 100-seed weight. Without reciprocal crosses, it was hard to say, despite the fact that the cytoplasmic effect of the female parent might have dominated the large seed size, since most of the progeny produced heavier seeds than those of the smallest-seeded parent CA 2969 (Figure 2). In the present study, seed size was not divided into classes as small, medium and large according to preference. In inheritance studies on seed size in domesticated chickpeas, inheritance was reported as being monogenic [13], digenic [12,13,32,73] and polygenic [27,29–31,33]. The main reason for these differences stems from the phenotypic preference used by researchers. Although maternal genetic effects were not found on the inheritance of seed size in the domesticated chickpea by Upadhyaya et al. [12], it was suggested that the female parent should have a large seed size to increase seed size in the chickpea [74]. Small seed size in chickpeas was dominant over large seed size and controlled by two genes [32,73]. By contrast, normal seed size was dominant over small seed size [12]. In interspecific reciprocal backcrosses, Ceylan et al. [74] indicated that seed size and yield components could be improved by using the domesticated chickpea as female. As a polygenic trait, seed size was governed by both additive and dominant genes [14,27], as in the present study (Figure 3).

5. Conclusions

In conclusion, the inheritance pattern of the extra-large-seeded trait in the domesticated chickpea was polygenic and controlled by partial dominant alleles or might have been affected by the female parent. Not only the 100-seed weight but also the days to the first flowering, days to 50% flowering, plant height, first pod height, number of pods and seeds per plant and seed yield per plant in the F_2 and F_3 populations were found as transgressive segregations, revealing that the seed size in chickpeas could be improved by crossing suitable parents. Double-podded progeny had higher seed yields and numbers of pods and seeds per plant than those of single-podded ones, while single-podded progeny had larger seed sizes than those of double-podded progeny. Progeny having imparipinnate leaves attained higher seed yields than those of progeny having unifoliolate leaves. Extra-large seeds of ≥55 g and two pods per axil traits were assembled in a considerable number of progeny in F_2 and F_3. With the epistatic effect of the double-podded gene in certain genetic backgrounds, three pods per axil were present in some progeny. Progeny having imparipinnate leaves double or more pods should be considered in breeding programs because these progeny had higher numbers of pods and seeds per plant and seed yields than their counterparts. Extra-large-seeded and double-podded traits can be integrated with intraspecific crossing when suitable "kabuli" chickpeas are crossed.

Supplementary Materials: The following are available online at http://www.mdpi.com/2073-4395/10/6/901/s1. Table S1: Monthly rainfall, relative humidity and extreme maximum and minimum temperatures during progeny growing seasons from F_1 to F_3.

Author Contributions: Conceptualization, C.T.; methodology, K.G.K. and T.E.; software, H.S.; validation, T.E.; formal analysis, H.S.; investigation, K.G.K. and T.E.; resources, C.T.; data curation, D.S.; writing—original draft preparation, C.T.; writing—review and editing, D.S., B.A., K.A. and M.C.; visualization, H.S.; supervision, C.T.; project administration, C.T.; funding acquisition, C.T. All authors have read and agreed to the published version of the manuscript.

Funding: This research was funded by TUBITAK, grant number 2190111.

Acknowledgments: We are thankful to Fred J. Muehlbauer (Washington State University, Pullman, WA) and J. Gil (Universidad de Cordoba, Cordoba, Spain) for providing seeds of chickpeas. C.T. thanks Mehmet Tekin and Alper Adak.

Conflicts of Interest: The authors declare no conflict of interest.

References

1. Toker, C.; Yadav, S.S. Legumes cultivars for stress environments. In *Climate Change and Management of Cool Season Grain Legume Crops*; Yadav, S.S., McNeil, D.L., Redden, R., Patil, S.A., Eds.; Springer: Dordrecht, The Netherlands, 2011; pp. 351–376.
2. Devasirvatham, V.; Tan, D.K.Y. Impact of high temperature and drought stresses on chickpea production. *Agronomy* **2018**, *8*, 145. [CrossRef]
3. Gaur, P.M.; Samineni, S.; Thudi, M.; Tripathi, S.; Sajja, S.B.; Jayalakshmi, V.; Mannur, D.M.; Vijayakumar, A.G.; Ganga Rao, N.V.; Ojiewo, C.; et al. Integrated breeding approaches for improving drought and heat adaptation in chickpea (*Cicer arietinum* L.). *Plant Breed.* **2019**, *138*, 389–400. [CrossRef]
4. Ahmad, M.; Naseer, I.; Hussain, A.; Mumtaz, M.Z.; Mustafa, A.; Hilger, T.H.; Zahir, Z.A.; Xu, M. Appraising Endophyte–Plant symbiosis for improved growth, nodulation, nitrogen fixation and abiotic stress tolerance: An experimental investigation with chickpea (*Cicer arietinum* L.). *Agronomy* **2019**, *9*, 621. [CrossRef]
5. Latati, M.; Dokukin, P.; Aouiche, A.; Rebouh, N.Y.; Takouachet, R.; Hafnaoui, E.; Hamdani, F.Z.; Bacha, F.; Ounane, S.M. Species interactions improve above-ground biomass and land use efficiency in intercropped wheat and chickpea under low soil inputs. *Agronomy* **2019**, *9*, 765. [CrossRef]
6. Soe, K.M.; Htwe, A.Z.; Moe, K.; Tomomi, A.; Yamakawa, T. Diversity and effectivity of indigenous *Mesorhizobium* strains for chickpea (*Cicer arietinum* L.) in Myanmar. *Agronomy* **2020**, *10*, 287. [CrossRef]
7. Food and Agriculture Organization Statistical Databases (FAOSTAT). Available online: http://www.fao.org/faostat/en/#data/QC (accessed on 13 May 2020).
8. Van der Maessen, L.J.G. *Cicer* L. a Monograph of the Genus, with Special Reference to the Chickpea (*Cicer arietinum* L.), Its Ecology and Cultivation. Ph.D. Thesis, Veenman, Utrecht, The Nederland, 1972.
9. Muehlbauer, F.J.; Singh, K.B. Genetics of chickpea. In *The Chickpea*; Saxena, M.C., Singh, K.B., Eds.; CAB Int.: Wallingford, CT, USA, 1987; pp. 99–125.
10. Regan, K.; MacLeod, B.; Siddique, K. *Production packages for kabuli chickpea in Western Australia (No. 117)*; Farm Note; The University of Western Australia: Perth, Australia, 2006.
11. Abbo, S.; Grusak, M.A.; Tzuk, T.; Reifen, R. Genetic control of seed weight and calcium concentration in chickpea seed. *Plant Breed.* **2000**, *119*, 427–431. [CrossRef]
12. Upadhyaya, H.D.; Kumar, S.; Gowda, C.L.L.; Singh, S. Two major genes for seed size in chickpea (*Cicer arietinum* L.). *Euphytica* **2006**, *147*, 311–315. [CrossRef]
13. Sundaram, P.; Samineni, S.; Sajja, S.B.; Roy, C.; Singh, S.P.; Joshi, P.; Gaur, P.M. Inheritance and relationships of flowering time and seed size in kabuli chickpea. *Euphytica* **2019**, *215*, 144. [CrossRef]
14. Bicer, B.T.; Sakar, D. Heritability and gene effects for yield and yield components in chickpea. *Hereditas* **2008**, *145*, 220–224. [CrossRef]
15. Bicer, B.T. The effect of seed size on yield and yield components of chickpea and lentil. *Afr. J. Biotechnol.* **2009**, *8*.
16. Singh, K.B. Chickpea breeding. In *The Chickpea*; Saxena, M.C., Singh, K.B., Eds.; CAB Int.: Wallingford, CT, USA, 1987; pp. 127–162.

17. Singh, B.D.; Jaiswal, H.K.; Singh, R.M.; Singh, A.K. Isolation of early-flowering recombinants from the interspecific cross between *Cicer arietinum* and *Cicer reticulatum*. *Int. Chickpea Newslett.* **1984**, *11*, 14–16.

18. Pundir, R.P.S.; Reddy, K.N.; Mengesha, M.H. *ICRISAT Chickpea Germplasm Catalog: Evaluation and Analysis*; ICRISAT: Patancheru, India, 1988.

19. Toker, C. Estimate of heritabilities and genotype by environment interactions for 100-seed weight, days to flowering and plant height in kabuli chick-peas (*Cicer arietinum* L.). *Turk. J. Field Crop.* **1998**, *3*, 16–20.

20. Koseoglu, K.; Adak, A.; Sari, D.; Sari, H.; Ceylan, F.O.; Toker, C. Transgressive segregations for yield criteria in reciprocal interspecific crosses between *Cicer arietinum* L. and *C. reticulatum* Ladiz. *Euphytica* **2017**, *213*, 116. [CrossRef]

21. Dusunceli, F.; Wood, J.A.; Gupta, A.; Yadav, M.; Yadav, S.S. International trade. In *Chickpea Breeding and Management*; Yadav, S.S., Redden, R.J., Chen, W., Sharma, B., Eds.; CAB Int.: Wallingford, CT, USA, 2007; pp. 555–575.

22. Gaur, P.M.; Singh, M.K.; Samineni, S.; Sajja, S.B.; Jukanti, A.K.; Kamatam, S.; Varshney, R.K. Inheritance of protein content and its relationships with seed size, grain yield and other traits in chickpea. *Euphytica* **2016**, *209*, 253–260. [CrossRef]

23. Muehlbauer, F.J.; Sarker, A. Economic Importance of Chickpea: Production, Value, and World Trade. In *The Chickpea Genome*; Varshney, R.K., Thudi, M., Muehlbauer, F.J., Eds.; Springer: Cham, Switzerland, 2017; pp. 5–12.

24. Toker, C.; Lluch, C.; Tejera, N.A.; Serraj, R.; Siddique, K.H.M. Abiotic stresses. In *Chickpea Breeding and Management*; Yadav, S.S., Redden, R.J., Chen, W., Sharma, B., Eds.; CAB International: Wallingford, CT, USA, 2007; pp. 474–496.

25. Balasubrahmanyan, R. The association of size and colour in gram (*Cicer arietinum* L.). *Curr. Sci.* **1950**, *19*, 246–247. [PubMed]

26. Malhotra, R.S.; Singh, K.B. Detection of epistasis in chickpea. *Euphytica* **1989**, *40*, 169–172.

27. Singh, O.; Gowda, C.L.L.; Sethi, S.C.; Dasgupta, T.; Smithson, J.B. Genetic analysis of agronomic characters in chickpea. *Theor. Appl. Genet.* **1992**, *83*, 956–962. [CrossRef]

28. Singh, O.; Gowda, C.L.L.; Sethi, S.C.; Dasgupta, T.; Kumar, J.; Smithson, J.B. Genetic analysis of agronomic characters in chickpea. II. Estimates of genetic variances from line× tester mating designs. *Theor. Appl. Genet.* **1993**, *85*, 1010–1016. [CrossRef]

29. Kumar, S.; Singh, O. Inheritance of seed size in chickpea. *J. Genet. Breed.* **1995**, *49*, 99–103.

30. Malhotra, R.S.; Bejiga, G.; Singh, K.B. Inheritance of seed size in chickpea. *J. Genet. Breed.* **1997**, *51*, 45–50.

31. Hovav, R.; Upadhyaya, K.C.; Beharav, A.; Abbo, S. Major flowering time gene and polygene effects on chickpea seed weight. *Plant Breed.* **2003**, *122*, 539–541. [CrossRef]

32. Hossain, S.; Ford, R.; McNeil, D.; Pittock, C.; Panozzo, J.F. Inheritance of Seed Size in Chickpea (*Cicer arietinum* L.) and Identification of QTL Based on 100-seed Weight and Seed Size Index. *Aust. J. Crop Sci.* **2010**, *4*, 126.

33. Sharma, S.; Upadhyaya, H.D.; Gowda, C.L.L.; Kumar, S.; Singh, S. Genetic analysis for seed size in three crosses of chickpea (*Cicer arietinum* L.). *Can. J. Plant Sci.* **2013**, *93*, 387–395. [CrossRef]

34. Cobos, M.J.; Rubio, J.; Fernández-Romero, M.D.; Garza, R.; Moreno, M.T.; Millán, T.; Gil, J. Genetic analysis of seed size, yield and days to flowering in a chickpea recombinant inbred line population derived from a Kabuli × Desi cross. *Ann. Appl. Biol.* **2007**, *151*, 33–42. [CrossRef]

35. Verma, S.; Gupta, S.; Bandhiwal, N.; Kumar, T.; Bharadwaj, C.; Bhatia, S. High-density linkage map construction and mapping of seed trait QTLs in chickpea (*Cicer arietinum* L.) using genotyping-by-sequencing (GBS). *Sci. Rep.* **2015**, *5*, 17512. [CrossRef] [PubMed]

36. Singh, V.K.; Khan, A.W.; Jaganathan, D.; Thudi, M.; Roorkiwal, M.; Takagi, H.; Sutton, T. QTL-seq for rapid identification of candidate genes for 100-seed weight and root/total plant dry weight ratio under rainfed conditions in chickpea. *Plant Biotechnol. J.* **2016**, *14*, 2110–2119. [CrossRef] [PubMed]

37. Lal, D.; Ravikumar, R.L. Development of Genetic Linkage Map and Identification of QTLs for Agronomic Traits in Chickpea (*Cicer arietinum* L.). *Int. J. Curr. Microbiol. App. Sci.* **2018**, *7*, 66–77. [CrossRef]

38. Sheldrake, A.R.; Saxena, N.P.; Krishnamurthy, L. The expression and influence on yield of the 'double-podded ' character in chickpeas (*Cicer arietinum* L.). *Field Crops Res.* **1978**, *1*, 243–253. [CrossRef]

39. Kumar, J.; Srivastava, R.K.; Ganesh, M. Penetrance and expressivity of the gene for double podding in chickpea. *J. Hered.* **2000**, *91*, 234–236. [CrossRef]

40. Yasar, M.; Ceylan, F.O.; Ikten, C.; Toker, C. Comparison of expressivity and penetrance of the double podding trait and yield components based on reciprocal crosses of kabuli and desi chickpeas (*Cicer arietinum* L.). *Euphytica* **2014**, *196*, 331–339. [CrossRef]

41. Adak, A.; Sari, D.; Sari, H.; Toker, C. Gene effects of *Cicer reticulatum* on qualitative and quantitative traits in the cultivated chickpea. *Plant Breed.* **2017**, *136*, 939–947. [CrossRef]

42. Rubio, J.; Moreno, M.T.; Cubero, J.I.; Gil, J. Effect of the gene for double pod in chickpea on yield, yield components and stability of yield. *Plant Breed.* **1998**, *117*, 585–587. [CrossRef]

43. Rubio, J.; Flores, F.; Moreno, M.T.; Cubero, J.I.; Gil, J. Effects of the erect/bushy habit, single/double pod and late/early flowering genes on yield and seed size and their stability in chickpea. *Field Crops Res.* **2004**, *90*, 255–262. [CrossRef]

44. Srinivasan, S.; Gaur, P.M.; Chaturvedi, S.K.; Rao, B.V. Allelic relationships of genes controlling number of flowers per axis in chickpea. *Euphytica* **2006**, *152*, 331–337. [CrossRef]

45. Rajesh, P.N.; Tullu, A.; Gil, J.; Gupta, V.; Ranjekar, P.; Muehlbauer, F. Identification of an STMS marker for the double-podding gene in chickpea. *Theor. Appl. Genet.* **2002**, *105*, 604–607. [CrossRef]

46. Cho, S.; Kumar, J.; Shultz, J.L.; Anupama, K.; Tefera, F.; Muehlbauer, F.J. Mapping genes for double podding and other morphological traits in chickpea. *Euphytica* **2002**, *128*, 285–292. [CrossRef]

47. Radhika, P.; Gowda, S.J.M.; Kadoo, N.Y.; Mhase, L.B.; Jamadagni, B.M.; Sainani, M.N.; Gupta, V.S. Development of an integrated intraspecific map of chickpea (*Cicer arietinum* L.) using two recombinant inbred line populations. *Theor. Appl. Genet.* **2007**, *115*, 209–216. [CrossRef]

48. Taran, B.; Warkentin, T.D.; Vandenberg, A. Fast track genetic improvement of ascochyta blight resistance and double podding in chickpea by marker-assisted backcrossing. *Theor. Appl. Genet.* **2013**, *126*, 1639–1647. [CrossRef]

49. Ali, L.; Deokar, A.; Caballo, C.; Tar'an, B.; Gil, J.; Chen, W.; Millan, L.; Rubio, J. Fine mapping for double podding gene in chickpea. *Theor. Appl. Genet.* **2016**, *129*, 77–86. [CrossRef]

50. Singh, M.; Rani, S.; Malhotra, N.; Katna, G.; Sarker, A. Transgressive segregations for agronomic improvement using interspecific crosses between *C. arietinum* L. x *C. reticulatum* Ladiz. and *C. arietinum* L. x *C. echinospermum* Davis species. *PLoS ONE* **2018**, *13*, e0203082. [CrossRef]

51. De Vicente, M.C.; Tanksley, S.D. QTL analysis of transgressive segregation in an interspecific tomato cross. *Genetics* **1993**, *134*, 585–596.

52. Gaur, P.M.; Pande, S.; Upadhyaya, H.D.; Rao, B.V. Extra-large kabuli chickpea with high resistance to fusarium wilt. *J. Agric. Res.* **2006**, *2*, 1–2.

53. Bonfil, D.J.; Goren, O.; Mufradi, I.; Lichtenzveig, J.; Abbo, S. Development of early-flowering kabuli chickpea with compound and simple leaves. *Plant Breed.* **2007**, *126*, 125–129. [CrossRef]

54. Gil, J.; Castro, P.; Millan, T.; Madrid, E.; Rubio, J. Development of new kabuli large-seeded chickpea materials with resistance to Ascochyta blight. *Crop Pasture Sci.* **2017**, *68*, 967–972. [CrossRef]

55. Muehlbauer, F.J.; Temple, S.R.; Chen, W. Registration of Sierra chickpea. *Crop Sci.* **2004**, *44*, 1864–1865. [CrossRef]

56. Warkentin, T.; Banniza, S.; Vandenberg, A. CDC Frontier kabuli chickpea. *Can. J. Plant Sci.* **2005**, *85*, 909–910. [CrossRef]

57. Gan, Y.T.; Siddique, K.H.M.; MacLeod, W.J.; Jayakumar, P. Management options for minimizing the damage by ascochyta blight (*Ascochyta rabiei*) in chickpea (*Cicer arietinum* L.). *Field Crops Res.* **2006**, *97*, 121–134. [CrossRef]

58. Gowda, C.L.L.; Upadhyaya, H.D.; Dronavalli, N.; Singh, S. Identification of large-seeded high-yielding stable kabuli chickpea germplasm lines for use in crop improvement. *Crop Sci.* **2011**, *51*, 198–209. [CrossRef]

59. Taran, B.; Bandara, M.; Warkentin, T.; Banniza, S.; Vandenberg, A. CDC Orion kabuli chickpea. *Can. J. Plant Sci.* **2011**, *91*, 355–356. [CrossRef]

60. Vandemark, G.; Guy, S.O.; Chen, W.; McPhee, K.; Pfaff, J.; Lauver, M.; Muehlbauer, F.J. Registration of 'Nash'chickpea. *J. Plant Regist.* **2015**, *9*, 275–278. [CrossRef]

61. Urrea, C.A.; Muehlbauer, F.J.; Harveson, R.M. Registration of 'New Hope'Chickpea Cultivar with Enhanced Resistance to Ascochyta Blight. *J. Plant Regist.* **2017**, *11*, 107–111. [CrossRef]

62. Vandemark, G.; Nelson, H.; Chen, W.; McPhee, K.; Muehlbauer, F. Registration of 'Royal' Chickpea. *J. Plant Regist.* **2019**, *13*, 123–127. [CrossRef]

63. Rubio, J.; Moreno, M.T.; Martinez, C.; Gil, J. Registration of CA2969, an Ascochyta blight resistant and double-podded chickpea germplasm. (Registrations of Germplasm). *Crop Sci.* **2003**, *43*, 1567–1569. [CrossRef]

64. Auckland, A.K.; Van der Maesen, L.J.G. Chickpea. In *Hybridization of Crop Plants*; Fehr, W.R., Hadley, H.H., Eds.; American Society of Agronomy and Crop Science Society of America: Madison, WI, USA, 1980; pp. 249–259.

65. MINITAB 2014 Inc. *Minitab Statistical Software, Release 17 for Windows*; MINITAB 2014 Inc.: State College, PA, USA, 2014.

66. Poehlman, J.M.; Sleper, D.A. *Breeding Field Crops*; Iowa State University Press: Ames, IA, USA, 1995; p. 494.

67. Steel, R.G.D.; Torrie, J.H. *Principles and Procedures of Statistics: A Biometrical Approach*; McGraw-Hill: New York, NY, USA, 1980.

68. Rao, N.K.; Pundir, R.P.S.; Van Der Maesen, L.J.G. Inheritance of some qualitative characters in chickpea (*Cicer arietinum* L.). *Proc. Plant Sci.* **1980**, *89*, 497–503.

69. Pundir, R.P.S.; Mengesha, M.H.; Reddy, K.N. Leaf types and their genetics in chickpea (*Cicer arietinum* L.). *Euphytica* **1990**, *45*, 197–200.

70. Danehloueipour, N.; Clarke, H.J.; Yan, G.; Khan, T.N.; Siddique, K.H.M. Leaf type is not associated with ascochyta blight disease in chickpea (*Cicer arietinum* L.). *Euphytica* **2008**, *162*, 281–289. [CrossRef]

71. Toker, C.; Ceylan, F.O.; Inci, N.E.; Yildirim, T.; Cagirgan, M.I. Inheritance of leaf shape in the cultivated chickpea (*Cicer arietinum* L.). *Turk Field Crops* **2012**, *17*, 16–18.

72. Abbo, S.; Goren, O.; Saranga, Y.; Langensiepen, M.; Bonfil, D. Leaf shape × sowing density interaction affects chickpea grain yield. *Plant Breed.* **2013**, *132*, 200–204. [CrossRef]

73. Upadhyaya, H.D.; Sharma, S.; Gowda, C.L. Major genes with additive effects for seed size in kabuli chickpea (*Cicer arietinum* L.). *J. Genet.* **2011**, *90*, 479–482. [CrossRef]

74. Ceylan, F.O.; Adak, A.; Sari, D.; Sari, H.; Toker, C. Unveiling of suppressed genes in interspecific and backcross populations derived from mutants of *Cicer* species. *Crop Pasture Sci.* **2019**, *70*, 254–262. [CrossRef]

 *agronomy*

Review

Past, Present and Future Perspectives on Groundnut Breeding in Burkina Faso

Moumouni Konate [1,*] **, Jacob Sanou** [1]**, Amos Miningou** [2]**, David Kalule Okello** [3] **, Haile Desmae** [4]**, Paspuleti Janila** [5] **and Rita H. Mumm** [6,*]

1   INERA, DRREA-Ouest, 01 BP 910 Bobo Dioulasso 01, Burkina Faso; jacobsanou17@gmail.com
2   INERA, CREAF, 01 BP 476 Ouagadougou 01, Burkina Faso; miningou_amos@yahoo.fr
3   NaSARRI, Uganda National Groundnuts Improvement, P.O. Box 56, Soroti, Uganda; kod143@gmail.com
4   ICRISAT-Mali, BP 320 Bamako, Mali; h.desmae@cgiar.org
5   ICRISAT-India, Patancheru, Andhra Pradesh 502324, India; p.janila@cgiar.org
6   Department of Crop Sciences and the Illinois Plant Breeding Center, University of Illinois, 1102 South Goodwin Avenue, Urbana, IL 61801, USA
*   Correspondence: mouni.konate@gmail.com (M.K.); ritamumm@illinois.edu (R.H.M.); Tel.: +226-700-495-04 (M.K.); +1-217-244-9497 (R.H.M.)

Received: 23 December 2019; Accepted: 10 March 2020; Published: 14 May 2020

Abstract: Groundnut (*Arachis hypogaea* L.) is a major food and cash crop in Burkina Faso. Due to the growing demand for raw oilseeds, there is an increasing interest in groundnut production from traditional rain-fed areas to irrigated environments. However, despite implementation of many initiatives in the past to increase groundnut productivity and production, the groundnut industry still struggles to prosper due to the fact of several constraints including minimal development research and fluctuating markets. Yield penalty due to the presence of drought and biotic stresses continue to be a major drawback for groundnut production. This review traces progress in the groundnut breeding that started in Burkina Faso before the country's political independence in 1960 through to present times. Up to the 1980s, groundnut improvement was led by international research institutions such as IRHO (Institute of Oils and Oleaginous Research) and ICRISAT (International Crops Research Institute for the Semi-Arid Tropics). However, international breeding initiatives were not sufficient to establish a robust domestic groundnut breeding programme. This review also provides essential information about opportunities and challenges for groundnut research in Burkina Faso, emphasising the need for institutional attention to genetic improvement of the crop.

Keywords: peanut; plant breeding; research; funding; genomics; INERA; cultivar; selection; *Arachis hypogaea*

1. Introduction: The Importance of Groundnut in Burkina Faso

Groundnut (*Arachis hypogaea* L.), also known as peanut, is a self-pollinated crop and allotetraploid ($2n = 4x = 40$) with a genome size of 2.54 Gb [1] and which belongs to the Fabaceae family [2,3]. Groundnut is an important food cop worldwide with an annual production of over 47 million tons on near 28 million hectares in 2017, according to the last statistics of Food and Agriculture Organisation (FAO) [4]. The crop's cultivation, processing and trade significantly impacts the socio-economic development of a large number of developing and least developed countries [5]. Approximately 60% of the world's production comes from Asia, whereas Africa accounts for 26%. In 2017, groundnut productivity was the lowest in Africa (839.6 kg/ha) compared to the rest of the world (1685.6 kg/ha) [4]. Burkina Faso (formerly known as Upper Volta) contributes 1–1.5% to the global groundnut production with approximately 400,000 tons per year [4].

In Burkina Faso, groundnut contributes significantly to both food security and poverty alleviation [6]. As in many low-income countries in Africa and Asia, the crop is primarily grown for subsistence by smallholder farmers [7], predominantly women [5,8,9], under rain-fed and low input conditions [10]. Groundnut can account for up to 50% of cash income while providing many benefits including a food rich in digestible proteins, high-quality oils, and many functional compounds and elements such as iron and zinc which are important to nutrition and health, especially in children [11–13]. As a legume crop, groundnut cultivation improves soil fertility and productivity by fixing atmospheric nitrogen [14]. Additionally, the plant's haulm and by-products have value as a feed for livestock [15,16]. The groundnut value chain in Burkina Faso employs a significant number of people and contributes substantially to the economy [6,17] and to family wellbeing [5,18]. Clearly, groundnut improvement has a direct positive impact on the nutritional and economic status of smallholder farmers.

Despite the importance and benefits of groundnut for farmers and consumers in Burkina Faso, the production of this legume has been unsteady for the last 20 years or so [4]. Groundnut yields have remained low (~800 kg/ha) in sharp contrast with the crop's potential which can provide up to 5000 kg/ha in intensive agriculture systems like in the USA [4,19,20]. This low level of productivity is attributable to several constraining factors including diseases and pests, erratic rainfall, drought, poor soils, market instability, and lack of locally adapted high-yielding varieties [19,21]. The highest production in Burkina Faso was 519,345 tonnes in 2016 [4], more due to the extension of cultivated land than to an increase in crop productivity (Figure 1). The stagnation of domestic production has been exacerbated by an unreliable seed supply system for groundnut [19] and weak organisation of the groundnut industry which has left a current gap in processing capacity.

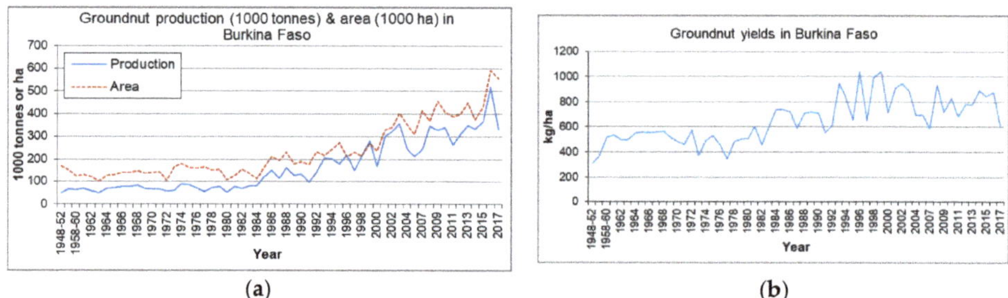

(a) (b)

Figure 1. Evolution of groundnut (**a**) production (1000 tonnes; shelled) and area (1000 ha); (**b**) yield (kg/ha) in Burkina Faso between 1948 and 2017 [4,22,23].

Groundnut breeding in Burkina Faso has been tightly correlated with activities in the crop's value chain which drive the whole groundnut industry including the research and development [6]. For more than a decade now, no major action plan has been established to develop the groundnut industry, especially after the 2008 food crisis [24]. To this extent, the focus of breeding efforts at INERA (Institute of Environment and Agriculture Research) was directed to the main staple food crops (i.e., maize, pearl millet, sorghum, rice) overlooking groundnut, which is often considered a cash crop, and thereby hampering groundnut cultivar development.

At present, information about progress and the current state of groundnut breeding in Burkina Faso is patchy. Most research results are confined to annual reports of individual projects with little published in international journals [25]. To our knowledge, publications in recent years have focused mainly on yield evaluation [26] and disease of local and exotic varieties for early leaf spot [27,28]. Earlier research activities involved the evaluation of resistance to foliar diseases in lines introduced from ICRISAT (International Crops Research Institute for the Semi-Arid Tropics) or the USA through the Peanut CRSP (Peanut Collaborative Research Support Program) [29–31] and similar development programmes.

In this paper, research on groundnut improvement in Burkina Faso is reviewed with an outlook for future breeding strategies. It appears that groundnut research started under colonial projects before the country's independence in 1960. However, it took nearly three decades before the national agriculture research programme emerged. Between 1980 and 2010, groundnut research was conducted by nationals with minimal access to technical and financial capacities. The most consistent support was from the USAID-funded Peanut CRSP. This was followed by years of resource scarcity, before groundnut research was rekindled with ICRISAT-led Gates-funded projects (Tropical Legumes) and local initiatives. Today, modern technologies offer the opportunity to advance and deliver improved groundnut varieties that meet farmer, consumer, processor, and export demands.

2. History of Groundnut Cultivation in Burkina Faso

Intensive groundnut production started in West Africa during the colonial period in the 1900s, providing raw material for the French oil factories as well a source of revenue [32], stimulating groundnut commercialization throughout the whole of Western Africa [33]. Subsequently, migrant groundnut farming arose as a labour system in West Africa associated with cash-cropping which drew thousands of young men from Burkina Faso and elsewhere (e.g., Mali, Guinea, Senegal, Mauritania) towards the Gambia River basin (Gambia), a hot-spot of groundnut cultivation at the time [32,33]. Although farmers had been growing the crop in many places in Burkina Faso, migrants returning home from the Gambia River basin were most likely the first to advance groundnut cultivation in the country. The swift widespread adoption of groundnut was probably due to the fact of its similarity to the West African native Bambara groundnut (*Vigna subterranea*) [33] to which farmers were already accustomed [23,33].

After successful promotion of groundnut cultivation in the region of Bobo-Dioulasso by the colonial administration in the early 1920s, crushing machines became operative in Ouagadougou and Banfora in 1922 and 1928, respectively [34]. Cultivation of groundnut extended rapidly around the country after 1936 [35,36], prompting the installation of the main factory for oil processing in 1941 in Bobo-Dioulasso [35]. Post-independence political unrest, drought waves, epidemics of groundnut rosette disease and groundnut price instability in global markets [22,32,33,37,38] resulted in low production levels through the early 1980s (Figure 1a). Production increased from 1985 due to the expanded acreage (Figure 1a) for cultivation of the crop [39] and significant governmental support to the groundnut sector by the SOFIVAR (Groundnut Funding and Extension Society) [37]. However, production has been erratic, and crop yields on average remained very low (<600 kg/ha) until the mid-1980s when, for the first time, yields reached >700 kg/ha (Figure 1b). The best yields (800–1000 kg/ha) were achieved between the mid-1990s and 2000, driven by market opportunities and national support provided to the groundnut sector [37,40]. Yields have fluctuated over the years since then (Figure 1b), highlighting the need for improved locally adapted varieties. The trend of an increase in groundnut production has been more due to the increase in area harvested than productivity improvement.

3. Groundnut Research in Burkina Faso

Groundnut research in Burkina Faso started before the country's autonomy, with French projects targeting the crop production for export. The French National Institute for Colonial Agriculture (called CNRS post-1939) was created to conduct development research to increase revenues from agricultural production in the French colonies of Africa [41,42]. The hardship of World War II further prompted the creation of specialised research institutions, such as IRHO (Institute of Oils and Oleaginous Research), to increase food and oilseed production in Africa [42,43]. Subsequently, IRHO, having merged with GERDAT (Study and Research Group for the Development of Tropical Agriculture), has conducted research on groundnut with a focus in the main production area in southwestern regions of Burkina Faso [36]. The research objectives were to increase yields while developing resistance to rosette disease which could totally decimate production during years of disease outbreaks [36]. This prompted IRHO, based at Niangoloko, to undertake groundnut improvement research to develop cultivars resistant to rosette disease and adapted to southwestern regions of the country [36]. Resistant cultivars were identified in 1952 around the Burkina/Côte-d'Ivoire border and used in breeding crosses at Bambey Agricultural Research Centre in Senegal [44]. By 1959, about 20 resistant cultivars with limited infection rate (≤6%) and better productivity (20–35% yield increase compared with local varieties) were introduced in Burkina Faso [36]. Additionally, experiments conducted between 1955 and 1963 resulted in some varieties with potential yields about three metric tonnes per hectare and showed that (1) groundnut densities between 111,100 and 133,300 plants/ha are optimal for top yields, and (2) phosphorus is the most limiting nutrient for groundnut in soils of western Burkina Faso [36]. The creation of ICRISAT in 1972 and nomination of groundnut as a mandate crop in 1976 [45] added momentum to research on groundnut [3,46]. Major constraints to production were identified [47], including pests [39,48], diseases [10,49], drought and aflatoxins [50–53].

Until the inception of PNRA in the mid-1980s, agricultural research was administered by the Ministry of Rural Development with little contribution of national scientists. Major research projects on groundnuts were implemented by international institutions [23,31,43,54] with the research station of IRHO at Niangoloko contributing somewhat but at a far lower level [32]. Varietal creation was limited at Niangoloko, as it was used primarily as a testing site, while breeding populations were developed at the research centre in Bambey, Senegal [44]. Apart from a few French agronomists, no Burkinabe conducted research in Niangoloko, probably because groundnut was not a priority for the government [54]. Government investment in agricultural extension prevailed over research even before technologies were developed to support food production as suggested by donors [23]. Then, progressive departure of international researchers and lack of resources, equipment, and trained scientists halted research activity and led to a loss of valuable germplasm and breeding records [32,55]. Restructuring in the 1980 to 1985 timeframe gave rise to INERA which initiated breeding programmes in 1988 [56].

In this context, programmes supported by donors such as USAID (United States Agency for International Development), the European Union (EU) and the Bill and Melinda Gates Foundation (BMGF) were instrumental in re-establishing groundnut research and development in Burkina Faso [23]. The USAID-funded Collaborative Research Support Program (CRSP) stimulated and sustained research activities from 1975 to 2012 [29–31]. This Peanut CRSP network involved the Institute for the Sahel (INSAH), IRHO and ICRISAT. However, these activities focused primarily on pest and disease management and resistance to leaf spot diseases as well as testing of advanced lines for production efficiency [30]. These programmes resulted in improved groundnut varieties which still hold a large share of production in Burkina Faso today [22,25]. They enabled many works that would not be possible otherwise. In this research collaboration, cultivar development was conducted mainly by IRHO and ICRISAT [36,47,49,55], while work at INERA involved testing only. Such involvement of international research in national programmes has been at the core of crop breeding activities in West Africa [25]. The growing number of trained and qualified scientists in Burkina Faso and in the region opens opportunities for more research leadership within countries. However, breeding programmes continue to be handicapped by the lack of the resources needed to conduct research activities [25] and lack of research leadership to some extent. Consequently, most groundnut varieties cultivated in Burkina Faso today were developed before or shortly after 1960, some from Senegal as sharing cultivars has been common practice among countries in the sub-region [57–59]. Varieties such as 28–206, 59–426, 69–101 and Fleur 11 were introduced from Senegal to Burkina Faso [25]. To date, more than 20 varieties have been registered in the national catalogue of crop varieties. Some of these (e.g., TS 32-1, CN94 C, RMP 91, RMP 12, KH 241D) have been under cultivation for more than 50 years (Table 1). Cultivar development was stagnant in the country until 1990 when two varieties (SH 67A, SH 470P) were released (Table 1), and the average age of commercial groundnut varieties is about 30 years [55].

Table 1. Groundnut varieties released from the 1950s to date and registered in the national catalogue of plant varieties.

Variety/Line	Pedigree	Botanical Type	Cycle (Days)	Year	Institution/Origin	Reference
CN 94C	90 Saria/Tougan 1) F_6	Spanish	90	1966	IRHO Saria, Burkina Faso	[10,60]
Te. 3	Local pop. Burkina	Spanish	90	1958	IRHO Niangoloko, Burkina Faso	[10,25,58,60,61]
TS 32-1	Spanlex/Te. 3,	Spanish	90	1966		
KH 149A	GH 119-7.1II-III/91 Saria	Spanish	90	1964		
KH 241D	GH 1185.2 II/91 Saria	Spanish	90	1964		
QH 243 C	KH 184 A/424 A, F_7	Spanish	90	1971		
RMP 91	48–37/Mani Pintar, F_9	Virginia	135	1963		
RMP 12	1036/Mani Pintar, F_9	Virginia	135	1960		
SH 67A	QH 243C/ PI 1166	Spanish	90	1990	INERA Niangoloko, Burkina Faso	[60]
SH 470P	Flower 113/ QH 200A, F_7	Spanish	90	1990		
69–101	55–455 14/28–206, F_5-B_3	Virginia	120	1969	IRHO/CNRA Bambey, Senegal	[10,25,57,60,61]
59–426	NA	Virginia	120	1959		
ICGSE 104	NA (Segregating material ICRISAT)	Valencia	75–80	1990	INERA—ICRISAT	[60]
Fleur 11	Variety from China	Spanish	90	1990	CNRA Bambey Senegal	[60,61]
Nafa 1 (ICGV 01276)	ICGV 92069/ICGV 93184	Virginia	110	2018	INERA—ICRISAT	[62]
Lokre (ICGV 91328)	J11/U4-7-5	Spanish	90	2018		
Miou Pale (ICGV 93305)	Var 27/U4-7-5	Valencia	90	2018		
Touinware (ICGV-IS 13806)	ICGV 86124/ICG 7878	Spanish	90	2018		
Beeda (ICGV-IS 13830)	ICGV 86124/ICG 7878	Spanish	90	2018		
Soukeba (ICGV-IS 13912)	ICGV 86124/ICG 7878	Spanish	90	2018		
Kiema 1	Local pop. Burkina	Spanish	90	2018		

NA = not available; pop. = population.

The current increase in groundnut production has been almost entirely due to the fact of land expansion [4]. With the exhaustion of the country's arable lands, a future increase in production must come from yield improvement based on genetic improvement and appropriate management practices. It has been estimated that less than 25% of approximately 460,000 hectares of land cultivated yearly in groundnut were occupied with improved varieties [55,63]. Replacement of popular groundnut varieties (TS 32-1, SH470-P, CN94 C, Fleur 11) with more productive ones is sought. Only recently, in 2018, have seven new varieties developed by ICRISAT been tested, registered in the national seed catalogue [62], and subsequently released for cultivation in Burkina Faso (Table 1). However, extension efforts are required to facilitate farmers' awareness and adoption of these varieties.

4. Research Resources

In the early 2000s, Burkina Faso began investing more in agricultural research capacity. As of January 2019, more than 65% of scientists hold a doctorate degree (HRM, personal communication), compared to less than 50% before 2000. Currently, the number of scientists at INERA alone reaches over 300, not counting scientists at university-based agricultural research centres. However, the increase in the number of researchers did not go along with the increase in research capacity [64]. Lack of resources coupled with poor competitiveness of local salaries with that of international positions motivated the departure of many scientists to CGIARs, NGOs or Western countries. National research programs struggle to keep up with evolving breeding methods and required infrastructures and equipment [64], limiting their effectiveness.

Furthermore, expenditures dedicated to research and development have been irregular [38] or reduced, sometimes due to the fact of political turmoil [23]. Since transitioning from PNRA, INERA has been chiefly financed through World Bank loans supporting three main projects [38,65,66]. What is more, of the proportion of funding dedicated to agriculture research, little goes to groundnut, making this crop less attractive to researchers. The consequence of resource limitation was attenuated by regional research initiatives through WECARD (West and Central African Council for Agricultural Research and Development) and ICRISAT. Regional efforts essentially guard against unnecessary replication of research in countries with similar agro-ecologies in the region and have the advantage of mobilising more donors [64]. International initiatives such as the Peanut CRSP enabled the implementation of many research activities on groundnut in the country for over three decades in collaboration with the University of Ouagadougou [23,31]. With the close of the World Bank-funded project PNDSA (National Project for Development of Agricultural Sector) in 2003, there was no funding at all for groundnut research at INERA until 2012, when groundnut development research started almost afresh with the second phase of Tropical Legumes project sponsored by the Gates Foundation. This project provided substantial support to the breeding programme in Burkina Faso, allowing INERA to reflect on product targets, breeding objectives to achieve these targets, and breeding process modernisation [67]. This funding had the merit of rekindling groundnut breeding in Burkina Faso, although there is still a need to develop clear and specific improvement goals, build technical capacity, and secure long-term funding.

In a nutshell, public sector groundnut breeding has not made satisfactory progress in Burkina Faso as in most sub-Saharan countries; no increase in genetic gain has been recorded in the last 30 years [68,69]. While technology can be accessed, especially through outsourcing, the primary challenge for groundnut improvement in Burkina Faso is the capacity to assemble relevant technological options to create an optimized varietal development pipeline for groundnut [68]. The national breeding programme needs to keep up with the evolving breeding methods through improved research leadership and technical expertise [68,70]. Therefore, a workforce of researchers able to apply advances in breeding methods, approaches and tools for cultivar development is needed. In West Africa, the University of Ghana/West Africa Centre for Crop Improvement has stepped up to provide such training at the postgraduate level with support from AGRA (Alliance for a Green Revolution in Africa) and others [71]. Other universities in Africa are also rising to meet this challenge of training the next generation of plant breeders. Furthermore, professional development programs, such as the African Plant Breeding Academy, coordinated by the University of California at Davis in the USA, offer continuing education to African plant breeders in the use of genomics-assisted selection and ways to optimize the breeding pipeline [72]. However, the question is, "Will these programs be enough to create the workforce needed for Burkina Faso?"

5. Production Environment

Groundnut was the top cash crop in Burkina Faso before cotton until 1977 [73]. Since then, groundnut production has primarily served domestic needs as a food crop [22,73]. Of the yearly production of more than 350,000 tonnes [4], only about 2% on average is exported [71] and this to other countries in West Africa [6]. A sharp increase in groundnut prices in early 1990s boosted production and exports towards Europe, at least for a few years [37,74]. However, the increased production and export levels were not sustained due to the decline in groundnut demand in Europe [73,75] and lack of adequate support to farmers [74]. Nevertheless, domestic demand for groundnut has been increasing in recent years to complement cotton seed for an increase of oil production which covers only approximately 30% of needs at the moment [76]. Currently, although groundnut production meets the demand for household consumption as food, little surplus is available for the processing market [22,23,73]. The development of productive cultivars for the farming system upstream is necessary to meet growing demands and sustain a stable value chain.

The distribution of groundnut cultivation in Burkina Faso (Figure 2) indicates Centre-Ouest as the top groundnut producing region, followed by the Boucle du Mouhoun region. Production is mid-level in Hauts-Bassins, Centre-Sud, Centre-Est and Est regions. This distribution has remained consistent since the 1960s [36], reflecting minimal efforts to expand groundnut to new areas. Nevertheless, Burkina Faso is among the top ten groundnut producing countries in Africa, based on area harvested (Table 2). Moreover, an estimation of country production proportionally to the population size shows that the importance of groundnut production in Burkina Faso is similar to that of Nigeria, the top producer in Africa [4]. Groundnut has great economic potential for the country [74], if only more political support were provided to increase investment in the sector.

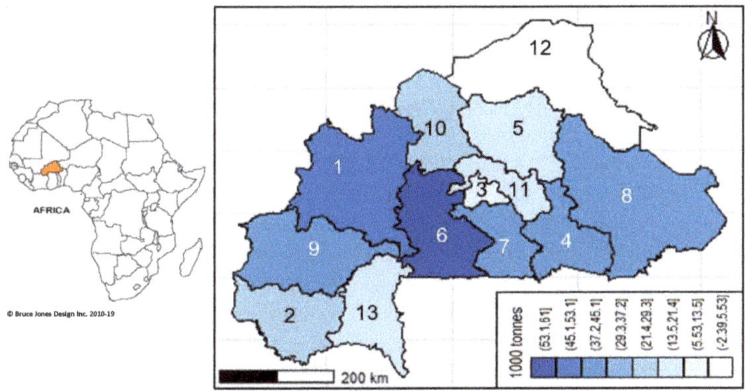

Figure 2. Map of groundnut production per administrative region in Burkina Faso, adapted based on the average of the last five years (2013–2017) [4]. 1: Boucle-du-Mouhoun; 2: Cascades; 3: Centre; 4: Centre-Est; 5: Centre-Nord; 6: Centre-Ouest; 7: Centre-Sud; 8: Est; 9: Hauts-Bassins; 10: Nord; 11: Plateau-Central; 12: Sahel; 13: Sud-Ouest. The choropleth map was drawn using R package GADMTools with breaks option "sd" (standard deviation).

Table 2. Top 10 groundnut producing countries based on average area harvested in the last five years (2013–2017) [4].

Rank	Country	Area (ha)	Production (Tonnes)	Yield (kg/ha)
1	Nigeria	2,766,845.8	3,068,586.8	1110.4
2	Sudan	2,027,954.4	1,629,402.2	797.8
3	UR * Tanzania	1,208,903.0	1,285,027.0	1052.8
4	Senegal	950,149.6	806,165.4	843.2
5	Niger	779,283.6	417,776.0	537.4
6	Chad	760,472.6	843,546.2	1117.1
7	Guinea	553,012.0	469,918.2	887.5
8	DRC *	492,000.0	370,447.4	753.6
9	Burkina Faso	480,635.4	380,894.2	799.8
10	Cameroon	439,308.4	610,196.2	1386.4

* UR Tanzania = United Republic of Tanzania; DRC = Democratic Republic of the Congo.

6. Constraints to Groundnut Productivity

The difference between groundnut potential yield and actual yields in farmer fields, referred to as yield gap [77,78], reaches over 50% Burkina Faso [4,62], due to the fact of several biotic and abiotic constraints [10,14,28,39,57]. With rain-fed systems, up to 50% of the yield potential can be compromised by moisture stress due to the fact of inconsistent rainfall [78]. Identifying ways to manage constraints that undermine the crop productivity and widen yield gap is key to developing an effective groundnut improvement programme. Desmae et al. [47] summarized the main traits of breeding interest identified in recent years: drought tolerance; resistance to rosette, foliar diseases (leaf spots and rust) and aflatoxin; and quality traits such as high oil content, especially with high proportion of high oleic acid. Potential sources were highlighted to improve groundnut varieties for these traits [47]. Not discounting the effects of G × E (interactions between genotype and environment), which can lead to inconsistent trait expression, these resources could be important assets to the groundnut improvement programme in Burkina Faso.

6.1. Abiotic Constraints

In Burkina Faso where rain-fed agriculture is predominant, rainfall patterns represent the most significant climatic factor affecting groundnut production. Low, erratic rainfall and increasing periods among rains render groundnut cultivation subject to substantial yield losses [79,80]. A strategy to cope with drought stress is to develop short-duration varieties to escape end-of-season drought as well drought tolerant varieties that hold up under conditions of low soil moisture [81].

Another constraint to realize yield potential is low levels of inputs in managing the crop. In Burkina Faso, groundnut is grown mostly under subsistence agriculture by smallholder farmers [74]. Fertiliser use for all crops since 2000 has averaged 11.1 kg/ha which is shockingly inadequate [64]. Appropriate use of fertilisers can lead to a 45% increase in groundnut yields [37]. However, farmers are incentivised to apply fertilisers and improve soil conditions only when there are market prospects [40]. Yet, critical elements, such as phosphorus and calcium, prove to be the top limiting nutrients for groundnut production, especially in the western regions of the country [36]. Although this issue can be overcome by applying appropriate chemical fertilisers, these are often out of reach for most smallholder farmers. Only 4% and 16% of farmers use chemical fertilisers or compost, respectively, in groundnut production, resulting in very low yields [37,74]. Therefore, it is advisable to develop cultivars that withstand the deficiency of both calcium and P to keep a good level of crop productivity. Soils in Burkina Faso present optimum pH between 6.0 and 6.5 for groundnut growth [82] which typically results in adequate availability of calcium and manganese [83]. Nevertheless, acidic pH should be taken into consideration in the breeding programme to anticipate the growing soil acidification in some areas in the country [84].

6.2. Biotic Constraints

As in most tropical regions of the world, diseases are major constraints to groundnut production in Burkina Faso [25]. Problems with foliar diseases including rust have been longstanding [85] with persistent occurrence in Burkina Faso [28]. Rosette disease, which affects the leaves and stem, is common in Western Burkina Faso [57,86,87] and is transmitted by the vector *Aphis craccivora* Koch [88] in a persistent manner [89]. Also, peanut clump disease, common in West Africa [90], has been reported in the country [57,86] and needs to be monitored. These diseases can cause important losses in groundnut production if not controlled [90]. Often foliar diseases occur simultaneously and collectively can cause from 24% up to 70% yield loss, following severe defoliation [10,25].

Groundnut productivity can also be reduced by soil pests [91]. Taxa associated with groundnut damage with high economic impact include termites (Isoptera), millipedes (Diplopoda) and scarabaeid larvae (Coleoptera) usually referred to as white grubs [39]. Species *Trochalus* sp., *Microtermes lepidus* Sjöstedt and *M. parvulus* Sjöstedt have been reported in Burkina Faso [92]; however, little is known about the economic importance of these pests at present. Additionally, the current erratic weather pattern can cause pests profiles to change thereby necessitating frequent nationwide surveys to document key pests associated with groundnut productivity. Studies at ICRISAT identified sources of resistance to these pests which hold promise for improving crop productivity [39,48]. Genetic resistance could be a key element of a broader strategy for effective pest control and control of pest-induced diseases, integrating use of pest-resistant varieties, cultural practices to minimize insect populations and bio-insecticides.

7. Suggested Foci for Groundnut Improvement in Burkina Faso

In principle, the target traits for groundnut improvement depend on farmers' needs, consumer and market demands and processing requirements [70]. The most pressing need in Burkina Faso is for groundnut varieties with high yield potential that also possess tolerance to major biotic and abiotic yield-reducing factors. Closing the yield gap is the main focus besides improving yield per se as in most of the developing countries [93]. Improved varieties must be able to thrive under minimal management conditions as farmers often simply cannot afford inputs such as pesticides and chemical fertilizers [85]. Furthermore, improved varieties must be developed to meet demands of the value chain [94], based on regular consultations with key stakeholders including both women and men farmers, marketers, processors and consumers [68,95].

7.1. Elements to Consider in Cultivar Development

In Burkina Faso as elsewhere in West Africa, market desirable traits in groundnut include high seed yield, high oil/high oleic oil content in the seeds and resistance to aflatoxins for food safety [75,96]. There is now a call for groundnuts specifically developed for end-use application: cooking oil, confectionary and peanut butter [47,97,98]. High oil content groundnut varieties are currently in high demand to supply oil-crushing factories. Recent studies have demonstrated the possibility to raise oil content to as much as 55% of seed composition, presenting up to 80% oleic [99,100]. Oil quality in terms of high proportion of oleic acid is desirable to increase product shelf life [101] and provide many health benefits to consumers [102–104]. The challenge is to build on these advances in groundnut improvement for oil content and quality [105] to put these traits together with other desirable agronomic traits (i.e., yield and disease resistance) in an ideal cultivar for stakeholders of the value chain [95].

The cornerstone and highest priority trait in crop breeding is the yield. Pod number per plant, shelling percentage, proportion of mature kernels per pod and seed weight are important parameters contributing to groundnut yield [70]. Other traits to consider include early maturity, ease in harvesting (peg strength) and shelling, kernel size, shape and colour, fresh seed dormancy and blanching ability [3,47,70]. Additionally, reticulation (venation and ridging visible on the pod), beak (appendage of the tip of the indehiscent pod) and constriction of pods are traits that provide not only varietal specifications, but also reflect market preferences [106]. For instance, slight pod constriction is preferred in the market, as it prevents flattened kernels, whereas pods with prominent reticulation or deep constriction tend to carry soil on them, thus reducing the market value [106].

To sustain productivity, groundnut resistance to biotic and abiotic stresses must be improved. To this end, ICRISAT has identified and developed sources of key traits, including resistance to early leaf spot (caused by *Cercospora arachidicola*), late leaf spot (caused by *Phaeoisariopsis personata*), rust (caused by *Puccinia arachidis*) and aflatoxins [10,27,28,30,47,53]. Additionally, significant progress has been achieved at ICRISAT in developing drought tolerant [107,108] and early maturing cultivars [47]. These sources can be utilised by Burkina-based breeding programmes to develop improved locally adapted germplasm.

7.2. Exploring Novel Industrial Uses of Groundnuts

Groundnut is considered as a "smart food", that is, a food that is highly nutritious, resilient to climate change with relatively low carbon and water footprints; as such, it has the potential to alleviate poverty [109]. Having high protein content and a healthy oil profile and serving as a source of key micronutrients including magnesium, groundnut has been used to make ready-to-use therapeutic food [110], used by UNICEF to treat acute malnutrition among children, women and men in developing countries [111,112]. Acute malnutrition affects near 500,000 children in Burkina Faso [112], resulting in 24.4% underweight and 10.2% mortality among children under five [113]. More generally, 25% of the population (~5 million people) are affected by hidden hunger [114,115]. To add more to the health benefit of groundnut, improvement for nutritional traits, i.e., bio-fortification, such as iron and zinc, must be on the breeding agenda [109,116]. This is important to reduce the prevalence of anaemia (>40%) among preschool-age children [113].

Besides, groundnut haulms can be used for livestock feed [117] thus giving additional value to the crop. Groundnut haulms are protein-rich and easy for animals to digest [16]. The need for feed has been increasing in recent years due to the drastic reduction of pastureland and the development of suburban farming in towns [118]. Therefore, animal feed production and market are promising sector of domestic economy [119], especially during the dry season when fresh grazing is not available [17]. However, livestock feed is rarely a production objective per se in subsistence farming. The development of dual-purpose varieties offering both high kernel yield and aboveground biomass could offer new opportunities to expand the groundnut value chain.

7.3. Broadening Genetic Base of Breeding Population

Cultivated groundnut is said to have a relatively narrow genetic base globally [1,120,121], perhaps due to the polyploidisation [122]. Therefore, useful genetic variability must be created through judicious choice of parents in creating new breeding populations for crop improvement [123]. The nature and magnitude of genetic variability present in the breeding population and the extent to which the trait is heritable are key to success of the crop improvement programme [123]. Pre-breeding activities deploying strategic crossing among cultivated varieties and also between cultivars and wild groundnuts [122,124–126] has enlarged the crop base genetic diversity. Interestingly, accessions and advanced breeding lines which are stored in gene banks across the globe [93] are abundant and accessible through appropriate legal procedures [127]. These resources constitute invaluable material for national and international breeding programmes.

8. Modernization Is Needed to Maximise Genetic Gain in Developing Varieties that Meet Stakeholder Demands

In principle, plant breeding is implemented through three basic steps, viz. (1) crossing choice individuals with traits of interest to create breeding populations with useful genetic variation, (2) identification and selection of progeny from the breeding crosses having outstanding performance aligned with the product target, and (3) development of stable new cultivars from selected progeny [128]. The success of this process can be measured by estimating the rate of genetic gain over time, using the so-called breeder's equation [129,130]: $\Delta G = (h^2\, \sigma_p\, i)/L$. The estimate of the rate of genetic gain (ΔG) is a product of the narrow sense heritability for the trait under selection (h^2), the standard deviation of the phenotypic variance of the trait (σ_p), and the selection intensity (i), divided by the length of time to complete a full breeding cycle (L). As such, it is also a function of selection accuracy (h; the square root of narrow sense heritability) and the additive genetic variation within the population (σ_a^2; a component of h^2). Each of these parameters can and should be manipulated in the breeding programme to maximise genetic gain in achieving the product target [68]. Such a strategy implies increasing heritability, selection accuracy, selection intensity and the speed of the breeding cycle and effectively exploiting genetic variation [128]. Modern breeding approaches, technologies and tools offer the means to increase the rate of genetic gain to effectively and efficiently reach product targets and thus get improved varieties out to farmers faster.

8.1. Modern Approaches, Technologies and Tools to Benefit Choice of Parents and Creation of Breeding Populations

Choice of parents is one of the most critical decisions to achieving success in cultivar development. Firstly, parental lines must represent viable sources of the suite of traits defined in the product target. Crossing of parents offers the opportunity for genetic recombination to result in new combinations of favourable alleles in the offspring. Ultimately, a potential new cultivar must contain favourable alleles for all the traits of interest. Genomics can aid in identifying lines with favourable alleles to employ as parents. For example, GWAS (genome-wide association studies) can be conducted to characterize germplasm, identify new sources of favourable alleles, and tag genes to be tracked through the breeding process. Genomic approaches using GEBV (genomic estimated breeding values) can be used to leverage genetic information as well as phenotypic information collected from prospective parent lines and their relatives to guide the breeder in choosing parents. Once crosses are made, mating designs and tailored breeding approaches can be deployed to maximize seed returns and accelerate progress to homozygosity [131]. Technologies such as doubled haploidy has been used to create "instant inbreeds" in some crops, including groundnut [132,133] which offers advantages in testing by cutting "noise" due to the segregation that is present in early generations.

To create useful genetic variation, technologies such as mutation, transformation, and gene editing can be deployed. Mutation breeding involves irradiating seed with gamma rays or using chemical mutagens like ethyl methane sulphonate, diethylsulfate or sodium azide [134–136] to evoke changes in the DNA. Successful cases of mutation breeding have been reported extensively for the improvement of important traits including groundnut yield [137,138], allergen reduction [135], and oleic acid content in the oil profile [139]. Therefore, mutation breeding can be a useful breeding approach, especially with genetic improvement of crops having narrow genetic base such as groundnut [135,140].

Groundnut improvement for tolerance to some of the biotic and abiotic stresses can be difficult, either due to the complex genetic control of that trait or absence of resistant sources. For instance, it has been difficult to develop resistance to *Aspergillus flavus* infection and aflatoxin production in groundnut in a sustainable manner [141,142]. Similar issues observed with groundnut response to other stress contexts such as drought and virus attacks have warranted alternative approaches to conventional breeding. In such conditions, genetic transformation presents great potential in groundnut improvement to utilize genes from other species [45,70,143,144]. Likewise, the difficulties of plant regeneration by tissue culture techniques and selection of transgenic events [144–146] are being overcome by recent advances in groundnut transformation process [147]. To date, at least a dozen of successful groundnut transformations have been reported in the literature [69]. Recently, agrobacterium-mediated transformation and groundnut tissue culture techniques were refined for optimum use which enabled development of genetically modified groundnut that was near-immune to aflatoxin contamination [147].

Furthermore, gene editing has shown great promise in creating new allelic variants using various technologies such as zinc finger nucleases (ZFNs), transcription activator-like effector nucleases (TALENs) and clustered regularly interspaced short palindromic repeats (CRISPR) [148–151]. Gene editing can result in gene modification (e.g., single base change), gene silencing (i.e., knockout), or gene insertion (i.e., knock-in) [150]. Recent studies have shown that multiple genetic changes can be performed in concert [152], suggesting the potential to utilize CRISPR to generate new genetic diversity for quantitative traits. For example, work by Campa et al. [153] demonstrated, in mammalians, simultaneous editing of up to 25 target sites using CRISPR in conjunction with nuclease Cas12A. Thus, genetic engineering is a powerful tool to achieve groundnut improvement for difficult traits, but also any other traits of interest [144,145,154]. However, public reluctance to consume food made from genetically modified plants [155–157] and complex regulation processes [150,158–160] could preclude the application of genetic engineering approaches in groundnut breeding. Although genetically modified crops and those derived from any gene editing pipelines are not banned per se in Burkina Faso, they are subject to stringent regulatory law before experimentation, trials and commercial release [161]. Public education and government support for appropriate review and regulation of genetically engineered products is key to overcoming potential obstacles.

8.2. Modern Approaches, Technologies and Tools to Benefit Evaluation and Selection

The rate of genetic gain toward product targets can be increased by increasing selection intensity. However, this can cripple progress if the selection intensity is so high as to effectively eliminate genetic variation. This potentially negative effect can be managed by increasing population size, that is, creating more progeny from each breeding cross. To deal with a larger number of individuals to test for each trait specified in the product target, technologies that can screen more individuals in less time with fewer resources are needed. Mechanization can help to manage activities such as planting, harvesting and threshing. In addition, technologies to facilitate high-throughput phenotyping can be used to increase efficiency in screening. Thus, the near infrared spectroscopy (NIRS) provides a robust, quick, cost-effective and non-destructive phenotyping for groundnut seed oil content and fatty acid profile [162,163]. Additionally, modern phenotyping facilities such as that at ICRISAT offer the possibility to dissect physiological factors with tight correlation with traits of interest (transpiration efficiency and drought tolerance, for instance) [164]. Likewise, advanced experimental designs, including randomised blocks, variable incomplete blocks, factorial, lattice, row-column and partial replicated designs [165] are useful in effectively partitioning genetic variation from environmental variation, G × E and error which can increase the accuracy of selection. Here again, technologies can come into play. For example, laser levelling of fields has been used to create more uniform fields for testing [166] which has been shown to result in more precise data upon which to make selection decisions [167,168]. However, cost may be a limitation for the use of this technology in resource poor breeding programmes [168].

To reduce the length of the breeding cycle, various approaches and technologies are available. Off-season nurseries offer the opportunity for more generations per year, cutting the overall time to complete the breeding cycle [68,163]. The term "speed breeding" has been coined to describe approaches and technologies to shorten the life cycle of the plant in generations where selection is not exercised [169,170]. For example, O'Connor et al. [171] were able to cycle 145 day groundnut lines at the rate of three generations per year to advance inbreeding from F2 to F4 based on controlled greenhouse conditions of optimal temperature and continuous light. In addition, marker-assisted selection, involving "tagged" genes of interest, can be utilized to cut the length of the breeding cycle. Individuals can be evaluated and selected based on genotype alone, eliminating wait times until the trait is manifested phenotypically and, in many cases, reducing field and labour resources required for phenotypic evaluation.

Advanced molecular approaches such as genomic selection utilize dense genome marker coverage to estimate genetic potential. Genomic selection facilitates faster identification of lines to serve as parents in the next breeding cycle, shortening cycle time [131]. In addition, genomic selection can be used to "predict" performance, replacing preliminary testing, as a means to advance progeny with greater genetic promise to advanced testing stages [131,172]. Furthermore, it can offer higher predictive ability when associated with modelling G × E interaction [173]. Estimated gain from genomic selection can be as much as 5 fold that of conventional breeding [174–176]. Although uptake of genomics-assisted breeding in Burkina Faso has been extremely low, as in many developing countries [69], due to the lack of human and infrastructure resources [70], nevertheless, opportunity for breeding programmes to embrace genomics-assisted breeding exists, using, where applicable, data available in partner institutions like ICRISAT where historical data on the performance of about 340 advanced breeding lines have been compiled [70].

8.3. Modern Approaches, Technologies and Tools to Support Commercialization and Release of Improved Cultivars

Modern approaches and technologies are available to support production of volumes of quality seed for distribution of new improved groundnut cultivars to farmers. DNA fingerprinting technologies ensure seed authenticity and purity [177]. Seed quality can be preserved using seed storage technologies, along with monitoring of relative humidity and seed moisture content with electronic meters or indicator papers [178]. Preserving seed quality is crucial for both the cultivar development process and the conservation of germplasm.

Data management tools that track materials through the entire breeding process facilitate traceability and document pedigree information. For example, the Breeding Management System

(BMS) [179] provides a comprehensive suite of mutually compatible software applications that work together to help breeders manage germplasm and collect, store and analyse their research data [180]. The BMS manages breeding data across all phases of the crop improvement cycle, keeping a safe, standardized and centralized record of data from one generation to the next in order to facilitate more economical and accelerated cultivar development. Such a system is not only crucial and foundational to breeding teams in their quest for selection accuracy, it supports breeding operations, resource allocation and data analyses [181], and provides support services at every step of the breeding process, all the way through to cultivar release. Such tools are essential to integrate and support all aspects of the breeding pipeline and to integrating efforts across team members.

A number of analytical and decision support tools for genomics-assisted breeding are freely available [182]. For example, CIMMYT offers several software options to facilitate various specialized analyses [183]. Other resources are available in the scientific literature to facilitate bioinformatics aspects of managing DNA-sequence data (e.g., GAPIT; Genome Association and Prediction Integrated Tool [184]). Use of publicly available tools such as these satisfies needs while avoiding license fees.

9. Research Challenges in Burkina Faso

Important breeding programmes are conducted across the world to improve groundnut as a multipurpose oilseed legume crop. Achievements in groundnut genomic research will affect cultivar development worldwide [105,185]. Collegial initiatives, such as the International Peanut Genome Consortium, resulted in major knowledge about the groundnut genome, which aided the deployment of molecular markers in breeding projects [1,70,99,163,182]. In addition to yield increase, cultivar development was geared towards traits as resistance to drought, aflatoxin resistance and foliar diseases, and seed quality and nutrient content [105,186,187]. The ICRISAT Headquarters (India) and its derivative research centres in Africa have been leading groundnut research programmes for decades, based on consumer and farmer preferences in semi-arid regions [186], thus providing some of the most significant impact on the crop production in Africa and Asia [186]. ICRISAT together with national partners have released near 200 improved ICRISAT-bred cultivars in 36 countries since 1986 [105]. In China, the world leading groundnut producer country, the breeding efforts using conventional and advanced methods, resulted in the release of more than 400 varieties in about 70 years [105]. In the USA, several university-based research teams (Georgia, Texas, and Auburn) conduct state-of-the-art research and breeding for traits of commercial importance [135,154,188,189]. Some of these research programmes overlap with breeding activities in national agricultural research systems (NARS) across Africa.

Although noticeable progress has been recorded in recent years [186], groundnut research has been comparatively trivial in many African NARS [69]. In Burkina Faso and other countries in West Africa, groundnut breeding operates to satisfy uneven environments and diverse stakeholders with low uptake of agricultural technologies [55,190]. To succeed cultivar development in such a context, design and implementation of trials must take into consideration the huge gap that often exists between optimum on-station conditions and irregular farmer field settings. This may result in increased costs due to the need for high numbers of field trials.

Additionally, the implementation of a modern breeding programme requires expertise, infrastructure, equipment, all of which requires a higher level of investment. To assess the benefit and ultimate value of implementing a new approach, technology, or tool, a cost-benefit analysis can be performed to provide justification for the additional expenditures. Furthermore, with or without further investment, other factors can go a long way to build in greater efficiencies in cultivar development: outsourcing some activities requiring special equipment or expertise (e.g., genotyping), establishing research networks to better leverage available resources and data, and forging partnerships with the private sector.

The lack of sustained funding and over-dependency on donors [64] restrains possibilities to implement long term view and renders the programme vulnerable to funding inconsistencies and abrupt changes in research agenda and vision. Therefore, efforts must be put into igniting government commitment to research for food and nutritional security in the country. Policy makers and those in the groundnut value chain must be made aware of possibilities and challenges if groundnut production is to impact national nutrition and trade and draw support from the private sector.

Ultimately, private sector intervention is probably the way forward to dependably invest substantial funding in the crop breeding and bring better governance in the breeding programme. The groundnut industry could be inspired by successful examples of private agricultural research in Burkina Faso and elsewhere, driven by cash crops such as cotton, banana and oil palm [64,191,192].

10. Conclusions

Groundnut production and genetic improvement in Burkina Faso has stagnated for too long. In the absence of a strong national program, research in the country has centred on evaluations of lines developed at international research institutes and programmes such as IRHO, ICRISAT and Peanut CRSP. However, lines from international institutions may have limited alignment with domestic product targets and fail to deliver adaptation under local conditions required for high yield. For best adaptability of cultivars to local conditions and national stakeholder needs, a strong national breeding programme built on the foundation of local germplasm collection must be the driver.

Most of the issues discussed in this review are applicable to many other national agricultural research programmes in sub-Saharan or West Africa. The scientific strength of breeding programmes requires expertise in plant breeding and genetics (i.e., at least two full-time PhD scientists per crop species [55]) as well as support in related disciplines important for groundnut improvement, viz. entomology, agronomy, weed science, pathology [55,69]. We contend that it is possible to significantly increase groundnut production and productivity through dedication of a strong local breeding programme which takes advantage of improved lines from international research institutions and modern breeding approaches, technologies and tools to develop locally adapted, high-yielding varieties with desirable traits. To this extent, strategies to accelerate genetic gain need to be adopted, along with gender integration in the entire crop development and value chain. Building technical and infrastructure capacities of the national breeding programme is needed to achieve such a research level and to expedite delivery of improved groundnut varieties to modern and smallholder farmers.

Author Contributions: M.K. wrote the first draft and incorporated the inputs from the co-authors and comments of editors. J.S., A.M., D.K.O., H.D., P.J. and R.H.M. reviewed the first draft and made inputs to improve the manuscript. All authors have read and agreed to the published version of the manuscript.

Funding: This research received no external funding and the APC was funded by authors.

Acknowledgments: This review was facilitated by INERA (Institute of Environment and Agriculture Research, Burkina Faso), AfPBA-AOCC (African Plant Breeding Academy and African Orphan Crops Consortium) and ICRISAT (International Crops Research Institute for the Semi-Arid Tropics).

Conflicts of Interest: The authors declare no conflict of interest.

References

1. Bertioli, D.J.; Jenkins, J.; Clevenger, J.; Dudchenko, O.; Gao, D.; Seijo, G.; Leal-Bertioli, S.C.M.; Ren, L.; Farmer, A.D.; Pandey, M.K.; et al. The genome sequence of segmental allotetraploid peanut arachis hypogaea. *Nat. Genet.* **2019**, *51*, 877–884. [CrossRef] [PubMed]
2. Valls, J.F.; Simpson, C.E. New species of arachis (leguminosae) from brazil, paraguay and bolivia. *Bonplandia* **2005**, *14*, 35–63. [CrossRef]
3. Stalker, H.T. Peanut (*Arachis hypogaea* L.). *Field Crops Res.* **1997**, *53*, 205–217. [CrossRef]
4. FAO Food and Agricultural Commodities Production. Available online: http://www.fao.org/faostat/en/#data/QC (accessed on 30 March 2020).
5. Tyroler, C. *Gender Considerations for Researchers Working in Groundnuts*; USAID—Feed The Future: Washington, DC, USA, 2018; p. 32.
6. Santara, I.; Mas Aparisi, A.; Balié, J. *Analyse des Incitations et Pénalisations Pour L'arachide au Burkina Faso*; FAO: Rome, Italy, 2013; p. 37.

7. Varshney, R.K.; Ribaut, J.-M.; Buckler, E.S.; Tuberosa, R.; Rafalski, J.A.; Langridge, P. Can genomics boost productivity of orphan crops? *Nat. Biotechnol.* **2012**, *30*, 1172–1176. [CrossRef]

8. Ndjeunga, J.; Ibro, A.; Cisse, Y.; Ben Ahmed, M.I.; Moutari, A.; Kodio, O.; Echekwu, C. *Characterizing Village Economies in Major Groundnut Producing Countries in West Africa: Cases of Mali, Niger and Nigeria*; ICRISAT: Hyderabad, India, 2010; p. 89.

9. Feldstein, H.S.; Butler Flora, C.; Poats, S.V. *Gender Variable in Agricultural Research*; International Development Research Centre: Ottawa, ON, Canada, 1989.

10. Subrahmanyam, P.; Bosc, J.P.; Hassane, H.; Smith, D.H.; Mounkaila, A.; Ndunguru, B.J.; Sankara, P. Les maladies de l'arachide au niger et au burkina faso. *Oléagineux* **1992**, *47*, 119–133.

11. Alper, C.M.; Mattes, R.D. Peanut consumption improves indices of cardiovascular disease risk in healthy adults. *J. Am. Coll. Nutr.* **2003**, *22*, 133–141. [CrossRef]

12. Francisco, M.L.D.L.; Resurreccion, A.V.A. Functional components in peanuts. *Crit. Rev. Food Sci. Nutr.* **2008**, *48*, 715–746. [CrossRef]

13. Jibrin, M.; Habu, S.; Echekwu, C.; Abdullahi, U.; Usman, I. Phenotypic and genotypic variance and heritability estimates for oil content and other agronomic traits in groundnut (*arachis hypogaea l.*). *Int. J. Sci. Res. Eng. Stud.* **2016**, *3*, 29–32.

14. Hamidou, F.; Harou, A.; Achirou, B.; Halilou, O.; Bakasso, Y. Nitrogen fixation by groundnut and cowpea for productivity improvement in drought conditions in the sahel. *Tropicultura* **2018**, *36*, 63–79.

15. Ranganayakulu, G.S.; Chandraobulreddy, P.; Thippeswamy, M.; Veeranagamallaiah, G.; Sudhakar, C. Identification of drought stress-responsive genes from drought-tolerant groundnut cultivar (*Arachis hypogaea L. Cv k-134*) through analysis of subtracted expressed sequence tags. *Acta Physiol. Plant.* **2012**, *34*, 361–377. [CrossRef]

16. Blummel, M.; Ratnakumar, P.; Vadez, V. Opportunities for exploiting variations in haulm fodder traits of intermittent drought tolerant lines in a reference collection of groundnut (*Arachis hypogaea L.*). *Field Crop. Res.* **2012**, *126*, 200–206. [CrossRef]

17. Desmae, H.; Sones, K. *Groundnut Cropping Guide*; Africa Soil Health Consortium: Nairobi, Kenya, 2017.

18. Njuki, J.; Kaaria, S.; Chamunorwa, A.; Chiuri, W. Linking smallholder farmers to markets, gender and intra-household dynamics: Does the choice of commodity matter? *Eur. J. Dev. Res.* **2011**, *23*, 426–443. [CrossRef]

19. Narh, S.; Boote, K.J.; Naab, J.B.; Abudulai, M.; M'Bi Bertin, Z.; Sankara, P.; Burow, M.D.; Tillman, B.L.; Brandenburg, R.L.; Jordan, D.L. Yield improvement and genotype × environment analyses of peanut cultivars in multilocation trials in west africa. *Crop Sci.* **2014**, *54*, 2413–2422. [CrossRef]

20. USDA. *Quick Stats*; US Department of Agriculture: Washington, DC, USA, 2019.

21. Zongo, A.; Khera, P.; Sawadogo, M.; Shasidhar, Y.; Sriswathi, M.; Vishwakarma, M.K.; Sankara, P.; Ntare, B.R.; Varshney, R.K.; Pandey, M.K.; et al. Ssr markers associated to early leaf spot disease resistance through selective genotyping and single marker analysis in groundnut (*Arachis hypogaea L.*). *Biotechnol. Rep.* **2017**, *15*, 132–137. [CrossRef]

22. Anonymous. Development of rainfed agriculture in the sahel: Overview and prospects. In Proceedings of the Fifth Conference of the Club du Sahel, Brussels, Belgium, 26–28 October 1983; Sahel, C.D., Ed.; CILSS: Ouagadougou, Burkina Faso, 1983.

23. Morris, W.H.M. *Production, commercialisation et exportation de l'arachide: Senegal, gambie, mali, burkina faso et niger*; Peanut CRSP: Griffins, GA, USA, 2020. Available online: https://pdf.usaid.gov/pdf_docs/PNABM077.pdf (accessed on 13 May 2020).

24. MAFAP. *Revue des Politiques Agricoles et Alimentaires au Burkina Faso*; SPAAA, Ed.; FAO: Rome, Italy, 2013.

25. Isleib, T.; Wynne, J.; Nigam, S. Groundnut breeding. In *The Groundnut Crop*; Springer: Berlin/Heidelberg, Germany, 1994; pp. 552–623.

26. Neya, F.B.; Sanon, E.; Koita, K.; Zagre, B.M.b.; Sankara, P. Diallel analysis of pod yield and 100 seeds weight in peanut (*Arachis hypogaea L.*) using griffing and hayman methods. *J. Appl. Biosci.* **2017**, *116*, 11619–11627. [CrossRef]

27. Zongo, A.; Konate, A.K.; Koïta, K.; Sawadogo, M.; Sankara, P.; Ntare, B.R.; Desmae, H. Diallel analysis of early leaf spot (*cercospora arachidicola hori*) disease resistance in groundnut. *Agronomy* **2019**, *9*, 15. [CrossRef]

28. Zongo, A.; Nana, A.; Sawadogo, M.; Konate, A.; Sankara, P.; Ntare, B.; Desmae, H. Variability and correlations among groundnut populations for early leaf spot, pod yield, and agronomic traits. *Agronomy* **2017**, *7*, 52. [CrossRef]

29. CRSP, P. *Annual Report of the Peanut Collaborative Research Support Program (Crsp)*; University of Georgia: Athens, GA, USA, 1983; p. 231.

30. Gapasin, D.; Cherry, J.; Gilbert, J.; Gibbons, R.; Hildebrand, G.; Nelson, D.; Valentine, H.; Williamson, H. *The Peanut Collaborative Research Support Program (Crsp)*; University of Georgia: Athens, GA, USA, 2005; p. 359.

31. Dalton, T.J.; Cardwell, K.; Katsvairo, T. *The Peanut Collaborative Research Support Program: A Report Submitted to the Bureau of Food Security, Usaid*; USAID: Washington, DC, USA, 2012; p. 56.

32. Swindell, K. Serawoollies, tillibunkas and strange farmers: The development of migrant groundnut farming along the gambia river, 1848–1895. *J. Afr. Hist.* **1980**, *21*, 93–104. [CrossRef]

33. Brooks, G.E. Peanuts and colonialism: Consequences of the commercialization of peanuts in west africa, 1830–1870. *J. Afr. Hist.* **1975**, *16*, 29–54. [CrossRef]

34. Anonymous. L'économie de la Haute Volta. Available online: https://clubjosephkizerbo.blog4ever.com/l-economie-de-la-haute-volta (accessed on 17 July 2019).

35. Anonymous. *Histoire du Burkina du 19e Siècle à Nos Jours*; Eurêka: Brussels, Belgium, 1996; p. 20.

36. IRHO. *L'arachide: Principaux Resultats Obtenus en Experimentation*; Institut de Recherches Pour les Huiles et Oleagineux (IRHO): Station de Niangoloko, Haute-Volta, 1964; p. 13.

37. Gbikpi, P. *L'agriculture Burkinabe*; Ministère de L'agriculture et des Ressources Animales: Luanda, Angola, 1996.

38. Stads, G.-J.; Boro, S.I. Indicateurs relatifs aux sciences et technologies agricoles, Burkina Faso. In *Abrégé l'ASTI*; INERA-IFPRI, Ed.; IFPRI: Washington, DC, USA, 2004; p. 10.

39. Umeh, V.; Youm, O.; Waliyar, F. Soil pests of groundnut in sub-Saharan Africa—A review. *Int. J. Trop. Insect Sci.* **2001**, *21*, 23–32. [CrossRef]

40. Barandao, A. Étude de Faisabilité Technico-Economique Pour L'implantation D'une Unité de Production de Plumpy'nut à Kongoussi. Ph.D. Thesis, EIER-ETSHER, Ouagadougou, Burkina Faso, 2006.

41. Volper, S.; Bichat, H. Des jardins d'essai au cirad: Une épopée scientifique française. In *Un Parcours Dans les Mondes de la Recherche Agronomique. L'Inra et le Cirad*; CNRS: Paris, France, 2014; Volume 3, pp. 113–124.

42. Tourte, R. *Histoire de la Recherche Agricole en Afrique Tropicale Francophone*; Organisation des Nations Unies pour L'alimentation et L'agriculture: Rome, Italy, 2005.

43. Surre, C. *L'institut de Recherches Pour les Huiles et Oléagineux: 1942–1984*; CIRAD: Paris, France, 1993.

44. Sauger, L.; Catherinet, M.; Durand, Y. Contribution a l'etude de la rosette chlorotique de l'arachide. *Bull. Agron. Minist. Fr. Outremer* **1954**, *13*, 163–180.

45. Nigam, S.; Dwivedi, S.; Gibbons, R. Groundnut breeding: Constraints, achievements and future possibilities. In *Plant Breeding Abstracts*; CAB International: Wallingford, UK, 1991; pp. 1127–1136.

46. Matlon, P.; Cantrell, R.; King, D.; Benoit-Cattin, M. *Coming Full Circle: Farmers' Participation in the Development of Technology*; IDRC: Ottawa, ON, Canada, 1984.

47. Desmae, H.; Janila, P.; Okori, P.; Pandey, M.K.; Motagi, B.N.; Monyo, E.; Mponda, O.; Okello, D.; Sako, D.; Echeckwu, C.; et al. Genetics, genomics and breeding of groundnut (*Arachis hypogaea* L.). *Plant Breed.* **2019**, *138*, 425–444. [CrossRef] [PubMed]

48. Amin, P.; Singh, K.; Dwivedi, S.; Rao, V. Sources of resistance to the jassid (empoasca kerri pruthi), thrips (frankliniella schultzei (trybom)) and termites (*odontotermes* sp.) in groundnut (*Arachis hypogaea* L.). *Peanut Sci.* **1985**, *12*, 58–60. [CrossRef]

49. ICRISAT. Resistance to Soil-borne Diseases of Legumes. In Proceedings of the Consultants' Group Discussion on the Resistance to Soil-Borne Diseases of Legumes, Patancheru, India, 8–11 January 1979; Nene, Y.L., Ed.; ICRISAT: Patancheru, India, 1979; p. 180.

50. Upadhyaya, H.D.; Nigam, S.N.; Mehan, V.K.; Lenne, J.M. Aflatoxin contamination of groundnut: Prospects for a genetic solution through conventional breeding. In Proceedings of the First Asia Working Group Meeting, Hanoi, Vietnam, 27–29 May 1996; Mehan, V.K., Gowda, C.L.L., Eds.; International Crops Research Institute for the Semi-Arid Tropics: Hanoi, Vietnam, 1997; pp. 81–85.

51. Upadhyaya, H.; Nigam, S.; Thakur, R. Genetic Enhancement for Resistance to Aflatoxin Contamination in Groundnut. In *Summary proceedings of the Seventh ICRISAT Regional Groundnut Meeting for Western and Central Africa*; ICRISAT: Cotonu, Benin, 2002; pp. 29–36.

52. Waliyar, F.; Bockelee-Morvan, A. Resistance of groundnut varieties to aspergillus flavus in senegal. In *International Workshop on Aflatoxin Contamination of Groundnut, Patancheru, Andhra Pradesh, India, 1989*; Hall, S.D., Ed.; ICRISAT: Andhra Pradesh, India, 1989; p. 426.

53. Nigam, S.N.; Waliyar, F.; Aruna, R.; Reddy, S.V.; Kumar, P.L.; Craufurd, P.Q.; Diallo, A.T.; Ntare, B.R.; Upadhyaya, H.D. Breeding peanut for resistance to aflatoxin contamination at icrisat. *Peanut Sci.* **2009**, *36*, 42–49. [CrossRef]

54. World-Bank. *Upper Volta—Agricultural Issues Study*; 3296; World Bank: Washington, DC, USA, 1982; p. 267.

55. Ndjeunga, J.; Mausch, K.; Simtowe, F. Assessing the effectiveness of agricultural r&d for groundnut, pearl millet, pigeonpea, and sorghum in west and central africa and east and southern africa. In *Crop Improvement, Adoption, and Impact of Improved Varieties in Food Crops in Sub-Saharan Africa*; Walker, T.S., Alwang, J., Eds.; CGIAR and CAB international: Wallingford, UK, 2015; chapter 7; pp. 123–147.

56. INERA. *Productions Scientifiques et Techniques des Chercheurs de L'inera (1990–2012)*; INERA: Ouagadougou, Burkina Faso, 2012; p. 97.

57. Germani, G. *Etude Nématologique de Deux Affections de L'arachide en Haute-Volta: La Chlorose et le Clump*; Office de la Recherche Scientifique et Technique D'outre—Mer (ORSTOM): Abidjan, Adiopodoumé, 1973; p. 31.

58. Picasso, C. Evolution des rendements et de ses composantes pour l'arachide et quelques cultures en rotation dans le sud du burkina faso. *Oléagineux* **1987**, *42*, 469–474.

59. Gillier, P.; Silvestre, P. *L'arachide*; Maisonneuve et Larose: Paris, France, 1969.

60. CNS. *Catalogue National des Especes et Varietes Agricoles du Burkina Faso*; Semences, C.N.S., Ed.; CNS: Ouagadougou, Burkina Faso, 2014; p. 81.

61. Mayeux, A.H.; F, W.; R, N.B. *Groundnut Germplasm Project*; ICRISAT: Andhra Pradesh, India, 2003; p. 78.

62. Anonymous. *Catalogue Régional des Espèces et Variétés Végétqles Cedeao-Uemoa-Cilss: Variétés Homologuées 2016–2018*; CORAF, Ed.; CORAF: Dakar, Senegal, 2019.

63. ASTI. *Cgiar's Diiva Project*; A Consolidated Database of Crop Varietal Releases, Adoption, and Research Capacity in Africa South of the Sahara; CGIAR: Montpellier, France, 2017.

64. Hollinger, F.; Staatz, J.M. *Croissance Agricole en Afrique de L'ouest: Facteurs Déterminants de Marché et de Politique*; L'organisation des Nations Unies Pour L'alimentation et L'agriculture: Rome, Italy, 2015.

65. Stads, G.-J.; Kaboré, S.S. *Burkina Faso: Évaluation de la Recherche Agricole*; IFPRI: Roma, Italy, 2010.

66. INERA. *Institut de L'environement et de Recherches Agricoles: Bilan de 10 Annees de Recherche 1988–1998*; CNRST/INERA: Stockholm, Sweden, 2000; p. 122.

67. Miningou, A.; Sawadogo, E.; Barry, S.; Traore, S.A. *Final Narrative Report on Groundnut 2018 TL III/Burkina Faso*; INERA: Stockholm, Sweden, 2019; p. 28.

68. Cobb, J.N.; Juma, R.U.; Biswas, P.S.; Arbelaez, J.D.; Rutkoski, J.; Atlin, G.; Hagen, T.; Quinn, M.; Ng, E.H. Enhancing the rate of genetic gain in public-sector plant breeding programs: Lessons from the breeder's equation. *Theor. Appl. Genet.* **2019**, *132*, 627–645. [CrossRef]

69. Abady, S.; Shimelis, H.; Janila, P.; Mashilo, J. Groundnut (*Arachis hypogaea* L.) improvement in sub-saharan africa: A review. *Acta Agric. Scand. Sect. B Soil Plant Sci.* **2019**, *69*, 528–545. [CrossRef]

70. Janila, P.; Variath, M.T.; Pandey, M.K.; Desmae, H.; Motagi, B.N.; Okori, P.; Manohar, S.S.; Rathnakumar, A.L.; Radhakrishnan, T.; Liao, B.; et al. Genomic tools in groundnut breeding program: Status and perspectives. *Front. Plant Sci.* **2016**, *7*, 289. [CrossRef]

71. Mumm, R.; Danquah, E. The state of soybean in africa: The african plant breeders of tomorrow. *Farmdoc Daily* **2019**, *9*, 15116–15120.

72. Mumm, R.H.; Howard-Yana, S.; Danquah, E.Y.; Van Deynze, A.; Edema, R.; Achigan-Dako, E.G.; Suza, W.P.; Madakadze, R.M. Aiming for excellence in training and sustaining african plant breeders. In *Plant and Animal Genome Meeting XXVIII*; PAG XXVIII: San Diego, CA, USA, 2020.

73. Kindo, Y.; Mas Aparisi, A. *Analyse des Incitations Par les Prix Pour Arachide au Burkina Faso Pour la Période 2005–2013*; FAO: Rome, Italy, 2015.

74. World-Bank. *Burkina Faso. Le Défi de la Diversification des Exportations Dans un Pays Enclavé: Étude Diagnostique sur L'intégration Commerciale Pour le Programme du Cadre Intégré 43134*; World Bank: Washington, DC, USA, 2007; p. 172.

75. FAO. *Worldwide Regulations for Mycotoxins in Food and Feed in 2003*; 92-5-105162-3; Food and Agriculture Organisation of the United Nations: Rome, Italy, 2004.

76. CCIB. *Donnees Ccib 2018*; Chambre de Commerce et d'Industrie du Burkina Faso: Ouagadougou, Burkina Faso, 2018.

77. van Ittersum, M.K.; Cassman, K.G.; Grassini, P.; Wolf, J.; Tittonell, P.; Hochman, Z. Yield gap analysis with local to global relevance—A review. *Field Crop. Res.* **2013**, *143*, 4–17. [CrossRef]

78. Lobell, D.B.; Cassman, K.G.; Field, C.B. Crop yield gaps: Their importance, magnitudes, and causes. *Annu. Rev. Environ. Resour.* **2009**, *34*, 179–204. [CrossRef]

79. KPR, V.; Sankar, G.M.; HP, S.; Balaguravaiah, D.; Padmalatha, Y. Modeling sustainability of crop yield in rainfed groundnut based on rainfall and land degradation. *Indian J. Dryland Agric. Res. Dev.* **2003**, *13*, 7–13.

80. Camberlin, P.; Diop, M. Inter-relationships between groundnut yield in senegal, interannual rainfall variability and sea-surface temperatures. *Theor. Appl. Climatol.* **1999**, *63*, 163–181. [CrossRef]

81. Janila, P.; Nigam, S.N. Phenotyping for groundnut (*Arachis hypogaea* L.) improvement. In *Phenotyping for Plant Breeding: Applications of Phenotyping Methods for Crop Improvement*; Panguluri, S.K., Kumar, A.A., Eds.; Springer: New York, NY, USA, 2013; pp. 129–167.

82. Stoop, W.A. Variations in soil properties along three toposequences in burkina faso and implications for the development of improved cropping systems. *Agric. Ecosyst. Environ.* **1987**, *19*, 241–264. [CrossRef]

83. Parker, M.B.; Walker, M.E. Soil ph and manganese effects on manganese nutrition of peanut1. *Agron. J.* **1986**, *78*, 614–620. [CrossRef]

84. Coulibaly, P.J.d.A.; Okae-Anti, D.; Ouattara, B.; Sawadogo, J.; Sedogo, M.P. Effect of Dry Cropping Season of Sorghum on Selected Physico-Chemical Properties in West Africa. *Int. J. Agric. Innov. Res.* **2018**, *7*, 192–196.

85. Gibbons, R.; Mertin, J. *Abstracts in English and French of the International Workshop on Groundnuts 13–17 October 1980*; ICRISAT: Andhra Pradesh, India, 1981.

86. Bonkoungou, S. Distribution Géographique et Quelques Aspects Ecologiques des Virus du Clump et de la Rosette de L'arachide au Burkina Faso. Ph.D. Thesis, Universite de Dschang, Dschang, Cameroun, 2001.

87. De Berchoux, C. La rosette de l'arachide en haute-volta comportement des lignées résistantes. *Oléagineux* **1960**, *15*, 229–233.

88. Dubern, J. *La Rosette Chlorotique de L'arachide: Contribution à L'étude de la Transmission Par Aphis Craccivora Koch*; ORSTOM: Adiopodoume, Cote d'Ivoire, 1977; p. 16.

89. Naidu, R.; Kimmins, F.; Deom, C.; Subrahmanyam, P.; Chiyembekeza, A.; Van der Merwe, P. Groundnut rossette: A virus disease affecting groundnut production in sub-saharan africa. *Plant Dis.* **1999**, *83*, 700–709. [CrossRef]

90. Nigam, S.; Prasada Rao, R.; Bhatnagar-Mathur, P.; Sharma, K. Genetic management of virus diseases in peanut. *Plant Breed. Rev.* **2012**, *36*, 293.

91. Wightman, J.; Rao, G.R. Groundnut pests. In *The Groundnut Crop*; Springer: Berlin/Heidelberg, Germany, 1994; pp. 395–479.

92. Umeh, V. Soil pests of groundnut in west africa–case study of mali, burkina-faso, niger, and nigeria. *Int. Arachis Newsl.* **1998**, *18*, 50–52.

93. Janila, P.; Nigam, S.N.; Pandey, M.K.; Nagesh, P.; Varshney, R. Groundnut improvement: Use of genetic and genomic tools. *Front. Plant Sci.* **2013**, *4*, 23. [CrossRef] [PubMed]

94. Ragot, M.; Bonierbale, M.; Weltzein, E. *From Market Demand to Breeding Decisions: A Framework: Working Paper 2*; CIP and CGIAR: Lima, Peru, 2018; p. 53.

95. Persley, G.J.; Anthony, V.M. *The Business of Plant Breeding: Market-Led Approaches to New Variety Design in Africa*; CABI: Wallingford, UK, 2017.

96. Dorner, J.W. Management and prevention of mycotoxins in peanuts. *Food Addit. Contam. Part A* **2008**, *25*, 203–208. [CrossRef] [PubMed]

97. Christèle, I.-V.; Laurencia, O.; Sylvie, A.; Hounhouigan, J.; Polycarpe, K.; Waliou, A.; Hama, F.B. *Traditional Recipes of Millet, Sorghum-and Maize-Based Dishes and Related Sauces Frequently Consumed by Young Children in Burkina Faso and Benin*; Wageningen University Publisher: Wageningen, The Netherlands, 2010.

98. Ayensu, D.A. *The Art of West African Cooking*, 1st ed.; Doubleday: New York, NY, USA, 1972; p. 145.

99. Chen, X.; Lu, Q.; Liu, H.; Zhang, J.; Hong, Y.; Lan, H.; Li, H.; Wang, J.; Liu, H.; Li, S.; et al. Sequencing of cultivated peanut, arachis hypogaea, yields insights into genome evolution and oil improvement. *Mol. Plant* **2019**, *12*, 920–934. [CrossRef]

100. Varshney, R.K. Exciting journey of 10 years from genomes to fields and markets: Some success stories of genomics-assisted breeding in chickpea, pigeonpea and groundnut. *Plant Sci.* **2016**, *242*, 98–107. [CrossRef]
101. Bolton, G.E.; Sanders, T.H. Effect of roasting oil composition on the stability of roasted high-oleic peanuts. *J. Am. Oil Chem. Soc.* **2002**, *79*, 129–132. [CrossRef]
102. Vassiliou, E.K.; Gonzalez, A.; Garcia, C.; Tadros, J.H.; Chakraborty, G.; Toney, J.H. Oleic acid and peanut oil high in oleic acid reverse the inhibitory effect of insulin production of the inflammatory cytokine tnf-α both in vitro and in vivo systems. *Lipids Health Dis.* **2009**, *8*, 25. [CrossRef]
103. O'Byrne, D.J.; Knauft, D.A.; Shireman, R.B. Low fat-monounsaturated rich diets containing high-oleic peanuts improve serum lipoprotein profiles. *Lipids* **1997**, *32*, 687–695. [CrossRef]
104. Rizzo, W.B.; Watkins, P.A.; Phillips, M.W.; Cranin, D.; Campbell, B.; Avigan, J. Adrenoleukodystrophy: Oleic acid lowers fibroblast saturated c22–26 fatty acids. *Neurology* **1986**, *36*, 357–361. [CrossRef]
105. Holbrook, C.C.; Burow, M.D.; Chen, C.Y.; Pandey, M.K.; Liu, L.; Chagoya, J.C.; Chu, Y.; Ozias-Akins, P. Chapter 4—Recent advances in peanut breeding and genetics. In *Peanuts*; Stalker, H.T., Wilson, R., Eds.; AOCS Press: Urbana, IL, USA, 2016; pp. 111–145.
106. Rao, V.R.; Murty, U. Botany—Morphology and anatomy. In *The Groundnut Crop*; Springer: Berlin/Heidelberg, Germany, 1994; pp. 43–95.
107. Nigam, S.N.; Chandra, S.; Sridevi, K.R.; Bhukta, M.; Reddy, A.G.S.; Rachaputi, N.R.; Wright, G.C.; Reddy, P.V.; Deshmukh, M.P.; Mathur, R.K.; et al. Efficiency of physiological trait-based and empirical selection approaches for drought tolerance in groundnut. *Ann. Appl. Biol.* **2005**, *146*, 433–439. [CrossRef]
108. Ntare, B.R.; Williams, J.H.; Dougbedji, F. Evaluation of groundnut genotypes for heat tolerance under field conditions in a sahelian environment using a simple physiological model for yield. *J. Agric. Sci.* **2001**, *136*, 81–88. [CrossRef]
109. ICRISAT. Smart Food. Available online: https://www.smartfood.org/why-smartfood-2/ (accessed on 25 August 2019).
110. Nutriset Plumpy'nut®et le Modèle Cmam: Comment la r&d de Nutriset a Contribué à Transformer le Traitement de la Malnutrition Aiguë Sévère. Available online: https://www.nutriset.fr/articles/fr/plumpynut-et-le-modele-cmam-contribution-de-la-recherche-et-developpement-de-nutriset (accessed on 25 July 2019).
111. Tidey, C. *Teaming up to Turn the Tide against Malnutrition in Niger*; UNICEF: New York, NY, USA, 2012.
112. Bloemen, S. *Des Solutions Locales Pour Répondre à Une Crise Alimentaire Régionale*; UNICEF: New York, NY, USA, 2013.
113. WHO. *Worldwide prevalence of anaemia 1993-2005: Who global database of anaemia*; World Health Organization: Geneva, Switzerland, 2008; p. 51.
114. Gödecke, T.; Stein, A.J.; Qaim, M. The global burden of chronic and hidden hunger: Trends and determinants. *Glob. Food Secur.* **2018**, *17*, 21–29. [CrossRef]
115. von Grebmer, K.; Saltzman, A.; Birol, E.; Wiesman, D.; Prasai, N.; Yin, S.; Yohannes, Y.; Menon, P.; Thompson, J.; Sonntag, A. *2014 Global Hunger Index: The Challenge of Hidden Hunger*; Welthungerhilfe, IFPRI, Concern Worldwide: Washington, DC, USA, 2014; p. 56.
116. Janila, P.; Nigam, S.N.; Abhishek, R.; Anil Kumar, V.; Manohar, S.S.; Venuprasad, R. Iron and zinc concentrations in peanut (*Arachis hypogaea* L.) seeds and their relationship with other nutritional and yield parameters. *J. Agric. Sci.* **2014**, *153*, 975–994. [CrossRef]
117. Cook, B.; Crosthwaite, I. Utilization of arachis species as forage. In *The Groundnut Crop*; Springer: Berlin/Heidelberg, Germany, 1994; pp. 624–663.
118. Orsini, F.; Kahane, R.; Nono-Womdim, R.; Gianquinto, G. Urban agriculture in the developing world: A review. *Agron. Sustain. Dev.* **2013**, *33*, 695–720. [CrossRef]
119. Ayantunde, A.A.; Blummel, M.; Grings, E.; Duncan, A.J. Price and quality of livestock feeds in suburban markets of west africa's sahel: Case study from bamako, mali. *Revue D'elevage Medecine Veterinaire Pays Tropicaux* **2014**, *67*, 13–21. [CrossRef]
120. Moretzsohn, M.C.; Gouvea, E.G.; Inglis, P.W.; Leal-Bertioli, S.C.; Valls, J.F.; Bertioli, D.J. A study of the relationships of cultivated peanut (*Arachis hypogaea*) and its most closely related wild species using intron sequences and microsatellite markers. *Ann. Bot.* **2012**, *111*, 113–126. [CrossRef]
121. Stalker, H.T.; Tallury, S.P.; Seijo, G.R.; Leal-Bertioli, S.C. Chapter 2—Biology, speciation, and utilization of peanut species. In *Peanuts*; Stalker, H.T., Wilson, R., Eds.; AOCS Press: Urbana, IL, USA, 2016; pp. 27–66.

122. Foncéka, D.; Hodo-Abalo, T.; Rivallan, R.; Faye, I.; Sall, M.N.; Ndoye, O.; Fávero, A.P.; Bertioli, D.J.; Glaszmann, J.-C.; Courtois, B.; et al. Genetic mapping of wild introgressions into cultivated peanut: A way toward enlarging the genetic basis of a recent allotetraploid. *BMC Plant Biol.* **2009**, *9*, 103. [CrossRef]

123. Knauft, D.A.; Wynne, J.C. Peanut breeding and genetics. In *Advances in Agronomy*; Sparks, D.L., Ed.; Academic Press: Cambridge, MA, USA, 1995; Volume 55, pp. 393–445.

124. Abdou, Y.A.-M.; Gregory, W.C.; Cooper, W.E. Sources and nature of resistance to cercospora arachidicola hori and cercosporidium personatum (beck & curtis) deighton in arachis species. *Peanut Sci.* **1974**, *1*, 6–11.

125. Mallikarjuna, N.; Senthilvel, S.; Hoisington, D. Development of new sources of tetraploid arachis to broaden the genetic base of cultivated groundnut (*Arachis hypogaea L.*). *Genet. Resour. Crop Evol.* **2011**, *58*, 889–907. [CrossRef]

126. Janila, P.; Ramaiah, V.; Rathore, A.; Rupakula, A.; Reddy, R.K.; Waliyar, F.; Nigam, S.N. Genetic analysis of resistance to late leaf spot in interspecific groundnuts. *Euphytica* **2013**, *193*, 13–25.

127. Correa, C.M. Considerations on the standard material transfer agreement under the fao treaty on plant genetic resources for food and agriculture. *J. World Intellect. Prop.* **2006**, *9*, 137–165. [CrossRef]

128. Moose, S.P.; Mumm, R.H. Molecular plant breeding as the foundation for 21st century crop improvement. *Plant Physiol.* **2008**, *147*, 969–977. [CrossRef]

129. Lush, J.L. *Animal Breeding Plans*; Read Books Ltd.: Redditch, UK, 1937.

130. Eberhart, S. Factors effecting efficiencies of breeding methods. *Afr. Soils* **1970**, *15*, 655–680.

131. Heffner, E.L.; Lorenz, A.J.; Jannink, J.-L.; Sorrells, M.E. Plant breeding with genomic selection: Gain per unit time and cost. *Crop Sci.* **2010**, *50*, 1681–1690. [CrossRef]

132. Bera, S.; Bhatt, O. In vitro callogenesis and plant regeneration from anther culture in groundnut (*Arachis hypogaea L.*). *J. Plant Genet. Resour.* **2007**, *20*, 118–121.

133. Maluszynski, M.; Kasha, K.J.; Szarejko, I. Published doubled haploid protocols in plant species. In *Doubled Haploid Production in Crop Plants: A Manual*; Maluszynski, M., Kasha, K.J., Forster, B.P., Szarejko, I., Eds.; Springer: Dordrecht, The Netherlands, 2003; pp. 309–335.

134. Gunasekaran, A.; Pavadai, P. Studies on induced physical and chemical mutagenesis in groundnut (*Arachis hypogia*). *Int. Lett. Nat. Sci.* **2015**, *8*, 25–35. [CrossRef]

135. Knoll, J.E.; Ramos, M.L.; Zeng, Y.; Holbrook, C.C.; Chow, M.; Chen, S.; Maleki, S.; Bhattacharya, A.; Ozias-Akins, P. Tilling for allergen reduction and improvement of quality traits in peanut (*Arachis hypogaea L.*). *BMC Plant Biol.* **2011**, *11*, 81. [CrossRef]

136. Mensah, J.; Obadoni, B. Effects of sodium azide on yield parameters of groundnut (*Arachis hypogaea L.*). *Afr. J. Biotechnol.* **2007**, *6*, 668–671.

137. Busolo-Bulafu, C.M. *Mutation Breeding of Groundnuts (Arachis hypogaea L.) in Uganda*; IAEA: Vienna, Austria, 1991.

138. Tshilenge-Lukanda, L.; Kalonji-Mbuyi, A.; Nkongolo, K.; Kizungu, R. Effect of gamma irradiation on morpho-agronomic characteristics of groundnut (*Arachis hypogaea L.*). *Am. J. Plant Sci.* **2013**, *4*, 2186–2192. [CrossRef]

139. Mondal, S.; Badigannavar, A. A narrow leaf groundnut mutant, TMV2-NLM has a g to a mutation in AHFAD2A gene for high oleate trait. *Indian J. Genet. Plant Breed.* **2013**, *73*, 105–109. [CrossRef]

140. Tshilenge-Lukanda, L.; Funny-Biola, C.; Tshiyoyi-Mpunga, A.; Mudibu, J.; Ngoie-Lubwika, M.; Mukendi-Tshibingu, R.; Kalonji-Mbuyi, A. Radio-sensitivity of some groundnut (*Arachis hypogaea L.*) genotypes to gamma irradiation: Indices for use as improvement. *Br. J. Biotechnol.* **2012**, *3*, 169–178. [CrossRef]

141. Fountain, J.C.; Khera, P.; Yang, L.; Nayak, S.N.; Scully, B.T.; Lee, R.D.; Chen, Z.-Y.; Kemerait, R.C.; Varshney, R.K.; Guo, B. Resistance to aspergillus flavus in maize and peanut: Molecular biology, breeding, environmental stress, and future perspectives. *Crop J.* **2015**, *3*, 229–237. [CrossRef]

142. Torres, A.M.; Barros, G.G.; Palacios, S.A.; Chulze, S.N.; Battilani, P. Review on pre and post-harvest management of peanuts to minimize aflatoxin contamination. *Food Res. Int.* **2014**, *62*, 11–19. [CrossRef]

143. Ozias-Akins, P.; Yang, H.; Gill, R.; Fan, H.; Lynch, R.E. Reduction of aflatoxin contamination in peanut: A genetic engineering approach. In *Crop Biotechnology*; American Chemical Society: New York, NY, USA, 2002; Volume 829, pp. 151–160.

144. Holbrook, C.C.; Ozias-Akins, P.; Chu, Y.; Guo, B. Impact of molecular genetic research on peanut cultivar development. *Agronomy* **2011**, *1*, 3–17. [CrossRef]

145. Bhatnagar-Mathur, P.; Sunkara, S.; Bhatnagar-Panwar, M.; Waliyar, F.; Sharma, K.K. Biotechnological advances for combating aspergillus flavus and aflatoxin contamination in crops. *Plant Sci.* **2015**, *234*, 119–132. [CrossRef]

146. Higgins, C.M.; Dietzgen, R.G. *Genetic Transformation, Regeneration and Analysis of Transgenic Peanut*; Australian Centre for International Agricultural Research: Canberra, Australia, 2000; p. 90.

147. Sharma, K.K.; Pothana, A.; Prasad, K.; Shah, D.; Kaur, J.; Bhatnagar, D.; Chen, Z.-Y.; Raruang, Y.; Cary, J.W.; Rajasekaran, K.; et al. Peanuts that keep aflatoxin at bay: A threshold that matters. *Plant Biotechnol. J.* **2018**, *16*, 1024–1033. [CrossRef]

148. van de Wiel, C.C.M.; Schaart, J.G.; Lotz, L.A.P.; Smulders, M.J.M. New traits in crops produced by genome editing techniques based on deletions. *Plant Biotechnol. Rep.* **2017**, *11*, 1–8. [CrossRef]

149. Mishra, R.; Zhao, K. Genome editing technologies and their applications in crop improvement. *Plant Biotechnol. Rep.* **2018**, *12*, 57–68. [CrossRef]

150. Kleter, G.A.; Kuiper, H.A.; Kok, E.J. Gene-edited crops: Towards a harmonized safety assessment. *Trends Biotechnol.* **2019**, *37*, 443–447. [CrossRef]

151. Subburaj, S.; Tu, L.; Jin, Y.-T.; Bae, S.; Seo, P.J.; Jung, Y.J.; Lee, G.-J. Targeted genome editing, an alternative tool for trait improvement in horticultural crops. *Hortic. Environ. Biotechnol.* **2016**, *57*, 531–543. [CrossRef]

152. Wolter, F.; Schindele, P.; Puchta, H. Plant breeding at the speed of light: The power of crispr/cas to generate directed genetic diversity at multiple sites. *BMC Plant Biol.* **2019**, *19*, 176. [CrossRef] [PubMed]

153. Campa, C.C.; Weisbach, N.R.; Santinha, A.J.; Incarnato, D.; Platt, R.J. Multiplexed genome engineering by cas12a and crispr arrays encoded on single transcripts. *Nat. Methods* **2019**, *16*, 887–893. [CrossRef] [PubMed]

154. Yuan, M.; Zhu, J.; Gong, L.; He, L.; Lee, C.; Han, S.; Chen, C.; He, G. Mutagenesis of FAD2 genes in peanut with CRISPR/Cas9 based gene editing. *BMC Biotechnol.* **2019**, *19*, 24. [CrossRef] [PubMed]

155. Ezezika, O.; Thomas, F.; Lavery, J.; Daar, A.; Singer, P. A social audit model for agro-biotechnology initiatives in developing countries: Accounting for ethical, social, cultural and commercialization issues. *J. Technol. Manag. Innov.* **2009**, *4*, 24–33. [CrossRef]

156. Savadori, L.; Savio, S.; Nicotra, E.; Rumiati, R.; Finucane, M.; Slovic, A.P. Expert and public perception of risk from biotechnology. *Risk Anal.* **2004**, *24*, 1289–1299. [CrossRef]

157. Assemblée-Nationale, A. *Régime de sécurité en matière de biotechnologie au burkina faso*; Assemblée-Nationale: Ouagadougou, Burkina Faso, 2006; p. 40.

158. Hill, R.A. Conceptualizing risk assessment methodology for genetically modified organisms. *Environ. Biosaf. Res.* **2005**, *4*, 67–70. [CrossRef]

159. Johnson, K.L.; Raybould, A.F.; Hudson, M.D.; Poppy, G.M. How does scientific risk assessment of GM crops fit within the wider risk analysis? *Trends Plant Sci.* **2007**, *12*, 1–5. [CrossRef]

160. Potrykus, I. Regulation must be revolutionized. *Nature* **2010**, *466*, 561. [CrossRef]

161. Assemblee Nationale, B.F. *Loi no 064–2012/an du 20 Décembre 2012 Portant Régime de Sécurité en Matière de Biotechnologie*; Presidence: Ouagadougou, Burkina Faso, 2012.

162. Sundaram, J.; Kandala, C.V.; Holser, R.A.; Butts, C.L.; Windham, W.R. Determination of in-shell peanut oil and fatty acid composition using near-infrared reflectance spectroscopy. *J. Am. Oil Chem. Soc.* **2010**, *87*, 1103–1114. [CrossRef]

163. Janila, P.; Pandey, M.K.; Shasidhar, Y.; Variath, M.T.; Sriswathi, M.; Khera, P.; Manohar, S.S.; Nagesh, P.; Vishwakarma, M.K.; Mishra, G.P.; et al. Molecular breeding for introgression of fatty acid desaturase mutant alleles (AHFAD2A and AHFAD2B) enhances oil quality in high and low oil containing peanut genotypes. *Plant Sci.* **2016**, *242*, 203–213. [CrossRef]

164. ICRISAT. Why G.E.M.S? Available online: http://gems.icrisat.org/ (accessed on 27 February 2020).

165. Hinkelmann, K.; Kempthorne, O. *Design and Analysis of Experiments Volume 2: Advanced Experimental Design*; John Wiley & Sons Inc.: Hoboken, NJ, USA, 2005.

166. Naresh, R.; Singh, S.; Misra, A.; Tomar, S.; Kumar, P.; Kumar, V.; Kumar, S. Evaluation of the laser leveled land leveling technology on crop yield and water use productivity in western uttar pradesh. *Afr. J. Agric. Res.* **2014**, *9*, 473–478.

167. Al-Naggar, A.; Abdalla, A.; Gohar, A.; Hafez, E. Breeding values of 254 maize (*Zea mays L.*) doubled haploid lines under drought conditions at flowering and grain filling. *J. Adv. Biol. Biotechnol.* **2016**, 1–15. [CrossRef]

168. Yost, M.; Sorensen, B.; Creech, E.; Allen, N.; Larsen, R.; Ramirez, R.; Ransom, C.; Reid, C.; Gale, J.; Kitchen, B. *Defense against Drought*; Utah State University Extension: Salt Lake City, UT, USA, 2019; p. 8.

169. Watson, A.; Ghosh, S.; Williams, M.J.; Cuddy, W.S.; Simmonds, J.; Rey, M.-D.; Asyraf Md Hatta, M.; Hinchliffe, A.; Steed, A.; Reynolds, D.; et al. Speed breeding is a powerful tool to accelerate crop research and breeding. *Nat. Plants* **2018**, *4*, 23–29. [CrossRef] [PubMed]

170. Chiurugwi, T.; Kemp, S.; Powell, W.; Hickey, L.T. Speed breeding orphan crops. *Theor. Appl. Genet.* **2019**, *132*, 607–616. [CrossRef]

171. O'Connor, D.J.; Wright, G.C.; Dieters, M.J.; George, D.L.; Hunter, M.N.; Tatnell, J.R.; Fleischfresser, D.B. Development and application of speed breeding technologies in a commercial peanut breeding program. *Peanut Sci.* **2013**, *40*, 107–114. [CrossRef]

172. Larkin, D.L.; Lozada, D.N.; Mason, R.E. Genomic selection—Considerations for successful implementation in wheat breeding programs. *Agronomy* **2019**, *9*, 479. [CrossRef]

173. Lado, B.; Barrios, P.G.; Quincke, M.; Silva, P.; Gutiérrez, L. Modeling genotype environment interaction for genomic selection with unbalanced data from a wheat breeding program. *Crop Sci.* **2016**, *56*, 2165–2179. [CrossRef]

174. Bassi, F.M.; Bentley, A.R.; Charmet, G.; Ortiz, R.; Crossa, J. Breeding schemes for the implementation of genomic selection in wheat (*Triticum* spp.). *Plant Sci.* **2016**, *242*, 23–36. [CrossRef]

175. Ben Hassen, M.; Bartholomé, J.; Valè, G.; Cao, T.-V.; Ahmadi, N. Genomic prediction accounting for genotype by environment interaction offers an effective framework for breeding simultaneously for adaptation to an abiotic stress and performance under normal cropping conditions in rice. *G3* **2018**, *8*, 2319–2332. [CrossRef]

176. Robertsen, C.D.; Hjortshøj, R.L.; Janss, L.L. Genomic selection in cereal breeding. *Agronomy* **2019**, *9*, 95. [CrossRef]

177. Mumm, R.H.; Walters, D.S. Quality control in the development of transgenic crop seed products research supported in part by exygen research, 30158 research drive, state college, PA 16801. *Crop Sci.* **2001**, *41*, 1381–1389. [CrossRef]

178. Bradford, K.J.; Dahal, P.; Bello, P. Using relative humidity indicator paper to measure seed and commodity moisture contents. *Agric. Environ. Lett.* **2016**, *1*. [CrossRef]

179. IBP. Your Partner for Modern Plant Breeding. Available online: https://www.integratedbreeding.net/ (accessed on 27 February 2020).

180. Rathore, A.; Singh, V.K.; Pandey, S.K.; Rao, C.S.; Thakur, V.; Pandey, M.K.; Anil Kumar, V.; Das, R.R. Current status and future prospects of next-generation data management and analytical decision support tools for enhancing genetic gains in crops. In *Plant Genetics and Molecular Biology*; Varshney, R.K., Pandey, M.K., Chitikineni, A., Eds.; Springer: Cham, Switzerland, 2018; pp. 277–292.

181. Cobb, J.N.; Biswas, P.S.; Platten, J.D. Back to the future: Revisiting mas as a tool for modern plant breeding. *Theor. Appl. Genet.* **2019**, *132*, 647–667. [CrossRef] [PubMed]

182. Varshney, R.K.; Singh, V.K.; Hickey, J.M.; Xun, X.; Marshall, D.F.; Wang, J.; Edwards, D.; Ribaut, J.-M. Analytical and decision support tools for genomics-assisted breeding. *Trends Plant Sci.* **2016**, *21*, 354–363. [CrossRef] [PubMed]

183. CYMMIT. CIMMYT Research Software. Available online: https://data.cimmyt.org/dataverse/cimmytswdvn (accessed on 27 February 2020).

184. Lipka, A.E.; Tian, F.; Wang, Q.; Peiffer, J.; Li, M.; Bradbury, P.J.; Gore, M.A.; Buckler, E.S.; Zhang, Z. Gapit: Genome association and prediction integrated tool. *Bioinformatics* **2012**, *28*, 2397–2399. [CrossRef] [PubMed]

185. Pandey, M.K.; Monyo, E.; Ozias-Akins, P.; Liang, X.; Guimarães, P.; Nigam, S.N.; Upadhyaya, H.D.; Janila, P.; Zhang, X.; Guo, B.; et al. Advances in arachis genomics for peanut improvement. *Biotechnol. Adv.* **2012**, *30*, 639–651. [CrossRef]

186. Ojiewo, C.O.; Janila, P.; Bhatnagar-Mathur, P.; Pandey, M.K.; Desmae, H.; Okori, P.; Mwololo, J.; Ajeigbe, H.; Njuguna-Mungai, E.; Muricho, G.; et al. Advances in crop improvement and delivery research for nutritional quality and health benefits of groundnut (*Arachis hypogaea* L.). *Front. Plant Sci.* **2020**, *11*, 29. [CrossRef]

187. Akram, N.A.; Shafiq, F.; Ashraf, M. Peanut (*Arachis hypogaea* L.): A prospective legume crop to offer multiple health benefits under changing climate. *Compr. Rev. Food Sci. Food Saf.* **2018**, *17*, 1325–1338. [CrossRef]

188. Chu, Y.; Wu, C.L.; Holbrook, C.C.; Tillman, B.L.; Person, G.; Ozias-Akins, P. Marker-assisted selection to pyramid nematode resistance and the high oleic trait in peanut. *Plant Genome* **2011**, *4*, 110–117. [CrossRef]

189. Wilson, J.N.; Chopra, R.; Baring, M.R.; Selvaraj, M.G.; Simpson, C.E.; Chagoya, J.; Burow, M.D. Advanced backcross quantitative trait loci (qtl) analysis of oil concentration and oil quality traits in peanut (*Arachis hypogaea* L.). *Trop. Plant Biol.* **2017**, *10*, 1–17. [CrossRef]

190. Ndjeunga, J.; Bantilan, M.C.S. Uptake of improved technologies in the semi-arid tropics of west africa: Why is agricultural transformation lagging behind? *Electron. J. Agric. Dev. Econ.* **2005**, *2*, 85–102.

191. Ruttan, V.W. Changing role of public and private sectors in agricultural research. *Science* **1982**, *216*, 23–29. [CrossRef] [PubMed]

192. Naseem, A.; Spielman, D.J.; Omamo, S.W. Private-sector investment in r&d: A review of policy options to promote its growth in developing-country agriculture. *Agribusiness* **2010**, *26*, 143–173.

 *agronomy*

Article

Ethyl Methyl Sulfonate-Induced Mutagenesis and Its Effects on Peanut Agronomic, Yield and Quality Traits

Tingting Chen [1], Luping Huang [1], Miaomiao Wang [1], Yang Huang [1], Ruier Zeng [1], Xinyue Wang [1], Leidi Wang [1], Shubo Wan [2,*] and Lei Zhang [1,*]

1   Guangdong Key Laboratory of Plant Molecular Breeding, State Key Laboratory for Conservation and Utilization of Subtropical Agro-Bioresources, South China Agricultural University, Guangzhou 510642, China; chentingting@scau.edu.cn (T.C.); lupinghuang2019@163.com (L.H.); 15802020854@163.com (M.W.); hy1294276320@163.com (Y.H.); ruierzeng@126.com (R.Z.); wangxinyuescau@163.com (X.W.); wangld@mail.iap.ac.cn (L.W.)
2   Biotechnology Research Center, Shandong Academy of Agricultural Science, Jinan 250100, China
*   Correspondence: wanshubo2016@163.com (S.W.); zhanglei@scau.edu.cn (L.Z.)

Received: 2 April 2020; Accepted: 30 April 2020; Published: 5 May 2020

Abstract: Peanut is an important oilseed and food crop worldwide; however, the development of new cultivars is limited by its remarkably low genetic variability. Therefore, in order to enhance peanut genetic variability, here, we treated two widely cultivated peanut genotypes, Huayu 22 and Yueyou 45, with different concentrations of the mutagen ethyl methyl sulfonate (EMS) for different durations. Based on median lethal dose (LD50) value, optimal EMS treatment concentrations for each duration were identified for each genotype. Mutants induced by EMS differed in various phenotypic traits, including plant height, number of branches, leaf characteristics, and yield and quality in plants of the M2 generation. Moreover, we identified potentially useful mutants associated with dwarfism, leaf color and shape, high oil and/or protein content, seed size and testa color, among individuals of the M2 generation. Mutations were stably inherited in M3-generation individuals. In addition to their contribution to the study and elucidation of the mechanisms underlying the regulation of the expression of some important agronomic traits, the mutants obtained in this study provide valuable germplasm resources for use in peanut improvement programs.

Keywords: *Arachis hypogaea* L.; EMS; mutant; dwarf; high protein; breeding

---

1. Introduction

Peanut (*Arachis hypogaea* L.) is one of the most important oil and protein crops in the world. In 2017, global peanut production reached 47.10 M tons in a total cultivated area of 27.95 M ha [1]. As peanut demand continues to increase, there is an urgent need to breed new peanut varieties with high yield and quality, in addition to resistance to biotic and abiotic stress factors [2]. However, cultivated peanut is an allotetraploid species that shows very low genetic variability through artificial selection over many decades [3]. Furthermore, peanut crop improvement should start with the formation of new germplasm that may then be used as source of highly desirable, outstanding performance traits.

Induced mutagenesis is one of the most important approaches for broadening crop genetic variability to overcome the limitations associated with a narrow genetic basis [4]. Induced mutants not only serve as an important functional genomics tool, but additionally, as intermediate material in crop breeding [5–8]. Thus, induced mutagenesis have also been used in peanut [9–11] to generate mutants for high-oleate content [12–15], color of the testa [16], photosynthetic activity [17], salinity resistance [18] and resistance to biotic stress [19], and pod development [20]. Among available mutagens, ethyl methyl sulfonate (EMS) is a potent and popular chemical mutagen that has been effectively used to induce a high-density of random irreversible point mutations uniformly distributed in the genome [21,22]. The EMS induces single nucleotide changes by alkylation of specific nucleotides, resulting in a high frequency of single nucleotide polymorphisms (SNPs) and insertion/deletions (InDels) in the genic sequences and coding sequence (CDS) region, which is considered as an excellent resource for targeting induced local lesions in genomes (TILLING) [23,24]. However, few studies have reported a mutant peanut variety library constructed by an optimal EMS treatment, because mutagenic efficiency depends on many different factors including, EMS concentration, treatment duration, peanut genotype, temperature, etc., [25,26]. To further improve the mutagenesis protocol for peanut, in this study, two different peanut genotypes were treated with different EMS concentrations for different time periods to optimize EMS treatment and identify stably-inherited mutations to improve peanut varieties.

2. Materials and Methods

2.1. Plant Materials

Two different *A. hypogaea* cultivars, namely, 'Huayu 22' (HY22) and 'Yueyou 45' (YY45), from North and South China, respectively, were used in this study. HY22 belongs to the common larger-sized kernel varieties with high quality and yield potential [27]. On the other hand, YY45 belongs to the Spanish type of peanut with smaller-sized kernels with high resistance to rust, leaf spot and bacterial wilt [28], used predominantly for peanut candy, salted peanuts, and peanut butter.

2.2. EMS Treatment

Sixteen treatments were tested at 25 °C, which consisted of four EMS (Sigma, St Louis, USA) concentrations (0.3%, 0.6%, 0.9% and 1.2%), each combined with four treatment duration times (1, 3, 5, and 7 h). Seeds soaked in 0.1 M phosphate buffer (pH 7.0) were used as the control treatment. Fifty seeds of each, HY22 and YY45, were subjected to each combination treatment and the control treatment. There are two replications for each treatment. Treated and control seeds were then thoroughly washed under running water for 3 h and transferred to pots filling wet sand and placed in an incubator for germination at 25 °C in the dark until radical emergence [29,30].

2.3. Generating Mutant Populations

All germinated and non-germinated seeds were sown in the field. The size of individual plots was 1.6 × 1.6 m. Planting distance between and within rows was 20 cm. Standard cultural practices were carried out thereafter. The experiment was laid in a complete randomized-block design at the Experimental Station of South China Agricultural University in Guangzhou (23°5' N, 113°23' E), Guangdong, China. M1 plants were self-pollinated to create M2 families. M2 seeds were harvested from 349 and 374 surviving M1 plants of HY22 and YY45, respectively. In the spring of 2019, 20 seeds from each M2 family were planted to grow an M3 generation, again, in a complete randomized-block design.

2.4. Phenotype

Mutants were detected by visual observation of the plants during the whole plant growing cycle in each generation. Visual phenotypic variation in growth performance and leaf morphology were recorded and photographs were taken to document the comparison between mutants and their control. After harvest, plant height, number of branches and total pod number per mutant or control were recorded. Pods were harvested from the parent plants and then dried. Total pod number (TPN) per plant was counted. The weight of twenty full pods of each plant was determined. After removal of pod shells, the weight of twenty full seeds sampled from each plant was determined. The 100-pod weight (HPW) and 100-kernel weight (HKW) were calculated as five times the weight of twenty full pods and twenty full seeds, respectively. Oil and protein from clean seeds were measured using a near-infrared reflectance (NIR) analyzer (DA 7250, Perten Instruments, Inc., Springfield, IL, United States) by the methods reported previously [31].

2.5. Statistical Analysis

All obtained data from the M1 generations were subjected to analysis of variance (ANOVA) using SPSS 16.0 software (SPSS, Chicago, IL, USA). The analyzed data were presented as means (±SD) of two replicates.

3. Results

3.1. Germination Rate and LD50 in the M1 Generation

Field germination rates of HY22 and YY45 in generation M1 were surveyed at 10 d after sowing. Figure 1 showed that there were 8 and 10 lethal treatments for HY22 (0.6% EMS for 5 or 7 h, 0.9% for 3, 5, or 7 h, and 1.2% EMS for 3, 5, or 7 h) and YY45 (0.6% and 0.9% EMS for 3, 5, or 7 h, and 1.2% for 1, 3, 5, or 7 h), respectively. Except those lethal treatments, the germination rates of the other EMS-treated seeds of both genotypes were significantly ($p < 0.05$) reduced by EMS treatment, except 0.3% EMS for 1 h in YY45, compared to the corresponding control (i.e., 79% for HY22 and 87% for YY45) (Figure 1). Moreover, under the same EMS concentration, different changes of germination rates were observed in different durations of both genotypes. In HY22, for 0.3% EMS concentration, no significant difference was found among durations of 1, 3, 5 h or 5, 7 h. However, significant difference was showed between 1, 3, and 7 h. For 0.6% EMS concentration, there was no significant difference between 1 h and 3 h. In YY45, for 0.3% EMS concentration, no significant difference was found between the durations of 1, 3 h and 5, 7 h. However, significant differences were showed between 1 and 5 h, 1 and 7 h, 3 and 5 h, and 3 and 7 h. In addition, the germination rates of EMS-treated YY45 seeds were higher than those of HY22 under 0.3% EMS for 1, 3, 5, or 7 h, and under 0.6% EMS for 1 h.

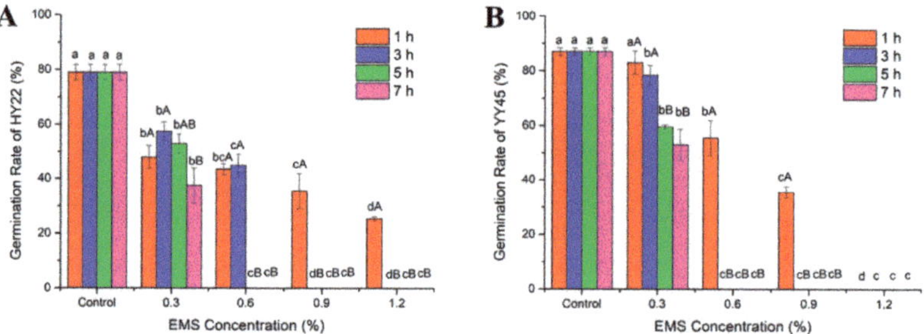

Figure 1. The germination rate of M1 generation HY22 and YY45 induced by different ethyl methyl sulfonate (EMS) treatments. (**A**) Germination rate of 'Huayu 22' (HY22) induced by different EMS treatment; (**B**) Germination rate of 'Yueyou 45' (YY45) induced by different EMS treatment. Note: Mean values within different EMS concentration followed by lower cases are significantly different at $p < 0.05$. Mean values within different duration followed by upper cases are significantly different at $p < 0.05$.

The median lethal dose (LD50) is usually used as a critical parameter for chemically induced mutagenesis. The LD50 values for mutagenesis of peanut with different EMS concentrations for each duration were estimated through linear regression based on lethal rate (Figure 2). Based on the lethal rate of HY22, LD50 values were calculated 0.51%, 1.41%, 0.27% and 0.19% for durations of 1, 3, 5 and 7 h, respectively. For YY45, LD50 values were estimated 0.63%, 0.40%, 0.33% and 0.31% for durations of 1, 3, 5 and 7 h, respectively (Figure 2, Supplementary Table S1). The data showed that the LD50 of HY22 were lower than those of YY45 for 1, 5 and 7 h, whereas it is converse for 3 h.

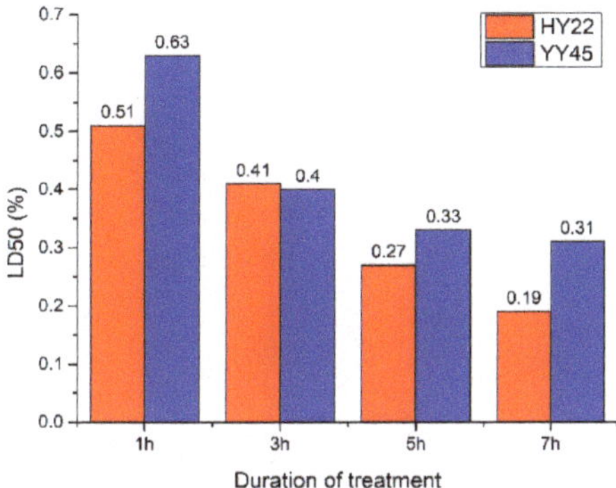

Figure 2. Median lethal dose (LD50) values of HY22 and YY45 with different EMS treatments.

3.2. Plant Height and Branch Number in Population M2

In order to analyze the agronomic traits in peanut, plant height and number of branches were recorded after harvest. For plant height, the peak of frequency (number of M2 families) in M2 populations of HY22 and YY45, were 25.49% (78) and 31.77% (115) at the groups of 40–44.99 cm and 35–39.99 cm, while the average value among HY22 and YY45 control plants was 43.4 and 38.5 cm, respectively. In M2 populations of HY22, the groups of 0–39.99 cm and 45–89.99 cm accounted for 27.78% (85) and 46.73% (143), respectively. As for YY45, the groups of 0–34.99 cm and 40-89.99 cm accounted for 24.03% (87) and 44.20% (160), respectively (Figure 3A).

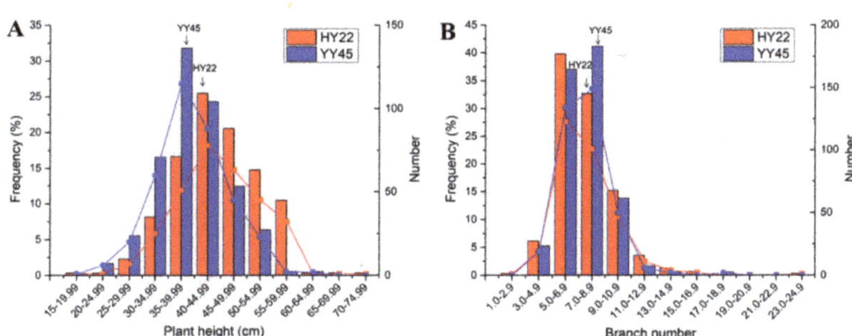

Figure 3. Frequency and number distribution of agronomic traits of M2 generation of HY22 and YY45 induced by EMS treatments. (**A**) Plant height; (**B**) Number of branches.

For the number of branches, the average of number of branches in HY22 and YY45 plants were 8.5 and 7.4, respectively (Figure 3B). The peak of frequency (number of M2 families) in M2 populations of HY22 and YY45, were 39.80% (123) and 41.16% (149) at the groups of 5.0–6.9 and 7.0–8.9. Except for groups including the control plants, there were 46.28% (143) families in the less-branches group and 21.04% (65) families in the more-branches group among HY22 plants in the M2 generation, whereas 42.26% (152) less-branches families and 16.57% (60) more-branches families were observed among YY45 plants of the same generation.

3.3. Plant Yield in the M2 Generation

In order to explore the effect of EMS treatment on yield, total pod number (TPN), hundred-pod weight (HPW) and hundred-kernel weight (HKW) per plant were studied in generation M2.

For TPN per plant, the peaks both appeared at the group of 10.0–14.9. TPN of 85.16% (264) of HY22 families were decreased, compared to the control (TPN: 24.9). Only 4.52% (14) of families were increased (Figure 4A). TPN in 45 YY45 M2 families was lower than that in the control (TPN: 11.4), accounting for 12.43%. TPN in 168 YY45 M2 families was higher than that in the control, accounting for 46.41% (Figure 4A).

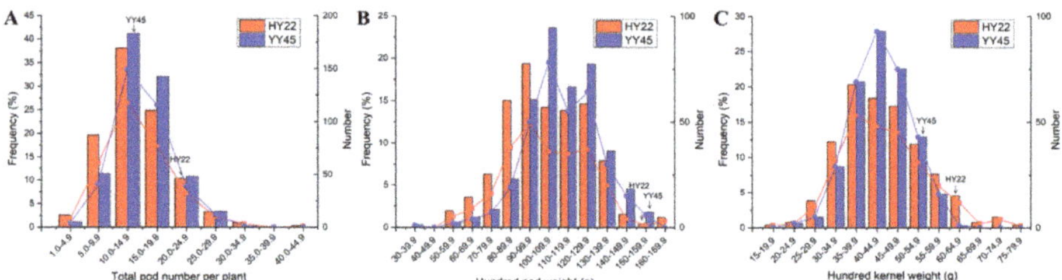

Figure 4. Frequency and number distribution of yield traits of M2 generation of HY22 and YY45 induced by EMS treatments. (**A**) Total pod number per plant; (**B**) Hundred pod weight; (**C**) Hundred kernel weight.

For HPW per plant, we observed the peak at the groups of 90–99.9 g and 100–109.99 g among the M2 generation of HY22 and YY45, whereas HPW of HY22 and YY45 control plants were 159 and 156 g, respectively. In the mutant populations of HY22 and YY45, HPW of 249 and 325 families, respectively, decreased compared to the corresponding control, and accounted for 98.42% and 98.19%, of the corresponding values, respectively. The data showed that HFW in M2 families of the two genotypes decreased relative to the corresponding WT (Figure 4B). For HKW per plant, the peaks of HY22 and YY45 appeared at the groups of 35–39.9 g and 40–44.9 g, which were lower than mean value in HY22 control plants (61 g) and YY45 (54 g). Similarly, in mutant populations of HY22 and YY45, HKW of 242 and 273 families, respectively, decreased compared to the corresponding control, and accounted for 92.72% and 81.98%, of the corresponding values, respectively (Figure 4C). Thus, it was verified that HKW in M2 families of the two genotypes decreased compared to the corresponding control.

3.4. Seed Quality in Plants of Generation M2

Data in Figure 5A show that the peak of oil content (OC) at the group of 50–51.99 in the M2 population of HY22, accounting for 29.27%, was lower than the HY22 control plants (mean OC: 55.57%). Compared with the control, low and high OC were observed in 256 (89.20) and 11 (3.81%) M2 families, respectively. In the M2 population of YY45, the peak appeared at the group of 48–49.99, accounting for 32.54%, which was lower than the control plants (mean OC: 51.16%). On the other hand, 237 (70.12%) and 44 (13.02%) M2 families had low and high OC compared to the control YY45 (Figure 5A). The data showed that OC in the mutant population of the two genotypes were decreased compared to their corresponding control.

It was found that the peak of protein content (PC) appeared at the group of 24–24.99 and 25–25.99 in the M2 population of HY22 and YY45, whereas PC of their control plants was 23.1% and 23.5%, respectively. In total, 164 and 65 families of HY22 had higher and lower PC than the control, accounting for 57.14% and 22.65%, respectively. Meanwhile, high PC was observed in 244 (73.05) M2 families of YY45, while lower PC was observed in 41 (12.28%) families (Figure 5B).

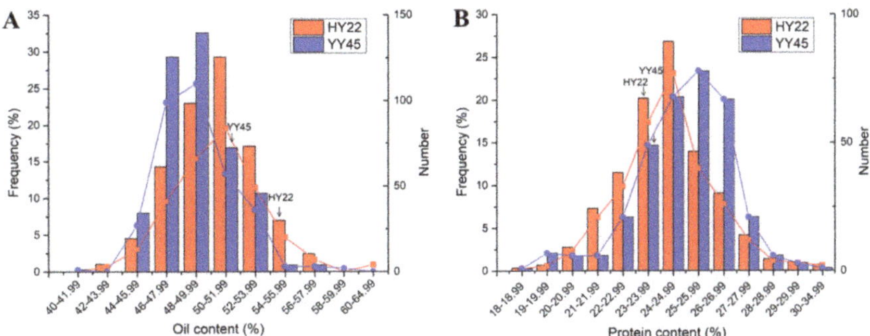

Figure 5. Frequency distribution of seed quality of M2 generation of HY22 and YY45 induced by EMS treatments. (**A**) Oil content; (**B**) Protein content.

3.5. Mutants

We identified several important mutations in the M2 generation that were stably inherited to the M3 generation. These included dwarfism, leaf color and shape, high oil and/or protein, seed size, and color of testa mutants.

3.5.1. Dwarf Mutants

Two mutants, HY-53 and HY-67, were dwarfs with height of 22.9 and 35.0 cm, and smaller leaves than those of the control. These traits were stably inherited in the M3 generation (Figure 6A,B). Mutant YY-18 was a dwarf with height of 21.0 cm, smaller and darker-green leaves than control YY45 (Figure 6C).

Figure 6. Mutants with dwarf and small-size leaves of HY22 and YY45. (**A** and **B**) Mutants of HY22; (**C**) Mutants of YY45. Red arrows indicate the corresponding control.

3.5.2. Leaf Color and Shape Mutants

We found nine leaf-shape mutants among HY22 plants of the M2 generation, including two large-leaf, four small-leaf, and two slender-leaf mutants and one wrinkled-leaf mutant. There were 13 leaf-color mutants: yellow (Figure 7A), dark-green, and pale-green (Figure 7B) observed in five, four, and one mutant, respectively; one more mutant had the leaves with a chlorine-like color in the veins (Figure 7C), while two more mutants showed mosaic leaves (Figure 7D,E).

There were five leaf-shape mutants in YY45 plants of the M2, including four small-leaf mutants and one rolled-leaf mutants. Additionally, we found nine leaf-color mutants; one yellow-leaf, three dark-green-leaf and one pale-green-leaf mutant (Figure 7F); lastly, there were four rolled-leaf mutants (Figure 7G).

Figure 7. Mutants of leaf color of HY22 and YY45. (**A–E**) Mutants of yellow, pale green, veinal chlorine and mosaic leaves of HY22; (**F,G**) Mutants of mosaic and pale green leaves of YY45. Red arrow indicates the corresponding control.

3.5.3. High oil and/or Protein Mutants

Table 1 showed there were one high-oil dwarf mutant, six high-protein dwarf mutants, and one high-protein yellow-leaf mutant among HY22 individuals in population M2. On the other hand, we observed one high-oil, small-leaf mutant, two high-oil, high-protein, dwarf mutants, two high-protein dwarf mutants, and one high-protein large-leaf mutant among YY45 plants in population M2.

Table 1. Mutants related to high oil and/or protein of HY22 and YY45 induced by EMS treatment.

	Mutant ID	Phenotype		Oil Content (%)	Protein Content (%)	Plant Height (cm)
	HY93	High oil	Dwarf	64.58	23.67	30.6
	HY100	High protein	Dwarf	51.82	32.95	34.2
	HY104	High protein	Dwarf	40.33	30.06	36.7
HY22	HY106	High protein	Dwarf	44.68	31.77	37.1
mutants	HY123	High protein	Dwarf	44.96	32.03	30.3
	HY125	High protein	Dwarf	50.20	32.40	34.3
	HY126	High protein	Dwarf	47.06	32.13	35.0
	HY131	High protein	Yellow leaves	45.01	30.72	40.2
	YY37	High oil	Small leaves	59.70	21.06	45.0
	YY38	High oil and protein	Dwarf	61.93	32.42	32.0
YY45	YY47	High oil and protein	Dwarf	61.17	30.12	25.6
mutants	YY62	High protein	Dwarf	44.86	30.66	34.8
	YY64	High protein	Dwarf	51.79	30.07	27.0
	YY66	High protein	Big leaves	45.74	30.54	41.9

3.5.4. Seed Size and Testa Color Mutants

The mutant in Figure 8B is shorter than the HY22 control (Figure 8A). The seed testa in the mutant in Figure 8C became wrinkled and its red color more intense than that of the control. Cracks were observed on the testa of the mutant in Figure 8E compared to the YY45 control (Figure 8D).

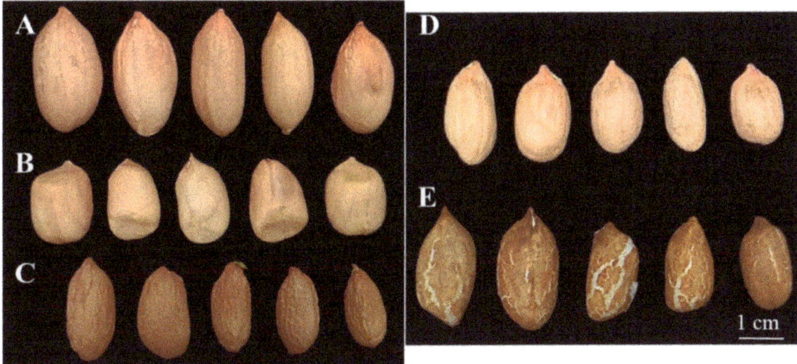

Figure 8. Seeds of mutants and their corresponding control. (**A**) The control HY22; (**B,C**) Mutants of HY22; (**D**) The control YY45; (**E**) Mutants of YY45.

4. Discussion

4.1. Optimal Treatment Conditions Based on LD50

Highly efficient mutagenesis is essential in mutagenesis-based breeding programs. Successful EMS-mediated mutagenesis depends on many factors, such as EMS concentration, treatment duration, and temperature, and others [4,32–34]. In the previous studies, several peanut genotypes, including Silihong, Baisha 1016 and Jinhua 8, were treated with different conditions. Then, researchers simply selected as optimal EMS treatment conditions closest to 50% of germination rate. However, few researchers have reported an analysis to determine proper EMS treatment conditions through linear regression for different peanut genotypes. In order to optimize EMS treatment conditions reasonably and correctly, here, we compared sixteen different treatments comprising four EMS concentrations and four treatment durations on two different peanut genotypes, HY22 and YY45. LD50 value for each duration was calculated based on lethal rate, which is as the optimal treatment conditions for each of the two genotypes studied.

Our results showed that EMS LD50 values for 1, 5 and 7 h in YY45 were higher than those in HY22, suggesting that the former is more resistant to EMS than the latter. Similarly, previous studies reported that sensitivity to chemical mutagens differs with genotype [35,36]. In addition, our results showed that 100-pod weight and 100-kernel weight decreased in the M2 population of both genotypes, as compared to the corresponding control. We inferred that this may be attributed to different physiological and biological processes related to yield, including enzyme activity and hormonal balance [37,38]. Furthermore, the toxic nature of the mutagen may reportedly damage cell constituents at the molecular level or alter enzyme activity [39,40].

In previous research, it is verified that polyploid species are tolerant to high densities of induced mutations compared to diploids [41], which supports the hypothesis that loss-of-function mutations in polyploids are masked by genetic redundancy among homoeologs [42]. Such redundancy reduces the probability of selection of favorable mutations induced by EMS [43], mainly producing point mutations (G to A and C to T) in many crop species [44]. Peanut (AABB) is a tetraploid crop with very similar A and B sub-genomes, which frequenctly have more than 98% DNA identify between corresponding genes [43]. In this study, we inferred that many potentially useful induced mutants were likely masked by functional redundancy among homeologs in the mutant population of both genotypes. TILLING (targeting induced local lesions in genome) approach provides a relatively simple strategy to identify mutants in a target sequence independently of its phenotypic effect [45]. Combined with the exome capture technique, which is a smaller, specific portion of a plant genome that can be captured for resequencing [46], the protein coding regions could be captured to discover SNPs [47] and catalog induced mutations [8,48]. These mutations can be combined to study gene function and to reveal previously hidden phenotypic variation [42].

4.2. Potentially Useful Mutants Associated Important Traits

Dwarf and chlorotic mutants are the most readily visible, usual mutant types. Dwarf mutants in peanut can provide useful insights for understanding the regulatory mechanisms of plant growth and development, whereby they can be of great assistance in breeding programs. In our study, several dwarf mutants were observed. Previous studies suggested that dwarfism might be due to abnormal biosynthesis of indole acetic acid (IAA), gibberellin (GA), brassinosteroid (BR) or strigolactones (SL) [49–55]. For example, the tryptophan-deficient dwarf1 (*tdd1*) rice mutant, which is embryonic-lethal because of a failure to develop most organs during embryogenesis, is caused by *TDD1*, which encodes a protein that functions upstream of Trp-dependent IAA biosynthesis [52]. Similarly, a rice GA-insensitive dwarf mutant showed a severe dwarf phenotype containing high concentrations of endogenous GA [56]. Furthermore, the Arabidopsis dwarf mutant, shrink1-D (*shk1-D*), is produced by the activation of the *CYP72C1* gene, which is a member of the cytochrome P450 monooxygenase gene family that regulates BR inactivation [57]. In turn, dwarf mutant dwarf11 (*d11*) rice bearing seeds of reduced length are controlled by the *D11* gene encoding a novel cytochrome P450 (CYP724B1), which showed homology to enzymes involved in BR biosynthesis [58]. Lastly, mutant dwarf 53 (*d53*) rice controlled by the *D53* gene encoding a substrate of the SCFD3 ubiquitination complex that functions as a repressor of SL signaling [59].

Several types of chlorosis, such as yellow, pale-green and vein-chlorosis, were identified in this study. In agreement with our results, previous studies found that a significant change in chlorophyll development always resulted in the variation of leaf color [60–63]. For example, chlorotic mutants defective in *ChlH*, *ChlD* or *ChlI* encoding the Mg-chelatase subunits have been identified in Arabidopsis, rice, barley, and tea [64–68]. The loss of chlorophyll resulted from either a reduction or an excess accumulation of ChlI [69]. In addition to genes encoding Mg-chelatase, other genes related to chlorotic leaves have also been identified, which influence the biosynthesis of chlorophyll and alter chlorophyll content. Firstly, OsYGL1, encoding a chlorophyll synthase responsible for catalyzing the esterification of chlorophyllide, was identified from a rice yellow-green leaf1 mutant that showed yellow-green leaves in young plants with reduced Chl synthesis, increased level of tetrapyrrole intermediates, and delayed chloroplast development [70]. Secondly, the *ylc1* mutant showed reduced levels of Chl and lutein in young leaves compared to those of the control and turned green gradually, approaching normal green at maximum tillering stage, which was controlled by the chloroplast-localized gene, *OsYLC1*, whose protein is involved in Chl and lutein accumulation and chloroplast development during early leaf development in rice [71]. Similarly, in this study, the Chl-deficient mutant in Figure 7A showed completely etiolated newly grown leaves, light-green on middle leaves and regular green on the bottom part at seedling stage, a pattern that warrants further study.

Three more mutant types associated with seed quality, including high-protein/dwarf, high-oil/dwarf, and high-oil/large-leaf mutant, were observed in our study (Table 1). A previous study showed that gene *FAD2* regulated high oleic-acid synthesis in peanut [14,72,73]. However, to date, only a few studies related to peanut seed-protein content have been reported. Mutations conferring different phenotypic variations are important for the functional analysis of corresponding genes; in addition, they offer an alternative plant material that peanut breeding programs can work with towards crop improvement for desirable traits. Therefore, in our future work, novel phenotypes resulting in phenotypic variation could be characterized using a combination of whole-genome resequencing, linkage maps and RNA-seq, providing a comprehensive picture of gene expression changes and newly introduced SNPs compared to wild-type, based on the reference genome of peanut. Once the gene function was identified in a mutant population by TILLING or separated population by genetic maps, the gene can be introgressed into breeding lines lacking that gene through crosses by the help of visible, biochemical, or molecular markers, i.e., molecular marker-assisted selection (MAS).

5. Conclusions

Based on LD50, here, we inferred the optimal EMS treatment conditions that will benefit future mutagenic research in peanut. Further, the effects of EMS treatment on growth, yield, oil, and seed quality were analyzed. Meanwhile, several mutants related to dwarfism, chlorosis, high oil and/or protein content, seed size and testa color were obtained. Our findings highlight the potential of EMS-induced mutant lines of HY22 and YY45; furthermore, the mutant lines selected in this study may be used as germplasm resources and breeding materials in peanut breeding programs.

Supplementary Materials: The following are available online at http://www.mdpi.com/2073-4395/10/5/655/s1. Table S1: LD50 calculation based on the lethal rate with different EMS treatments.

Author Contributions: T.C. did the experiment and wrote the manuscript. L.H., M.W. and Y.H. analyzed the date. R.Z., X.W. and L.W. did a part of experiment. L.Z. gave good suggestions on the manuscript. S.W. and L.Z. revised the manuscript. All authors have read and agreed to the published version of the manuscript.

Funding: This study was supported by the National Key R&D Program of China (2018YFD1000906), the Key Science and Technology Planning Project of Guangdong Province (2019B020214003), and the Guangdong Technical System of Peanut and Soybean Industry (2019KJ136-05).

Acknowledgments: We are grateful to the personnel of Shandong Academy of Agricultural Science and Guangdong Academy of Agricultural Science for providing the seed of Huayu 22 and Yueyou 45, respectively. We are also grateful to editor and the anonymous reviewers.

Conflicts of Interest: The authors declare no conflict of interest.

References

1. FAOSTAT. Food and Agriculture Organization of the United Nations. Available online: http://wwwfaoorg/ (accessed on 30 January 2020).
2. Pandey, M.K.; Monyo, E.; Ozias-Akins, P.; Liang, X.; Guimarães, P.; Nigam, S.N.; Upadhyaya, H.D.; Janila, P.; Zhang, X.; Guo, B. Advances in *Arachis* genomics for peanut improvement. *Biotechnol. Adv.* **2012**, *30*, 639–651. [CrossRef] [PubMed]
3. de Carvalho Moretzsohn, M.; Hopkins, M.S.; Mitchell, S.E.; Kresovich, S.; Valls, J.F.M.; Ferreira, M.E. Genetic diversity of peanut (*Arachis hypogaea* L.) and its wild relatives based on the analysis of hypervariable regions of the genome. *BMC Plant Biol.* **2004**, *4*, 11.
4. Asif, A.; Khalil Ansari, M.Y. Generation of mutant lines of *Nigella sativa* L. by induced mutagenesis for improved seed yield. *Ind. Crop. Prod.* **2019**, *139*, 111552. [CrossRef]
5. Menda, N.; Semel, Y.; Peled, D.; Eshed, Y.; Zamir, D. In silico screening of a saturated mutation library of tomato. *Plant J.* **2004**, *38*, 861–872. [CrossRef] [PubMed]
6. Galpaz, N.; Burger, Y.; Lavee, T.; Tzuri, G.; Sherman, A.; Melamed, T.; Eshed, R.; Meir, A.; Portnoy, V.; Bar, E. Genetic and chemical characterization of an EMS induced mutation in Cucumis melo *CRTISO* gene. *Arch. Biochem. Biophys.* **2013**, *539*, 117–125. [CrossRef] [PubMed]
7. Tsuda, M.; Kaga, A.; Anai, T.; Shimizu, T.; Sayama, T.; Takagi, K.; Machita, K.; Watanabe, S.; Nishimura, M.; Yamada, N. Construction of a high-density mutant library in soybean and development of a mutant retrieval method using amplicon sequencing. *BMC Genom.* **2015**, *16*, 1014. [CrossRef]
8. Henry, I.M.; Nagalakshmi, U.; Lieberman, M.C.; Ngo, K.J.; Krasileva, K.V.; Vasquez-Gross, H.; Akhunova, A.; Akhunov, E.; Dubcovsky, J.; Tai, T.H. Efficient genome-wide detection and cataloging of EMS-induced mutations using exome capture and next-generation sequencing. *Plant Cell* **2014**, *26*, 1382–1397. [CrossRef]
9. Branch, W.D. Variability among advanced gamma-irradiation induced large-seeded mutant breeding lines in the 'Georgia Browne' peanut cultivar. *Plant Breed.* **2002**, *121*, 275–277. [CrossRef]
10. Nadaf, H.; Kaveri, S.; Madhusudan, K.; Motagi, B. Induced genetic variability for yield and yield components in peanut (*Arachis hypogaea* L.). In *Induced Plant Mutations in the Genomics Era*; FAO: Rome, Italy, 2009; pp. 346–348.
11. Wang, J.; Qiao, L.; Zhao, L.; Wang, P.; Guo, B.; Liu, L.; Sui, J. Performance of peanut mutants and their offspring generated from mixed high-energy particle field radiation and tissue culture. *Genet. Mol. Res.* **2015**, *14*, 10837–10848. [CrossRef]

12. Bera, S.K.; Kamdar, J.H.; Kasundra, S.V.; Dash, P.; Maurya, A.K.; Jasani, M.D.; Chandrashekar, A.B.; Manivannan, N.; Vasanthi, R.P.; Dobariya, K.L.; et al. Improving oil quality by altering levels of fatty acids through marker-assisted selection of *ahfad2* alleles in peanut (*Arachis hypogaea* L.). *Euphytica* **2018**, *214*, 162. [CrossRef]

13. Mondal, S.; Badigannavar, A.M.; D'Souza, S.F. Induced variability for fatty acid profile and molecular characterization of high oleate mutant in cultivated groundnut (*Arachis hypogaea* L.). *Plant Breed.* **2011**, *130*, 242–247. [CrossRef]

14. Patel, M.; Jung, S.; Moore, K.; Powell, G.; Ainsworth, C.; Abbott, A. High-oleate peanut mutants result from a MITE insertion into the *FAD2* gene. *Theor. Appl. Genet.* **2004**, *108*, 1492–1502. [CrossRef] [PubMed]

15. Tang, Y.; Wang, X.; Wu, Q.; Fang, C.; Guan, S.; Yang, W.; Wang, C.T.; Wang, P. Identification of differentially expressed genes from developing seeds of a normal oil peanut cultivar and its high oil EMS mutant. *Res. Crop.* **2013**, *14*, 511–516.

16. Wan, L.; Li, B.; Pandey, M.K.; Wu, Y.; Lei, Y.; Yan, L.; Dai, X.; Jiang, H.; Zhang, J.; Wei, G. Transcriptome analysis of a new peanut seed coat mutant for the physiological regulatory mechanism involved in seed coat cracking and pigmentation. *Front. Plant Sci.* **2016**, *7*, 1491. [CrossRef]

17. Alberte, R.S.; Hesketh, J.D.; Kirby, J.S. Comparisons of photosynthetic activity and lamellar characteristics of virescent and normal green peanut leaves. *Z. Für Pflanzenphysiol.* **1976**, *77*, 152–159. [CrossRef]

18. Sui, J.; Jiang, D.; Zhang, D.; Song, X.; Wang, J.; Zhao, M.; Qiao, L. The salinity responsive mechanism of a hydroxyproline-tolerant mutant of peanut based on digital gene expression profiling analysis. *PLoS ONE* **2016**, *11*, e0162556. [CrossRef]

19. Gowda, M.; Bhat, R.; Motagi, B.; Sujay, V.; Kumari, V.; Sujatha, B. Association of high-frequency origin of late leaf spot resistant mutants with *AhMITE1* transposition in peanut. *Plant Breed.* **2010**, *129*, 567–569.

20. Wan, L.; Li, B.; Lei, Y.; Yan, L.; Ren, X.; Chen, Y.; Dai, X.; Jiang, H.; Zhang, J.; Guo, W. Mutant transcriptome sequencing provides insights into pod development in peanut (*Arachis hypogaea* L.). *Front. Plant Sci.* **2017**, *8*, 1900. [CrossRef]

21. Maluszynski, M.; Szarejko, I.; Barriga, P.; Balcerzyk, A. Heterosis in crop mutant crosses and production of high yielding lines using doubled haploid systems. *Euphytica* **2001**, *120*, 387–398. [CrossRef]

22. Greene, E.A.; Codomo, C.A.; Taylor, N.E.; Henikoff, J.G.; Till, B.J.; Reynolds, S.H.; Enns, L.C.; Burtner, C.; Johnson, J.E.; Odden, A.R. Spectrum of chemically induced mutations from a large-scale reverse-genetic screen in Arabidopsis. *Genetics* **2003**, *164*, 731–740.

23. Till, B.J.; Cooper, J.; Tai, T.H.; Colowit, P.; Greene, E.A.; Henikoff, S.; Comai, L. Discovery of chemically induced mutations in rice by TILLING. *BMC Plant Biol.* **2007**, *7*, 19. [CrossRef] [PubMed]

24. Sevanthi, A.; Kandwal, P.; Kale, P.B.; Prakash, C.; Ramkumar, M.; Yadav, N.; Mahato, A.K.; Sureshkumar, V.; Behera, M.; Deshmukh, R.K. Whole genome characterization of a few EMS-induced mutants of upland rice variety Nagina 22 reveals a staggeringly high frequency of SNPs which show high phenotypic plasticity towards the wild-type. *Front. Plant Sci.* **2018**, *9*, 1179. [CrossRef]

25. Roychowdhury, R.; Tah, J. Genetic variability study for yield and associated quantitative characters in mutant genotypes of *Dianthus caryophyllus* L. *Int. J. Biosci.* **2011**, *1*, 38–44.

26. Kodym, A.; Afza, R. Physical and chemical mutagenesis. In *Plant Functional Genomics*; Springer: Berlin/Heidelberg, Germany, 2003; pp. 189–203.

27. Zheng, Y.; Zheng, Y.; Wu, Z.; Sun, X.; Wang, C. Study on physiological characteristics and supporting techniques of Huayu 22 peanut. *Asian Agric. Res.* **2015**, *7*, 74–82.

28. Li, S.; Liang, X.; Zhou, G. Breeding of a new peanut variety Yueyou 45. *Guangdong Agric. Sci.* **2010**, *11*, 19–20.

29. Arisha, M.H.; Shah, S.N.; Gong, Z.; Jing, H.; Li, C.; Zhang, H. Ethyl methane sulfonate induced mutations in M2 generation and physiological variations in M1 generation of peppers (*Capsicum annuum* L.). *Front. Plant Sci.* **2015**, *6*, 399. [CrossRef]

30. Wang, L.; Zhang, B.; Li, J.; Yang, X.; Ren, Z. Ethyl methanesulfonate (EMS)-mediated mutagenesis of cucumber (*Cucumis sativus* L.). *Agric. Sci.* **2014**, *2014*, 48085.

31. Weng, Y.; Shi, A.; Ravelombola, W.S.; Yang, W.; Qin, J.; Motes, D.; Moseley, D.O.; Chen, P. A rapid method for measuring seed protein content in cowpea (*Vigna unguiculata* (L.) Walp). *Am. J. Plant Sci.* **2017**, *8*, 2387. [CrossRef]

32. Arisha, M.; Liang, B.; Shah, S.M.; Gong, Z.; Li, D. Kill curve analysis and response of first generation *Capsicum annuum* L. B12 cultivar to ethyl methane sulfonate. *Genet. Mol. Res.* **2014**, *13*, 10049–10061. [CrossRef]

33. Shah, S.; Gong, Z.; Arisha, M.; Khan, A.; Tian, S. Effect of ethyl methyl sulfonate concentration and different treatment conditions on germination and seedling growth of the cucumber cultivar Chinese long (9930). *Genet. Mol. Res.* **2015**, *14*, 2440–2449. [CrossRef]

34. Gnanamurthy, S.; Dhanavel, D. Effect of EMS on induced morphological mutants and chromosomal variation in Cowpea (*Vigna unguiculata* (L.) Walp). *Int. Lett. Nat. Sci.* **2014**, *17*, 33–43. [CrossRef]

35. Kumar, A.; Kumar, V.; Lal, S.K.; Jolly, M.; Sachdev, A. Influence of gamma rays and ethyl methane sulphonate (EMS) on the levels of phytic acid, raffinose family oligosaccharides and antioxidants in soybean seeds of different genotypes. *J. Plant Biochem. Biotechnol.* **2015**, *24*, 204–209. [CrossRef]

36. Ali, H.; Shah, T.M.; Iqbal, N.; Atta, B.M.; Haq, M.A. Mutagenic induction of double-podding trait in different genotypes of chickpea and their characterization by STMS marker. *Plant Breed.* **2010**, *129*, 116–119. [CrossRef]

37. Devi, A.S.; Mullainathan, L. Physical and chemical mutagenesis for improvement of chilli (*Capsicum annuum* L.). *World Appl. Sci. J.* **2011**, *15*, 108–113.

38. Borovsky, Y.; Tadmor, Y.; Bar, E.; Meir, A.; Lewinsohn, E.; Paran, I. Induced mutation in β-*CAROTENE HYDROXYLASE* results in accumulation of β-carotene and conversion of red to orange color in pepper fruit. *Theor. Appl. Genet.* **2013**, *126*, 557–565. [CrossRef]

39. Kumar, A.P.; Boualem, A.; Bhattacharya, A.; Parikh, S.; Desai, N.; Zambelli, A.; Leon, A.; Chatterjee, M.; Bendahmane, A. SMART–sunflower mutant population and reverse genetic tool for crop improvement. *BMC Plant Biol.* **2013**, *13*, 38. [CrossRef]

40. Khan, S.; Goyal, S. Improvement of mungbean varieties through induced mutations. *Afr. J. Plant Sci.* **2009**, *3*, 174–180.

41. Wang, T.L.; Uauy, C.; Robson, F.; Till, B. TILLING in extremis. *Plant Biotechnol. J.* **2012**, *10*, 761–772. [CrossRef]

42. Uauy, C.; Wulff, B.B.; Dubcovsky, J. Combining traditional mutagenesis with new high-throughput sequencing and genome editing to reveal hidden variation in polyploid wheat. *Annu. Rev. Genet.* **2017**, *51*, 435–454. [CrossRef]

43. Bertioli, D.J.; Jenkins, J.; Clevenger, J.; Dudchenko, O.; Gao, D.; Seijo, G.; Leal-Bertioli, S.C.; Ren, L.; Farmer, A.D.; Pandey, M.K. The genome sequence of segmental allotetraploid peanut *Arachis hypogaea*. *Nat. Genet.* **2019**, *51*, 877–884. [CrossRef]

44. Shirasawa, K.; Hirakawa, H.; Nunome, T.; Tabata, S.; Isobe, S. Genome-wide survey of artificial mutations induced by ethyl methanesulfonate and gamma rays in tomato. *Plant Biotechnol. J.* **2016**, *14*, 51–60. [CrossRef] [PubMed]

45. McCallum, C.M.; Comai, L.; Greene, E.A.; Henikoff, S. Targeted screening for induced mutations. *Nat. Biotechnol.* **2000**, *18*, 455–457. [CrossRef] [PubMed]

46. Krasileva, K.V.; Vasquez-Gross, H.A.; Howell, T.; Bailey, P.; Paraiso, F.; Clissold, L.; Simmonds, J.; Ramirez-Gonzalez, R.H.; Wang, X.; Borrill, P. Uncovering hidden variation in polyploid wheat. *Proc. Natl. Acad. Sci. USA* **2017**, *114*, E913–E921. [CrossRef] [PubMed]

47. Winfield, M.O.; Wilkinson, P.A.; Allen, A.M.; Barker, G.L.; Coghill, J.A.; Burridge, A.; Hall, A.; Brenchley, R.C.; D'Amore, R.; Hall, N. Targeted re-sequencing of the allohexaploid wheat exome. *Plant Biotechnol. J.* **2012**, *10*, 733–742. [CrossRef] [PubMed]

48. King, R.; Bird, N.; Ramirez-Gonzalez, R.; Coghill, J.A.; Patil, A.; Hassani-Pak, K.; Uauy, C.; Phillips, A.L. Mutation scanning in wheat by exon capture and next-generation sequencing. *PLoS ONE* **2015**, *10*, e0137549. [CrossRef]

49. Ashikari, M.; Wu, J.; Yano, M.; Sasaki, T.; Yoshimura, A. Rice gibberellin-insensitive dwarf mutant gene *Dwarf 1* encodes the α-subunit of GTP-binding protein. *Proc. Natl. Acad. Sci. USA* **1999**, *96*, 10284–10289. [CrossRef]

50. Ueguchi-Tanaka, M.; Fujisawa, Y.; Kobayashi, M.; Ashikari, M.; Iwasaki, Y.; Kitano, H.; Matsuoka, M. Rice dwarf mutant *d1*, which is defective in the α subunit of the heterotrimeric G protein, affects gibberellin signal transduction. *Proc. Natl. Acad. Sci. USA* **2000**, *97*, 11638–11643. [CrossRef]

51. Liu, C.; Wang, J.; Huang, T.; Wang, F.; Yuan, F.; Cheng, X.; Zhang, Y.; Shi, S.; Wu, J.; Liu, K. A missense mutation in the VHYNP motif of a DELLA protein causes a semi-dwarf mutant phenotype in Brassica napus. *Theor. Appl. Genet.* **2010**, *121*, 249–258. [CrossRef]

52. Sazuka, T.; Kamiya, N.; Nishimura, T.; Ohmae, K.; Sato, Y.; Imamura, K.; Nagato, Y.; Koshiba, T.; Nagamura, Y.; Ashikari, M. A rice *tryptophan deficient dwarf* mutant, *tdd1*, contains a reduced level of indole acetic acid and develops abnormal flowers and organless embryos. *Plant J.* **2009**, *60*, 227–241. [CrossRef]

53. Bishop, G.J. Brassinosteroid mutants of crops. *J. Plant Growth Regul.* **2003**, *22*, 325–335. [CrossRef]

54. Kwon, M.; Choe, S. Brassinosteroid biosynthesis anddwarf mutants. *J. Plant Biol.* **2005**, *48*, 1. [CrossRef]

55. Bennett, T.; Leyser, O. Strigolactone signalling: Standing on the shoulders of DWARFs. *Curr. Opin. Plant Biol.* **2014**, *22*, 7–13. [CrossRef] [PubMed]

56. Tanaka, N.; Matsuoka, M.; Kitano, H.; Asano, T.; Kaku, H.; Komatsu, S. *gid1*, a gibberellin-insensitive dwarf mutant, shows altered regulation of probenazole-inducible protein (PBZ1) in response to cold stress and pathogen attack. *Plant Cell Environ.* **2006**, *29*, 619–631. [CrossRef] [PubMed]

57. Takahashi, N.; Nakazawa, M.; Shibata, K.; Yokota, T.; Ishikawa, A.; Suzuki, K.; Kawashima, M.; Ichikawa, T.; Shimada, H.; Matsui, M. *shk1-D*, a dwarf Arabidopsis mutant caused by activation of the *CYP72C1* gene, has altered brassinosteroid levels. *Plant J.* **2005**, *42*, 13–22. [CrossRef]

58. Tanabe, S.; Ashikari, M.; Fujioka, S.; Takatsuto, S.; Yoshida, S.; Yano, M.; Yoshimura, A.; Kitano, H.; Matsuoka, M.; Fujisawa, Y. A novel cytochrome P450 is implicated in brassinosteroid biosynthesis via the characterization of a rice dwarf mutant, *dwarf11*, with reduced seed length. *Plant Cell* **2005**, *17*, 776–790. [CrossRef]

59. Jiang, L.; Liu, X.; Xiong, G.; Liu, H.; Chen, F.; Wang, L.; Meng, X.; Liu, G.; Yu, H.; Yuan, Y. DWARF 53 acts as a repressor of strigolactone signalling in rice. *Nature* **2013**, *504*, 401–405. [CrossRef]

60. Hooda, M.; Dhillon, R.; Bangarwa, K. Albinism in jojoba (Simmondsia chinensis). *Natl. J. Plant Improv.* **2004**, *1*, 69–70.

61. Chen, T.; Zhang, Y.; Zhao, L.; Zhu, Z.; Lin, J.; Zhang, S.; Wang, C. Physiological character and gene mapping in a new green-revertible albino mutant in rice. *J. Genet. Genom.* **2007**, *34*, 331–338. [CrossRef]

62. Pawar, N.; Pai, S.; Nimbalkar, M.; Kolar, F.; Dixit, G. Induction of chlorophyll mutants in zingiber officinale roscoe by gamma rays and EMS. *Emir. J. Food Agric.* **2010**, *22*, 406–411. [CrossRef]

63. Wang, Z.; Huang, Y.; Miao, Z.; Hu, Z.; Song, X.; Liu, L. Identification and characterization of *BGL11* (*t*), a novel gene regulating leaf-color mutation in rice (*Oryza sativa* L.). *Genes Genom.* **2013**, *35*, 491–499. [CrossRef]

64. Mochizuki, N.; Brusslan, J.A.; Larkin, R.; Nagatani, A.; Chory, J. *Arabidopsis genomes uncoupled 5* (*GUN5*) mutant reveals the involvement of Mg-chelatase H subunit in plastid-to-nucleus signal transduction. *Proc. Natl. Acad. Sci. USA* **2001**, *98*, 2053–2058. [CrossRef] [PubMed]

65. Rissler, H.M.; Collakova, E.; DellaPenna, D.; Whelan, J.; Pogson, B.J. Chlorophyll biosynthesis. Expression of a second *chl I* gene of magnesium chelatase in Arabidopsis supports only limited chlorophyll synthesis. *Plant Physiol.* **2002**, *128*, 770–779. [CrossRef] [PubMed]

66. Jensen, P.E.; Petersen, B.L.; Stummann, B.; Henningsen, K.; Willows, R.; Vothknecht, U.; Kannangara, C.; von Wettstein, D. Structural genes for Mg-chelatase subunits in barley: Xantha-f,-g and-h. *Mol. Gen. Genet. MGG* **1996**, *250*, 383–394. [PubMed]

67. Zhang, H.; Li, J.; Yoo, J.H.; Yoo, S.C.; Cho, S.H.; Koh, H.J.; Seo, H.S.; Paek, N.C. Rice *Chlorina-1* and *Chlorina-9* encode ChlD and ChlI subunits of Mg-chelatase, a key enzyme for chlorophyll synthesis and chloroplast development. *Plant Mol. Biol.* **2006**, *62*, 325–337. [CrossRef]

68. Jung, K.; Hur, J.; Ryu, C.; Choi, Y.; Chung, Y.; Miyao, A.; Hirochika, H.; An, G. Characterization of a rice chlorophyll-deficient mutant using the T-DNA gene-trap system. *Plant Cell Physiol.* **2003**, *44*, 463–472. [CrossRef]

69. Papenbrock, J.; Pfündel, E.; Mock, H.P.; Grimm, B. Decreased and increased expression of the subunit CHL I diminishes Mg chelatase activity and reduces chlorophyll synthesis in transgenic tobacco plants. *Plant J.* **2000**, *22*, 155–164. [CrossRef]

70. Wu, Z.; Zhang, X.; He, B.; Diao, L.; Sheng, S.; Wang, J.; Guo, X.; Su, N.; Wang, L.; Jiang, L.; et al. A chlorophyll-deficient rice mutant with impaired chlorophyllide Esterification in chlorophyll biosynthesis. *Plant Physiol.* **2007**, *145*, 29–40. [CrossRef]

71. Zhou, K.; Ren, Y.; Lv, J.; Wang, Y.; Liu, F.; Zhou, F.; Zhao, S.; Chen, S.; Peng, C.; Zhang, X.; et al. *Young Leaf Chlorosis 1*, a chloroplast-localized gene required for chlorophyll and lutein accumulation during early leaf development in rice. *Planta* **2013**, *237*, 279–292. [CrossRef]

72. Chen, Z.; Wang, M.L.; Barkley, N.A.; Pittman, R.N. A simple allele-specific PCR assay for detecting *FAD2* alleles in both A and B genomes of the cultivated peanut for high-oleate trait selection. *Plant Mol. Biol. Rep.* **2010**, *28*, 542–548. [CrossRef]

73. Wang, Y.; Zhang, X.; Zhao, Y.; Prakash, C.S.; He, G.; Yin, D. Insights into the novel members of the *FAD2* gene family involved in high-oleate fluxes in peanut. *Genome* **2015**, *58*, 375–383. [CrossRef]

Article

Construction of Soybean Mutant Diversity Pool (MDP) Lines and an Analysis of Their Genetic Relationships and Associations Using TRAP Markers

Dong-Gun Kim [1,2,†], Jae Il Lyu [1,†], Min-Kyu Lee [1,3], Jung Min Kim [1,3], Nguyen Ngoc Hung [1,3], Min Jeong Hong [1], Jin-Baek Kim [1], Chang-Hyu Bae [2,*] and Soon-Jae Kwon [1,*]

[1] Advanced Radiation Technology Institute, Korea Atomic Energy Research Institute, Jeongup 56212, Korea; dgkim@kaeri.re.kr (D.-G.K.); jaeil@kaeri.re.kr (J.I.L.); biolmk@kaeri.re.kr (M.-K.L.); jmkim0803@kaeri.re.kr (J.M.K.); nguyenhung@kaeri.re.kr (N.N.H.); hongmj@kaeri.re.kr (M.J.H.); jbkim74@kaeri.re.kr (J.-B.K.)

[2] Department of Life-resources, Graduate School, Sunchon National University, Suncheon 57922, Korea

[3] Division of Plant Biotechnology, College of Agriculture and Life Science, Chonnam National University, Gwangju 61186, Korea

* Correspondence: soonjaekwon@kaeri.re.kr (S.-J.K.); chbae@scnu.ac.kr (C.-H.B.); Tel.: +82-63-570-3312 (S.-J.K.); +82-61-750-3214 (C.-H.B.)

† These authors contributed equally to this work.

Received: 30 December 2019; Accepted: 6 February 2020; Published: 10 February 2020

Abstract: Mutation breeding is useful for improving agronomic characteristics of various crops. In this study, we conducted a genetic diversity and association analysis of soybean mutants to assess elite mutant lines. On the basis of phenotypic traits, we chose 208 soybean mutants as a mutant diversity pool (MDP). We then investigated the genetic diversity and inter-relationships of these MDP lines using target region amplification polymorphism (TRAP) markers. Among the different TRAP primer combinations, polymorphism levels and polymorphism information content (PIC) values averaged 59.71% and 0.15, respectively. Dendrogram and population structure analyses divided the MDP lines into four major groups. According to an analysis of molecular variance (AMOVA), the percentage of inter-population variation among mutants was 11.320 (20.6%), whereas mutant intra-population variation ranged from 0.231 (0.4%) to 14.324 (26.1%). Overall, intra-population genetic similarity was higher than that of inter-populations. In an analysis of the association between TRAP markers and agronomic traits using three different statistical approaches based on the single factor analysis (SFA), the Q general linear model (GLM), and the mixed linear model (Q+K MLM), we detected six significant marker–trait associations involving five phenotypic traits. Our results suggest that the MDP has great potential for soybean genetic resources and that TRAP markers are useful for the selection of soybean mutants for soybean mutation breeding.

Keywords: mutation breeding; soybean; mutant diversity pool (MDP); TRAP markers; association analysis

---

1. Introduction

Soybeans (*Glycine max* L.), used for food, livestock feed, and biofuel, is one of the most important agricultural crops worldwide. Soybeans are consumed directly by humans, especially in many Asian countries, in the form of traditional food products such as tofu, soy flour, and soymilk [1,2]. Soybean seeds are composed of 40%–42% protein, 18%–22% oil (85% unsaturated and 15% saturated fatty acids), 28% carbohydrates, and abundant quantities of other nutrients, such as phosphorus, calcium, iron, lysine, and vitamins A, B, and D [3]. In addition, soybeans play an important role in crop diversification and improve other crops through its addition of nitrogen to the soil during crop rotation [4].

Because the rate of spontaneous mutations in higher plants is quite low (10^{-5} to 10^{-8}) [5], physical and chemical mutagens can be used to induce mutations in cultivated plants [6]. Gamma radiation is a very effective tool to induce genetic variation in many plant characters, with the resulting changes dependent on the irradiation dose. Various plant organisms, such as seeds, pollen, whole plants, and embryoid bodies, can be irradiated [7]. Because gamma rays can also cause various types of DNA damage, including single- or double-strand breaks and substitutions [8,9], agronomic traits, such as flowering, maturation date, seed coat color, chloroplast number, and biomass yield, are frequently altered in soybean [10,11]. At present, 3200 mutant varieties of more than 210 plant species have been produced for commercial use. Approximately 170 mutant varieties of soybean, the second-most registered species after rice, are found in the FAO/IAEA Mutant Variety Database (http://mvd.iaea.org).

The use of molecular marker-based techniques in genetic studies, such as estimation of genetic diversity and population structure, has advanced remarkably in recent years. Among the different types of DNA markers, restriction fragment length polymorphisms (RFLPs), random amplified polymorphic DNAs (RAPDs), amplified fragment length polymorphisms (AFLPs), and inter-simple sequence repeats (ISSRs) have been extensively used in soybeans, each with their own advantages and limitations [12]. In addition, SNPs, which are widely distributed throughout genomes in both non-coding and coding regions, constitute the most abundant molecular markers recently used in plant genetic breeding [13], but their development is time-consuming and costly. The target region amplification polymorphism (TRAP) is a relatively new, simple, polymerase chain reaction (PCR)-based marker system that takes advantage of the available EST database sequence information to generate polymorphic markers targeting candidate gene [14]. Essentially, it derives an 18-mer primer from the EST sequence and pairs it with an arbitrary primer that targets the intron and/or exon region (AT- or GC-rich core). Because it can be used to generate markers for specific gene sequences, the TRAP technique is useful for genotyping germplasm and generating markers associated with desirable crop agronomic traits for marker-assisted breeding [15]. In recent years, the TRAP marker technique has been applied for genetic diversity analyses [16,17] and genetic mapping [18]. In addition, Im et al. [19] have developed a transposable element-based TRAP (TE-TRAP) marker system that is reportedly suitable for the mutation breeding of sorghum. Although TRAP markers have most commonly been used for genetic mapping and phylogenetic studies, they have also recently been applied to detect DNA mutations [20].

Rapid advances in the field of molecular biology and its allied sciences have led to the routine use of molecular markers, thereby providing plant breeders with a precise genetic-diversity analysis tool for plant improvement [21,22]. A combined molecular and morphological analysis is one of the most widely used approaches for the estimation of genetic distances within a group of genotypes, and molecular markers serve as an excellent tool for obtaining genetic information. Molecular markers are also of great value to plant breeders for assessment of genetic divergence among genotypes for various agronomic traits [23]. Another recent strategy for analyzing agronomic traits, association analysis based on molecular-marker linkage disequilibrium (LD), can reduce experimental time and costs. Association analysis has therefore been widely applied to study a variety of crops, such as rice [24], maize [25], and soybeans [26].

In this study, we constructed 208 mutant diversity pool (MDP) lines based on agronomic traits and investigated their genetic diversity and relationships using TRAP markers. Finally, we performed an association analysis between agronomic traits and polymorphic TRAP amplicons.

2. Materials and Methods

2.1. Plant Materials and Phenotypic Evaluation

A total of 1000 seeds each of one soybean landraces, KAS360-22, and six representative Korean soybean cultivars [27], 94Seori, BangSa (BS), PalDal (P), DanBaek (DB), DaePung (DP), and HwangKeum

(HK), were irradiated with 250 Gy of gamma rays using a ^{60}Co gamma-irradiator (150 TBq capacity; ACEL, Ottawa, ON, Canada) at the Korea Atomic Energy Research Institute in 2008. The irradiated M_1 and control (non-irradiated) seeds were immediately sown in the research field of the Advanced Radiation Technology Institute. To construct MDP lines, we sowed 1000 irradiated seeds of each of the seven soybean cultivars and harvested a total of 1695 $M_{1:2}$ individual seeds, only excluding those exhibiting growth aberrations, such as stunted growth, pollen sterility, and no germination due to the degree of radiosensitivity (Figure 1). Next, we generated 1695 individual gamma-irradiated mutants during M_1–M_5 generations by single-seed descent and then continued as bulks until M_{12} generation. In a first selection phase, we selected 523 mutant lines from the M_5 generation that possessed at least 30% superior agricultural characteristics related to various environmental factors, such as grain yield, growth type, and climate adaptability. In a second selection phase, we investigated the morphological phenotypes of the 523 mutant lines in the M_{12} generation to eliminate redundant phenotypes. Overall, we selected 208 genetically fixed mutant lines (201 mutants with their wild types), which we designated as the mutant diversity pool (MDP), and assessed the following four agronomic traits: days of flowering (DF), maturation date (MD), seed index (SI), node number (NN), and the following seven morphological traits: growth type (GT), flower color (FC), seed coat color (SCC), seed hilum color (SHC), stem anthocyanin (SA), plant height (PH), and ramification number (RN).

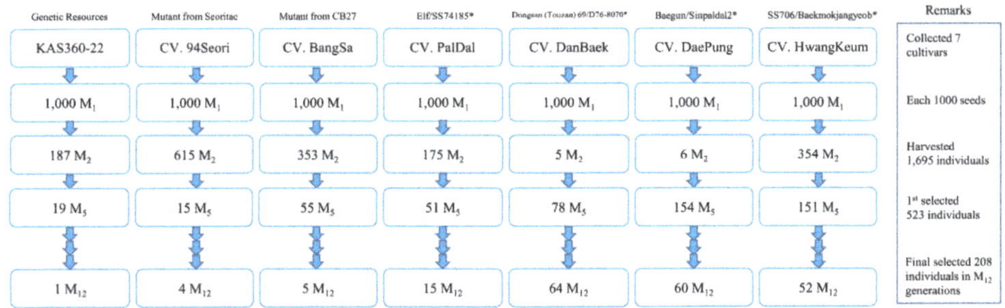

Figure 1. Schematic illustration of the breeding of M_1–M_{12} generations of 208 soybean mutant diversity pool (MDP) lines. * Information of cultivars was described in Lee et al. [27]

2.2. DNA Extraction

The control and treated seeds of the 208 genetically fixed mutant lines were immediately sown in 50-cell (5 × 10) vegetable nursery trays containing bio-bed soil (Dongbu Farm Hannong, Gimje, Korea) and then incubated in a greenhouse at 20 ± 5 °C under natural light for 1 month. Fresh leaf tissue from seedlings of each mutant line was collected and subjected to total genomic DNA extraction using a DNeasy 96 Plant kit (Qiagen, Leipzig, Germany) following the manufacturer's protocol. The extracted DNAs were stored at −20 °C until use. For PCR analysis, DNA concentrations were determined using a NanoDrop ND-1000 spectrophotometer (Thermo Fisher Scientific., Waltham, MA, USA) were and then adjusted to 10 ng/µl.

2.3. TRAP Analysis

Four fixed primers and four arbitrary primers were used to generate TRAP markers (Table 1). All four arbitrary primers and one of the fixed primers were designed from other studies of monocot plants [14,16]. The three other fixed primers (with 'MIR' prefixes) were designed based on *Arabidopsis thaliana* microRNA sequences [28] using the Primer3 program (http://frodo.wi.mit.edu/primer3/). PCR amplifications with 16 primer combinations were carried out on all DNA samples according to the protocol of Hu et al. [29] with slight modification. Briefly, reactions were performed in 20-µl volumes containing 2 µl genomic DNA (10 ng/µl), 1 µl fixed primer (10 pmol/µl), 1 µl of each arbitrary primer (10 pmol/µl), 0.8 µl of dNTPs (2.5 mM), 2.0 µl 10 × PCR buffer, and 0.3 µl Phoenix *Taq* DNA polymerase

(5 U/µl; cat no. Phoenix2013). DNA amplification was performed in a thermocycler (G Storm, UK) according to the following program: initial denaturation at 94 °C for 2 min, followed by 5 cycles of 94 °C for 45 s, 35 °C for 45 s, and 72 °C for 60 s, then 35 cycles of 94 °C for 45 s, 53 °C for 45 s, and 72 °C for 60 s, and a final extension at 72 °C for 7 min. The amplified products were analyzed separately using a fragment analyzer automated capillary electrophoresis instrument (FA; Advanced Analytical Technologies, Ankeny, USA), and the collected images were scored manually.

Table 1. List of soybean target region amplification polymorphism- (TRAP) marker primers.

Primer name	Sequence (5′–3′)
Fixed primers	
B14G14B	AAT CTC AAG GAC AAA AGG
MIR 156A	GAT CTC TTT GGC CTG TC
MIR 157B	GAT CAT TGT CCA GAT TC
MIR 159A	GAT CCT TGG TTC TTT GG
Arbitrary primers	
Sa4	TTA CCT TGG TCA TAC AAC ATT
Sa12	TTC TAG GTA ATC CAA CAA CA
Ga3	TCA TCT CAA ACC ATC TAC AC
Ga5	GGA ACC AAA CAC ATG AAG A

2.4. Data Analysis

TRAP marker alleles were scored as binary data, with '0' indicating the absence of a given allele and '1' indicating its presence. The binary data were entered into a Microsoft Excel spreadsheet, and genetic similarities were computed. Using the 0/1-matrix, we calculated gene diversity, percent polymorphism, polymorphism information content (PIC), and genetic distance with the genetic analysis package PowerMarker. A dendrogram was constructed according to the unweighted pair group method with arithmetic mean (UPGMA) algorithm based on the Nei distance method in PowerMarker v3.25 as well as the embedded MEGA7 program. To analyze population structure, a Bayesian population analysis was performed in STRUCTURE 2.3.4. A graphical determination of the optimal number of populations (K) was carried out using STRUCTURE HARVESTER (http://taylor0.biology.ucla.edu/structureHarvester/). The number of putative populations (K), assumed according to a set of allele frequencies at each locus, can provide the degree of admixture of mutant lines. Each individual was assigned to one or more populations based on membership (q) using a Q-matrix derived from STRUCTURE 2.3.4. Ideally, the average estimated log probability of the data Pr(x|k) should plateau at the most appropriate value of K. To determine the optimal K, we calculated the average probability of K for values of $K = 2 - 15$ based on 10 Markov chain Monte Carlo runs, each consisting of 10,000 burn-in (initiation) iterations followed by 100,000 iterations under a population admixture model. We then conducted an analysis of molecular variance (AMOVA) and calculated genetic distances to support the genetic diversity information. An AMOVA of 999 permutations was completed to assess inter- and intra-population variance (wild type and mutants) using GenALEx v6.501. Pairwise fixation index (Fst) values were also estimated by AMOVA. The hypothesis of an association of molecular markers with phenotypic data was tested using three different methods in the software program TASSEL 4. The first association method, a single factor analysis (SFA) of variance that does not consider population structure, was performed using each marker as the independent variable. The mean performance of each allelic class was compared using the general linear model (GLM) function in TASSEL. Next, a Q GLM method was carried out using the same software. This method applies population structure detected by STRUCTURE (Q matrix) as co-factors. To obtain an empirical threshold for marker significance and an experiment-wise P-value, 10,000 permutations of the data were performed. The final marker–trait association test, Q + K MLM, considers both the kinship matrix and the population structure Q-matrix. The K matrix of pairwise kinship coefficients for all pairs of lines was calculated from the TRAP data by TASSEL. Basic statistics and a correlation analysis of agronomic traits were performed using Microsoft Excel and Python v2.7.

3. Results

3.1. Phenotypic Analysis and Correlation Analysis

A summary of agronomic and morphological traits of the 208 MDP lines is shown in Table 2 and Supplementary Table S1. With respect to growth characteristics, a variety of phenotypes were observed. DF ranged from 42 (mutant numbers; S87, S88, and S149) to 64 (S138), while MD varied from 112 (S6) to 150 (S8). In addition, the seed index (SI) ranged from 7.9 (S13) to 28.8 (S7) g, and PH ranged from 23 (S14) to 92.2 (S76) cm. NN and RN varied from 8.4 (S14) to 24.6 (S78) and 2.6 (S81) to 8.8 (S12) cm, respectively. As shown in the histogram in Figure 2, the phenotypic values of these six quantitative traits in the 208 soybean MDP lines followed a Gaussian distribution. The SI was relatively well distributed, whereas DF and MD had dynamic distributions. Substantial variation was observed in agronomic traits between mutants and wild types of the 208 MDP lines (Figure 3, Figure S1). Compared with their wild types, P- and DB-derived mutants possessed a wider distribution of PH and NN phenotypes, whereas the phenotypes of the six quantitative traits of the HK-derived mutants were only slightly changed. DB-derived mutants, in particular, had mostly increased values of agronomic traits, such as DF, SI, PH, NN, and RN, compared with the wild type. In contrast, HK-derived mutants tended to have reduced values relative to the wild type, albeit only very slightly smaller (Figure 3). With regards to the five qualitative traits, altered phenotypes, such as changes in growth type and color-related traits, were confirmed in the MDP lines (Figure S1). The seven wild-type plants exhibited determinate growth, whereas 46 P-, DB-, and DP-derived mutants were indeterminate. In addition, changes were observed in color-related traits, including FC, SCC, SHC, and SA. These results indicate that MDP lines were successfully constructed through multiplex genetic and phenotypic mutation induced by gamma irradiation. In addition, we calculated the pairwise correlation coefficients of the 11 agronomic traits in the 208 soybean MDP lines. The strongest positive correlations were between PH and NN (0.912*), GT and NN (0.824*), and GT and PH (0.749*), while the most negative correlations were those between MD and SI (−0.512*) and between SI and NN (−0.357*) (Table 3).

Table 2. Summary of quantitative trait values in 208 soybean MDP lines.

Values	Agronomic Traits				Morphological Traits	
	Days of Flowering	Maturity Days	Seed Index (g)	Node Number (ea)	Plant Height (cm)	Ramification Number (ea)
Min	42	112	7.9	8.4	23.0	2.6
Mean	52.83	133.58	18.06	13.40	42.73	4.96
Max	64	150	28.8	24.6	92.2	8.8
Line No.						
Min	S87, S88, S149	S6	S13	S14	S14	S81
Max	S138	S8	S7	S78	S76	S12

Min, minimum; max, maximum; Line No., see Supplementary Table S1. Data were investigated in 2019 at the KAERI research field, Jeongup, Korea. Traits are all means of five biological replicates.

Table 3. Matrix of correlation coefficients between 11 agronomic traits in 208 soybean MDP lines.

	DF	MD	GT	FC	SCC	SHC	SI	SA	PH	NN	RN
DF (days of flowering)	–	0.419 *	0.266 *	0.436 *	0.038	0.060	−0.327 *	0.435 *	0.272 *	0.323 *	0.248 *
MD (Maturity days)		–	0.269 *	0.369 *	0.042	0.034	−0.512 *	0.491 *	0.407 *	0.381 *	0.224 *
GT (Growth type)			–	0.031	0.129 *	0.237 *	0.308 *	0.117	0.749 *	0.824 *	0.101
FC (Flower color)				–	0.301 *	0.280 *	0.369 *	0.685 *	0.058	0.003	0.099
SCC (Seed coat color)					–	0.354 *	0.071	0.243 *	0.166 *	0.134 *	0.068
SHC (Seed hilum color)						–	0.005	0.197 *	0.241 *	0.260 *	0.124 *
SI (Seed index)							–	0.554 *	−0.295 *	−0.357 *	−0.211 *
SA (Stem anthocyanin)								–	0.027	0.103	0.009
PH (Plant height)									–	0.912 *	0.177 *
NN (Node number)										–	0.202 *
RN (Ramification number)											–

* Significant at the 0.05 probability level.

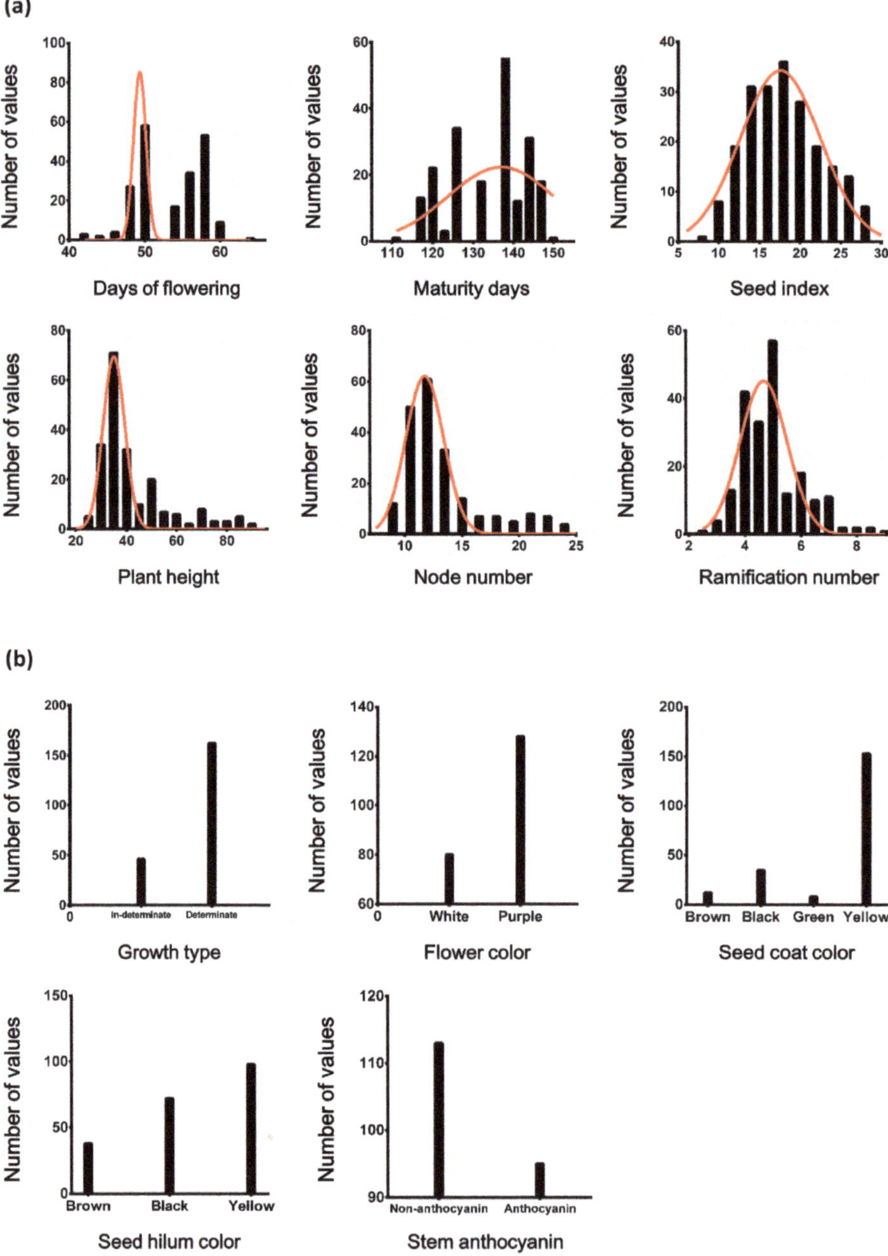

Figure 2. Distribution of agronomic traits among 208 soybean MDP lines. Data are presented for (**a**) six quantitative traits (DF, MD, SI, PH, NN, and RN) with gaussian fitting curve and (**b**) five qualitative traits (GT, FC, SCC, SHC, and SA).

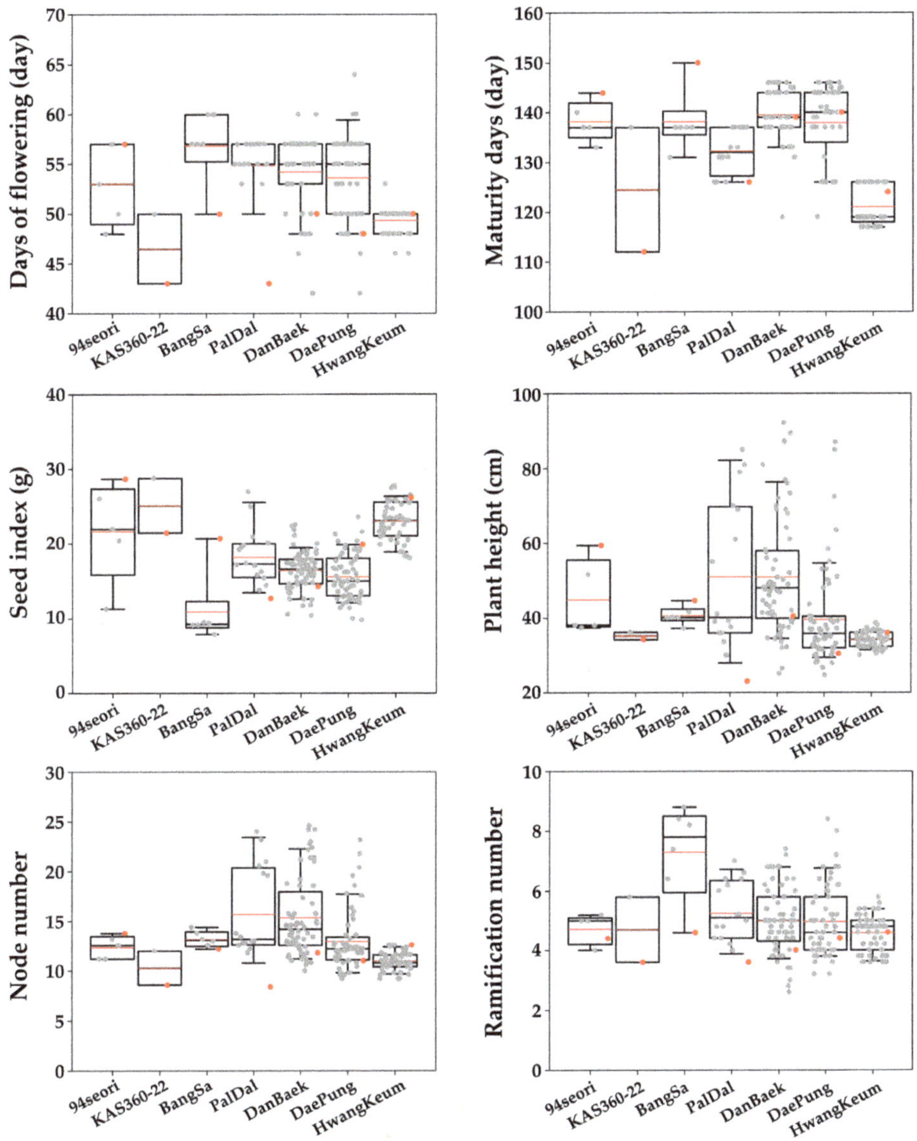

Figure 3. Changes in the phenotypes of six quantitative traits in 208 MDP lines. The box plots of phenotypic distributions in seven MDP lines and their wild types are shown. The data shown are the mean values of individual mutants (gray) and wild types (red).

3.2. TRAP Marker Polymorphism

A summary of the TRAP markers produced by 16 primer combinations (four fixed forward primers in combination with arbitrary reverse primers) across all 208 soybean mutants is given in Table 4. Sixteen primer combinations amplified a total of 551 fragments. The number of amplified fragments ranged from 25 (for primers MIR157B + Ga5) to 45 (for primers B14G14B + Ga5). A total of 551 amplicons were scored, of which 222 (40.29%) were monomorphic alleles and 329 (59.71%) were polymorphic. An average of 34.44 amplicons, 20.56 polymorphic, were scored per primer combination. The highest (84.00%) and lowest (32.35%) polymorphism levels were obtained with primers MIR157B + Ga5 and B14G14B + Ga3, respectively. PIC varied among the primer combinations, ranging from 0.07 (B14G14B + Sa12) to 0.23 (MIR157B + Sa4), with a mean value of 0.15.

Table 4. Summary of polymorphism of 16 TRAP marker sets in 208 soybean MDP lines.

Primer Combination	Total Number of Fragments	Polymorphic Fragments	Polymorphism (%)	PIC
B14G14B + Sa4	38	23	60.53	0.15
B14G14B + Sa12	38	17	44.74	0.07
B14G14B + Ga3	34	11	32.35	0.08
B14G14B + Ga5	45	29	64.44	0.16
MIR156A + Sa4	39	31	79.49	0.16
MIR156A + Sa12	37	22	59.46	0.16
MIR156A + Ga3	35	17	48.57	0.12
MIR156A + Ga5	35	25	71.43	0.20
MIR157B + Sa4	31	23	60.53	0.23
MIR157B + Sa12	31	19	61.29	0.16
MIR157B + Ga3	30	17	56.67	0.14
MIR157B + Ga5	25	21	84.00	0.21
MIR159A + Sa4	36	22	58.33	0.18
MIR159A + Sa12	37	18	48.65	0.10
MIR159A + Ga3	29	15	51.72	0.16
MIR159A + Ga5	31	20	64.52	0.16
Total	551	329	–	–
Average	34.44	20.56	59.71	0.15

3.3. Genetic Relationships and Population Structure of the 208 MDP Lines

A dendrogram was constructed to clarify genetic relationships of the 208 MDP lines. At a genetic distance of 0.097, the seven wild-type cultivars and their mutants could be divided into four major groups (Figure 4). Group I included five mutants with their wild types KAS360-22 and 94seori. Group II comprised 22 mutants originating from BS and P. Group III was made up of two subgroups: III-a, which mainly contained DP mutants and their wild type DP as well as some HK mutants, and III-b, which mainly included HK mutants and HK with a few DP mutants. Group IV was distinct from the other three groups and consisted of all 64 DB mutants with DB and DP mutants. We performed a population structure analysis with a predefined number of sub-populations (K) ranging from 2 to 15. The optimal K was determined using an ad-hoc statistic (ΔK), which was based on the rate of change in the log probability of the data between successive K-values (Figure S2). According to the analysis, the optimal K value was 4, which corresponded to a division of the genetic composition into four groups (Figure 4). Each accession was assigned to single or multiple membership depending on whether its genotype indicated admixture. The result of this analysis was consistent with the topology of the dendrogram.

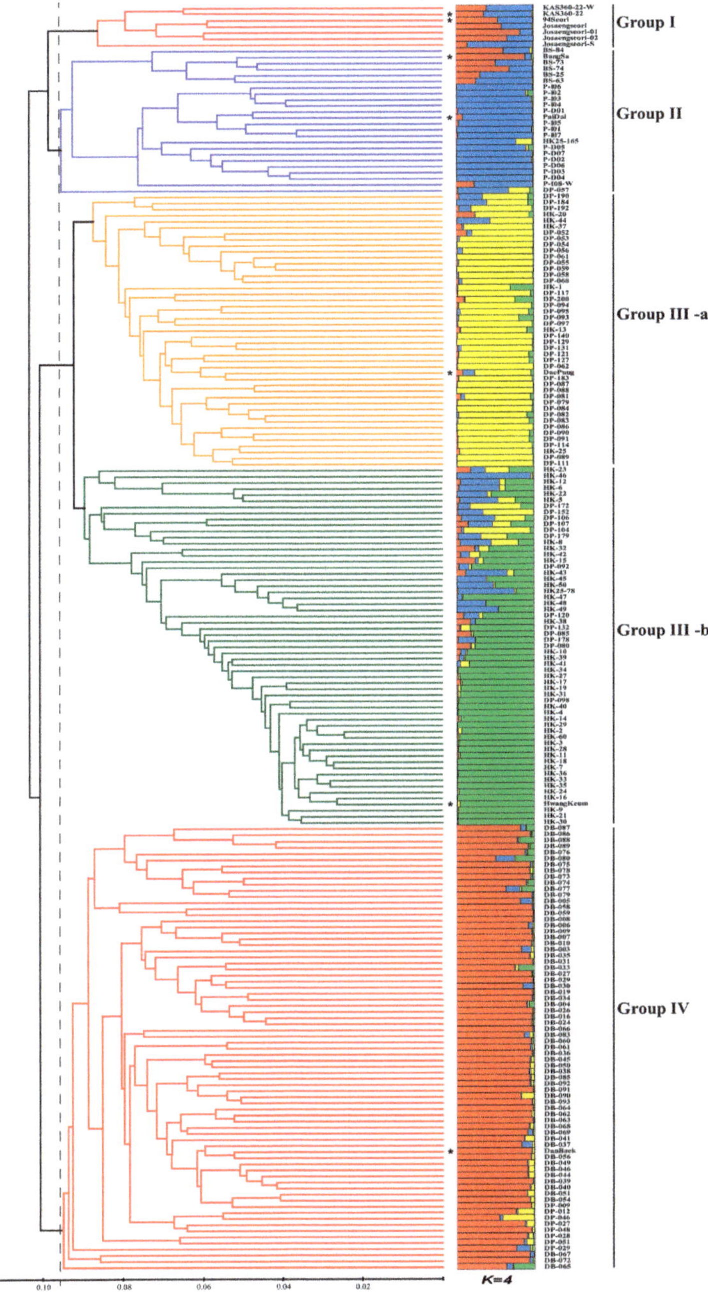

Figure 4. Dendrogram revealed by unweighted pair group method with arithmetic mean (UPGMA) cluster analysis and the population structure of 208 soybean MDP lines based on TRAP markers. * Indicated original cultivars.

3.4. AMOVA

An AMOVA of the 208 MDP lines based on TRAP markers was performed to analyze the distribution of inter- (among) and intra- (within) mutant population genetic diversity. According to the AMOVA, the estimated inter-mutant population variance was 11.320 (20.6%), while approximately 79.4% of the variation in all variance positions was attributed to intra-mutant population variance. These results indicate that the majority of the variance was intra-mutant populations. However, notwithstanding the estimation of variance in each intra-mutant population, the variation of intra-mutant population was mostly lower than inter-mutant population. The highest percentage of observed intra-mutant population genetic variation was that of DB populations (26.1%) and the lowest was intra KAS360-22 populations (0.4%). We also examined genetic differentiation between soybean MDP lines using Fst data estimated from pairwise comparisons. Fst values varied from 0.065 (KAS360-22 and 94seori) to 0.351 (HK and 94seori), with an average of 0.248 (Table 5).

Table 5. Analysis of molecular variance results and pairwise Fst values estimated from 208 soybean MDP lines.

Source	Est. Var.	Percentage of variation (%)
Inter-mutant pops	11.320	20.6
Intra 94Seori pops	0.820	1.5
Intra KAS360-22 pops	0.231	0.4
Intra BangSa pops	0.866	1.6
Intra PalDal pops	2.765	5.0
Intra DanBaek pops	14.324	26.1
Intra DaePung pops	14.252	26.0
Intra HwangKeum pops	10.342	18.8

Control	94Seori	KAS360-22	BangSa	PalDal	DanBaek	DaePung	HwangKeum
94Seori	–						
KAS360-22	0.065	–					
BangSa	0.273	0.243	–				
PalDal	0.312	0.270	0.285	–			
DanBaek	0.239	0.229	0.222	0.268	–		
DaePung	0.276	0.245	0.258	0.226	0.129	–	
HwangKeum	0.351	0.343	0.322	0.270	0.254	0.126	–

Significance ($p < 0.001$) was determined based on 1000 iterations.

3.5. Association Analysis

Associations between 329 polymorphic fragments (16 combinations of TRAP markers) and 11 phenotypic traits of 208 soybean MDP lines were analyzed by three methods: (1) SFA; (2) a structure association analysis using a general linear model where population membership served as covariates (Q GLM); and (3) a composite approach in which the average relationship was estimated by kinship (K) and implemented under a mixed linear model (Q + K MLM). The average level of significance of each marker–trait association was based on the three different analyses are shown in Table 6. A total of 27 significant ($p < 0.001$) marker–trait associations (SMTAs) were detected using the three methods. The 157B + Sa4 primer combination was significantly associated with four phenotypic traits (GT, SA, PH, and NN), while 156A + Sa12, 156A + Sa4, 157B + Ga5, 157B + Ga3, 159A + Sa12, 159A + Ga5, and 159A + Ga3 were associated with one trait each: FC, PH, FC, PH, SHC, SI, and RN, respectively. The lowest calculated P-value using SFA was for the association of 157B + Sa12_6 with the FC trait ($p = 1.44 \times 10^{-12}$, $R^2 = 0.216$). Under the Q GLM model, the lowest P-value of a SMTA was that of B14 + Sa4_18 with PH ($P = 2.46 \times 10^{-8}$, $R^2 = 0.119$). Under the Q+K MLM model, the lowest SMTA P-value was observed for the association of B14+Sa4_18 with PH ($P = 9.81 \times 10^{-7}$, $R^2 = 0.107$) (Table 6). Among the three different analytical approaches (SFA, Q GLM, and Q + K MLM), the highest total number of SMTAs (626) was detected using SFA, followed by the Q GLM approach (178). The lowest number of SMTAs (143) was detected using the Q+K MLM approach; this number corresponded to 22.8% and 80.3% of the total number of SMTAs detected with SFA and Q GLM, respectively (Table S2). Six SMTAs at $p < 0.0001$—two for GT (156A + Ga5_16 and 157B + Sa4_4), one for FC (157B + Sa12_6), one for SCC (157B + Sa12_20), one for PH (B14 + Sa4_18), and one for NN (B14 + Sa4_18)—were revealed by all three approaches when kinship and/or population structure was considered in this collection.

Table 6. Selection of 27 significant marker–trait associations (SMTA) markers based on three association analysis approaches.

Trait	Marker	SFA [a]	R^2	Q GLM [b]	R^2	Q + K MLM [c]	R^2	Average p-value
Maturity days	B14 + Ga5_28	**	0.073	**	0.044	*	0.044	*
Growth type	156A + Ga5_16	**	0.116	**	0.101	**	0.111	**
	157B + Sa4_4	**	0.125	**	0.102	**	0.102	**
Flower color	157B + Sa12_6	**	0.216	**	0.074	**	0.072	**
	156A + Sa12_17	**	0.159	*	0.057	*	0.057	*
	157B + Ga5_9	**	0.125	*	0.046	*	0.046	*
Seed coat color	157B + Sa12_20	**	0.083	**	0.083	**	0.086	**
Seed hilum color	159A + Sa12_28	*	0.055	*	0.058	*	0.063	*
	159A + Sa4_33	**	0.111	*	0.058	*	0.057	*
	157B + Sa12_20	*	0.066	*	0.053	*	0.056	*
Seed index	156A + Ga3_7	**	0.082	**	0.054	*	0.054	**
	159A + Sa4_32	**	0.096	*	0.050	*	0.050	*
	156A + Ga3_1	*	0.065	*	0.046	*	0.046	*
	159A + Ga5_2	**	0.139	*	0.044	*	0.044	*
Stem anthocyanin	157B + Sa4_7	**	0.079	**	0.047	*	0.044	**
Plant height	B14 + Sa4_18	**	0.152	**	0.119	**	0.107	**
	B14 + Ga5_30	**	0.173	**	0.069	*	0.059	**
	157B + Ga3_10	**	0.135	**	0.062	*	0.052	*
	157B + Sa4_4	**	0.072	*	0.053	*	0.051	*
	156A + Sa4_6	*	0.056	*	0.048	*	0.049	*
	156A + Ga5_13	**	0.100	**	0.063	*	0.049	*
Node number	B14 + Sa4_18	**	0.121	**	0.097	**	0.087	**
	B14 + Ga5_30	**	0.162	**	0.078	*	0.066	**
	157B + Sa4_4	**	0.083	*	0.062	*	0.055	*
	156A + Ga5_16	**	0.074	*	0.046	*	0.054	*
Ramification number	159A + Ga3_24	*	0.058	*	0.054	*	0.061	*
	159A + Sa4_32	**	0.080	**	0.072	*	0.058	*

[a] SFA: single factor analysis of variance. [b] Q GLM: general linear model using a Q population structure matrix. [c] Q + K MLM: mixed linear model using Q population structure and K kinship matrixes. * $p \leq 0.001$, ** $p \leq 0.0001$.

4. Discussion

In this study, we constructed an MDP from populations of 1695 gamma-irradiated mutants in two selection phases over M_1 to M_{12} generations; first, in the M_5 generation, we selected 523 mutant lines exhibiting at least 30% superior agricultural characteristics, and, second, we eliminated redundant morphological phenotypes in the M_{12} generation (Figure 1). Finally, we constructed 208 MDP lines and investigated 11 agronomic traits. Our collection strategy for selecting MDP lines differed in some respects from the general core-collection method. With the latter approach, a core collection assembled from an existing collection is chosen to represent the genetic and phenotypic diversity of the larger collection without overlapping phenotypes [30]. Such an approach has become accepted as an efficient tool for improving the conservation of many crops [31,32]. In our study, we similarly eliminated overlapping phenotypes from our collected MDP lines in the second selection phase, but we considered specific changed agronomic characteristics of individual mutants rather than their representation of the original populations.

Our examination of agronomic traits in the MDP lines revealed a variety of DF, MD, GT, FC, SA, PH, NN, and RN phenotypes as well as those related to seed traits, such as SCC, SHC, and SI (Table 2, Figure 2, Table S1). We also observed changes in phenotypes between MDP lines and their wild types (Figure 3, Table S3, Figure S1). The FAO/IAEA mutant variety database (MVD, http://mvd.iaea.org)

includes 174 publicly released soybean mutants. These mutants have various desirable agronomical and biochemical characteristics, such as an improved maturity date, yield, protein content, fatty acid content, and changed seed/stem color, with approximately 62% of released mutants mainly selected for their altered maturity dates and yields. In our phenotypic evaluation of the 208 MDP lines, we detected a wider variety of changes to the quantitative traits, including SI, PH, NN, and RN (Figure 3), as well as to the qualitative traits, such as FC, SCC, and SHC (Figure S1). According to our previous study, in addition, some of DB- and DP-derived mutants in the MDP lines had changed compositions of fatty acids, including linolenic acid and oleic acid [33]. Given all of these results, our MDP lines may be useful resources as a genetic diversity pool for soybean breeding.

To investigate genetic relationships among the 208 MDP lines, we evaluated DNA polymorphism patterns in these lines using TRAP markers. In the rapid, efficient PCR-based TRAP marker system, expressed sequence tag database information and bioinformatics tools are used to generate polymorphic markers around targeted candidate gene sequences. Previous studies of lettuce (*Lactuca sativa*) [34], sugarcane (*Saccharum officinarum*) [35], spinach (*Spinacia oleracea*) [29], geranium (*Pelargonium inquinans*) [36], sunflower (*Helianthus annuus*) [37], and faba beans (*Vicia faba*) [16] have demonstrated that TRAP markers are useful for assessing genetic diversity. Using this system in the present study, we PCR-amplified 551 fragments with 16 primer combinations and observed considerable variation in the percentage of polymorphic amplicons among primer pairs—from 32.35% to 84.00% (Table 4). In a study of faba beans, Kwon et al. [16] obtained 221 amplified fragments with 12 TRAP primer combinations and observed an average polymorphism rate of 55.2%. In the present study, we observed a polymorphism level of 59.7% among 551 amplified fragments. In contrast, a study of sugarcane detected a polymorphism rate of 74% from 925 amplified fragments [17], a level much higher than in the soybeans (*Glycine max*) and the faba beans. Compared with the results of previous studies of soybeans based on ISSR and RAPD techniques [38,39], the use of the TRAP system yielded more DNA fragments per primer combination. A previous AFLP analysis generated an average of 40 to 50 DNA fragments per primer pair [40,41], similar to the outcome of our TRAP analysis. The present results demonstrate that the TRAP marker system is a simple yet powerful technique for estimating soybean genetic diversity.

To reveal relationships among the 208 MDP lines, we constructed a UPGMA-based dendrogram using the TRAP marker data. On the basis of genetic distances, the 208 MDP lines clustered into four groups. An analysis of the population structure based on an ad-hoc statistic (ΔK) likewise divided the MDP lines into four groups. These results indicate that four genotype-based sub-populations are present in the 208 MDP lines (Figure 4) that largely correspond to their wild-type cultivars. As denoted by different colors, the main membership composition of the four groups and their subgroups is as follows: Group I including two wild types (94seori, and KAS360-22) possessed 52% red and 46% blue; Group II including BS and P was 80% blue; Group III-a including DP was 88% yellow; Group III-b including HK was 71% green; and Group IV including DB was 90% red. In a previous genetic diversity analysis based on 20 SSR markers, 91 Korean soybean cultivars were divided into seven groups at a genetic distance of 0.81. In that study, HK and P were clearly separated, but three cultivars (BS, DB, and DP) grouped together [42]. Using TRAP markers in the present study, we were able to better resolve groups of wild-type cultivars. In addition, we performed an AMOVA to separate the total molecular variance of the mutants into inter- and intra-population components (Table 5) and assessed their significance using permutational testing procedures. Overall, based on the dendrogram and population structure, 201 mutant lines grouped with their wild type except 29 (14%) mutant lines, including 22 DP- and 7 HK-mutants. Nevertheless, these mutant lines also possessed their genetic membership according to population structure. In AMOVA, all intra-mutant population also showed lower variation than inter-mutant population except for two populations, DB- and DP-, since DB- and DP- had most large mutant lines, 64 and 60, respectively. A similar result was described by Lee et al. [20]. Each of the ten wild types was clustered with their M_1 generation mutants by gamma radiation in faba bean. However, the genetic variation of the mutants is not much higher than among cultivars or accessions. Although TRAP markers have most commonly been used for genetic mapping and dendrogram studies, they have also recently been applied to detect DNA mutations. Because of their many advantages, including simplicity, reliability, moderate throughput, and ease of sequencing of selected bands, TRAP markers have been used widely in plants. For example, the TRAP system has been used to study genetic variability induced by gamma ray treatments in sugarcane [43] and sorghum (*Sorghum bicolor*) [19]. Lee et al. [20] recently exploited a TRAP marker to estimate the frequency of mutations induced by gamma rays in an M_1 generation of faba bean. The 242 amplified fragments obtained using eight primer combinations had an average polymorphism rate of 66.7%, which is higher than the percentage in our study because they used early generation. TRAP markers have several advantages over other types of markers: they are easy to use (like RAPDs), high in polymorphisms (like AFLPs), and their primers can be readily designed from known sequences of putative genes [44].

In association mapping, false discoveries are a major concern and can be partially attributed to spurious associations caused by population structure and unequal relatedness among individuals. Two major approaches, namely, GLM and MLM, are used to study marker–trait associations. The number of SMTAs detected by GLM is generally much higher than that revealed by MLM [45]. GLM-based studies of marker–trait associations consider only the Q matrix generated during the study of population structure. In contrast, MLM simultaneously accounts for both population structure and kinship (genetic relatedness among individuals) and is hence more reliable. In the present study, the GLM method (Q) uncovered 178 SMTAs between the 11 phenotypic traits and 27 TRAP markers. Using the MLM method (Q + K), 143 SMTAs involving 27 TRAP markers were identified (Table 6, Table S2). These results confirm a previous observation that the number of SMTAs estimated with GLM is higher than that uncovered with MLM [46,47]. Most interestingly, the three approaches considering kinship and/or population structure in the MDP collection in this study revealed six SMTAs at $p < 0.0001$ in all approach methods. These six SMTAs involved five agronomic traits: GT (2), FC (1), SCC (1), PH (1), and NN (1).

5. Conclusions

In this study, we successfully constructed soybean MDP lines and compared their agronomic traits. We also performed the first-ever study of genetic diversity and relationships using the TRAP marker system in soybean. To examine MDP genetic diversity and relationships, we performed dendrogram, population structure, and molecular variance analyses based on their TRAP genotypes. Finally, we uncovered six SMTAs ($p < 0.0001$) involved with TRAP genotypes and agronomic traits using three association mapping methods (SFA, Q GLM, and Q + K MLM). Our results can serve as a foundation for future research on genotype–phenotype interactions in large mutant populations.

Supplementary Materials: The following are available online at http://www.mdpi.com/2073-4395/10/2/253/s1, Figure S1: Changes in five qualitative-trait phenotypes of 208 MDP lines; Figure S2: ΔK values for different numbers of populations (K) assumed in the STRUCTURE analysis, Table S1: Agronomic characteristics of the 208 soybean MDP lines used in this study; Table S2: Comparison of the number of SMTAs ($p < 0.01$) based on three analytical approaches (SFA, Q GLM, and Q + K MLM); Table S3: Comparison of phenotypes and genotypes among seven original cultivars.

Author Contributions: Conceptualization, D.-G.K., J.I.L., and S.-J.K.; methodology, D.-G.K., J.I.L., and S.-J.K.; software, D.-G.K. M.-K.L.; validation, D.-G.K., J.-B.K., C.-H.B., and S.-J.K.; investigation, D.-G.K., J.M.K.; resources, D.-G.K., M.J.H.; data curation, D.-G.K., M.-K.L., N.N.H.; writing—original draft preparation, D.-G.K.; writing—review and editing, J.I.L., S.-J.K.; visualization, M.J.H., J.-B.K.; supervision, C.-H.B., S.-J.K.; funding acquisition, J.-B.K., S.-J.K. All authors have read and agreed to the published version of the manuscript.

Funding: This research was funded by the Radiation Technology R&D Program (NRF-2017M2A2A6A05018538) through the National Research Foundation of Korea funded by the Ministry of Science and ICT.

Acknowledgments: We thank Edanz Group (www.edanzediting.com/ac) for editing the English text of a draft of this manuscript.

Conflicts of Interest: The authors declare no conflict of interest.

References

1. Hartman, G.L.; West, E.D.; Herman, T.K. Crops that feed the World 2. Soybean—Worldwide production, use, and constraints caused by pathogens and pests. *Food Secur.* **2011**, *3*, 5–17. [CrossRef]
2. Qiu, L.-J.; Xing, L.-L.; Guo, Y.; Wang, J.; Jackson, S.A.; Chang, R.-Z. A platform for soybean molecular breeding: The utilization of core collections for food security. *Plant Mol. Biol.* **2013**, *83*, 41–50. [CrossRef] [PubMed]
3. Krober, O.A.; Cartter, J.L. Quantitative Interrelations of Protein and Nonprotein Constituents of Soybeans 1. *Crop Sci.* **1962**, *2*, 171–172. [CrossRef]
4. Singh, G. *The Soybean: Botany, Production and Uses*; CABI: Wallingford, UK, 2010.
5. Jiang, S.-Y.; Ramachandran, S. Natural and artificial mutants as valuable resources for functional genomics and molecular breeding. *Int. J. Biol. Sci.* **2010**, *6*, 228–251. [CrossRef] [PubMed]
6. Sharma, D.; Rana, D. Response of castor (Ricinus communis) genotypes to low doses of gamma irradiation. *Indian J. Agric. Sci.* **2007**, *77*, 467–469.
7. Ryu, J.; Im, S.; Kim, D.; Ahn, J.; Kim, J.; Kim, S.; Kang, S. Effects of Gamma-ray irradiation on growth characteristics and DNA damage in licorice (Glycyrrhiza uralensis). *Radiat. Indust* **2014**, *8*, 89–95.
8. Wallace, S.S. Biological consequences of free radical-damaged DNA bases. *Free Radic. Biol. Med.* **2002**, *33*, 1–14. [CrossRef]
9. Wu, J.; Morimyo, M.; Hongo, E.; Higashi, T.; Okamoto, M.; Kawano, A.; Ohmachi, Y. Radiation-induced germline mutations detected by a direct comparison of parents and first-generation offspring DNA sequences containing SNPs. *Mutat. Res. Fundam. Mol. Mech. Mutagenesis* **2006**, *596*, 1–11. [CrossRef]
10. Lee, K.J.; Kim, J.-B.; Kim, S.H.; Ha, B.-K.; Lee, B.-M.; Kang, S.-Y.; Kim, D.S. Alteration of seed storage protein composition in soybean [Glycine max (L.) Merrill] mutant lines induced by γ-irradiation mutagenesis. *J. Agric. Food Chem.* **2011**, *59*, 12405–12410. [CrossRef]
11. Ha, B.-K.; Lee, K.J.; Velusamy, V.; Kim, J.-B.; Kim, S.H.; Ahn, J.-W.; Kang, S.-Y.; Kim, D.S. Improvement of soybean through radiation-induced mutation breeding techniques in Korea. *Plant Genet. Resour.* **2014**, *12*, S54–S57. [CrossRef]
12. Powell, W.; Morgante, M.; Andre, C.; Hanafey, M.; Vogel, J.; Tingey, S.; Rafalski, A. The comparison of RFLP, RAPD, AFLP and SSR (microsatellite) markers for germplasm analysis. *Mol. Breed.* **1996**, *2*, 225–238. [CrossRef]
13. Agarwal, M.; Shrivastava, N.; Padh, H. Advances in molecular marker techniques and their applications in plant sciences. *Plant Cell Rep.* **2008**, *27*, 617–631. [CrossRef] [PubMed]
14. Hu, J.; Vick, B.A. Target region amplification polymorphism: A novel marker technique for plant genotyping. *Plant Mol. Biol. Report.* **2003**, *21*, 289–294. [CrossRef]
15. Hu, Z.; Zhang, H.; Kan, G.; Ma, D.; Zhang, D.; Shi, G.; Hong, D.; Zhang, G.; Yu, D. Determination of the genetic architecture of seed size and shape via linkage and association analysis in soybean (Glycine max L. Merr.). *Genetica* **2013**, *141*, 247–254. [CrossRef]
16. Kwon, S.-J.; Hu, J.; Coyne, C.J. Genetic diversity and relationship among faba bean (Vicia faba L.) germplasm entries as revealed by TRAP markers. *Plant Genet. Resour.* **2010**, *8*, 204–213. [CrossRef]
17. Devarumath, R.M.; Kalwade, S.B.; Bundock, P.; Eliott, F.G.; Henry, R. Independent target region amplification polymorphism and single-nucleotide polymorphism marker utility in genetic evaluation of sugarcane genotypes. *Plant Breed.* **2013**, *132*, 736–747. [CrossRef]
18. Chen, S.-S.; Chen, G.-Y.; Chen, H.; Wei, Y.-M.; Li, W.; Liu, Y.-X.; Liu, D.-C.; Lan, X.-J.; Zheng, Y.-L. Mapping stripe rust resistance gene YrSph derived from Tritium sphaerococcum Perc. with SSR, SRAP, and TRAP markers. *Euphytica* **2012**, *185*, 19–26. [CrossRef]
19. Im, S.; Kwon, S.; Ryu, J.; Jeong, S.; Kim, J.; Ahn, J.; Kim, S.; Jo, Y.; Choi, H.; Kang, S. Development of a transposon-based marker system for mutation breeding in sorghum (Sorghum bicolor L.). *Genet. Mol. Res. GMR* **2016**, *15*, gmr8713. [CrossRef]
20. Lee, M.-K.; Lyu, J.I.; Hong, M.J.; Kim, D.-G.; Kim, J.M.; Kim, J.-B.; Eom, S.H.; Ha, B.-K.; Kwon, S.-J. Utility of TRAP markers to determine indel mutation frequencies induced by gamma-ray irradiation of faba bean (Vicia faba L.) seeds. *Int. J. Radiat. Biol.* **2019**, *95*, 1160–1171. [CrossRef]
21. Andersen, J.R.; Lübberstedt, T. Functional markers in plants. *Trends Plant Sci.* **2003**, *8*, 554–560. [CrossRef]

22. Brar, D.S. Molecular marker assisted breeding. In *Molecular Techniques in Crop Improvement*; Springer: Berlin/Heidelberg, Germany, 2002; pp. 55–83.

23. Pawar, K.; Yadav, S.; Arman, M.; Singh, A. Assessment of divergence in soybean (Glycine max L. Merrill) germplasm for yield attributing traits. *IJSR* **2013**, *2*, 1–2. [CrossRef]

24. Borba, T.C.D.O.; Brondani, R.P.V.; Breseghello, F.; Coelho, A.S.G.; Mendonça, J.A.; Rangel, P.H.N.; Brondani, C. Association mapping for yield and grain quality traits in rice (Oryza sativa L.). *Genet. Mol. Biol.* **2010**, *33*, 515–524. [CrossRef] [PubMed]

25. Xue, Y.; Warburton, M.L.; Sawkins, M.; Zhang, X.; Setter, T.; Xu, Y.; Grudloyma, P.; Gethi, J.; Ribaut, J.-M.; Li, W. Genome-wide association analysis for nine agronomic traits in maize under well-watered and water-stressed conditions. *Theor. Appl. Genet.* **2013**, *126*, 2587–2596. [CrossRef] [PubMed]

26. Hu, Z.; Zhang, D.; Zhang, G.; Kan, G.; Hong, D.; Yu, D. Association mapping of yield-related traits and SSR markers in wild soybean (Glycine soja Sieb. and Zucc.). *Breed. Sci.* **2014**, *63*, 441–449. [CrossRef] [PubMed]

27. Lee, C.; Choi, M.-S.; Kim, H.-T.; Yun, H.-T.; Lee, B.; Chung, Y.-S.; Kim, R.W.; Choi, H.-K. Soybean [Glycine max (L.) Merrill]: Importance as a crop and pedigree reconstruction of Korean varieties. *Plant Breed. Biotechnol.* **2015**, *3*, 179–196. [CrossRef]

28. Maher, C.; Stein, L.; Ware, D. Evolution of Arabidopsis microRNA families through duplication events. *Genome Res.* **2006**, *16*, 510–519. [CrossRef]

29. Hu, J.; Mou, B.; Vick, B.A. Genetic diversity of 38 spinach (Spinacia oleracea L.) germplasm accessions and 10 commercial hybrids assessed by TRAP markers. *Genet. Resour. Crop Evol.* **2007**, *54*, 1667–1674. [CrossRef]

30. Frankel, O. Genetic perspectives of germplasm conservation. In *Genetic Manipulation: Impact on Man and Society*; Cambridge University Press: Cambridge, UK, 1984; pp. 161–170.

31. Balfourier, F.; Roussel, V.; Strelchenko, P.; Exbrayat-Vinson, F.; Sourdille, P.; Boutet, G.; Koenig, J.; Ravel, C.; Mitrofanova, O.; Beckert, M. A worldwide bread wheat core collection arrayed in a 384-well plate. *Theor. Appl. Genet.* **2007**, *114*, 1265–1275. [CrossRef]

32. Zhang, H.; Zhang, D.; Wang, M.; Sun, J.; Qi, Y.; Li, J.; Han, L.; Qiu, Z.; Tang, S.; Li, Z. A core collection and mini core collection of Oryza sativa L. in China. *Theor. Appl. Genet.* **2011**, *122*, 49–61. [CrossRef]

33. Hong, M.J.; Jang, Y.E.; Kim, D.G.; Kim, J.M.; Lee, M.K.; Kim, J.B.; Eom, S.H.; Ha, B.K.; Lyu, J.I.; Kwon, S.J. Selection of mutants with high linolenic acid contents and characterization of fatty acid desaturase 2 and 3 genes during seed development in soybean (Glycine max). *J. Sci. Food Agric.* **2019**, *99*, 5384–5391. [CrossRef]

34. Hu, J.; Ochoa, O.E.; Truco, M.J.; Vick, B.A. Application of the TRAP technique to lettuce (Lactuca sativa L.) genotyping. *Euphytica* **2005**, *144*, 225–235. [CrossRef]

35. Alwala, S.; Kimbeng, C.A.; Veremis, J.C.; Gravois, K.A. Linkage mapping and genome analysis in a Saccharum interspecific cross using AFLP, SRAP and TRAP markers. *Euphytica* **2008**, *164*, 37–51. [CrossRef]

36. Palumbo, R.; Hong, W.-F.; Wang, G.-L.; Hu, J.; Craig, R.; Locke, J.; Krause, C.; Tay, D. Target region amplification polymorphism (TRAP) as a tool for detecting genetic variation in the genus Pelargonium. *HortScience* **2007**, *42*, 1118–1123. [CrossRef]

37. Yue, B.; Cai, X.; Vick, B.A.; Hu, J. Genetic diversity and relationships among 177 public sunflower inbred lines assessed by TRAP markers. *Crop Sci.* **2009**, *49*, 1242–1249. [CrossRef]

38. Atak, Ç.; Alikamanoğlu, S.; Açık, L.; Canbolat, Y. Induced of plastid mutations in soybean plant (Glycine max L. Merrill) with gamma radiation and determination with RAPD. *Mutat. Res. Fundam. Mol. Mech. Mutagenesis* **2004**, *556*, 35–44. [CrossRef]

39. Jain, R.K.; Joshi, A.; Jain, D. Molecular Marker Based Genetic Diversity Analysis in Soybean Glycine max (L.) Merrill Genotypes. *Int. J. Curr. Microbiol. Appl. Sci* **2017**, *6*, 1034–1044. [CrossRef]

40. Maughan, P.; Maroof, M.S.; Buss, G.; Huestis, G. Amplified fragment length polymorphism (AFLP) in soybean: Species diversity, inheritance, and near-isogenic line analysis. *Theor. Appl. Genet.* **1996**, *93*, 392–401. [CrossRef]

41. Zargar, M.; Romanova, E.; Shmelkova, E.; Trifonova, A.; Kezimana, P. AFLP-analysis of genetic diversity in soybean [Glycine max (l.) Merr.] cultivars Russian and foreign selection. *Agron. Res.* **2017**, *15*, 2217–2225.

42. Kim, S.-H.; Jung, J.-W.; Moon, J.-K.; Woo, S.-H.; Cho, Y.-G.; Jong, S.-K.; Kim, H.-S. Genetic diversity and relationship by SSR markers of Korean soybean cultivars. *Korean J. Crop Sci.* **2006**, *51*, 248–258.

43. Khan, I.A.; Bibi, S.; Yasmin, S.; Khatri, A.; Seema, N.; Afghan, S. Genetic variability in mutated population of sugarcane clone NIA-98 through molecular markers (RAPD and TRAP). *Pak. J. Bot.* **2010**, *42*, 605–614.

44. Qiao, L.; Liu, H.; Sun, J.; Zhao, F.; Guo, B.; Weng, M.; Liu, T.; Dai, J.; Wang, B. Application of target region amplification polymorphism (TRAP) technique to Porphyra (Bangiales, Rhodophyta) fingerprinting. *Phycologia* **2007**, *46*, 450–455. [CrossRef]
45. Neumann, K.; Kobiljski, B.; Denčić, S.; Varshney, R.; Börner, A. Genome-wide association mapping: A case study in bread wheat (Triticum aestivum L.). *Mol. Breed.* **2011**, *27*, 37–58. [CrossRef]
46. Kwon, S.; Simko, I.; Hellier, B.; Mou, B.; Hu, J. Genome-wide association of 10 horticultural traits with expressed sequence tag-derived SNP markers in a collection of lettuce lines. *Crop J.* **2013**, *1*, 25–33. [CrossRef]
47. Kwon, S.-J.; Brown, A.F.; Hu, J.; McGee, R.; Watt, C.; Kisha, T.; Timmerman-Vaughan, G.; Grusak, M.; McPhee, K.E.; Coyne, C.J. Genetic diversity, population structure and genome-wide marker-trait association analysis emphasizing seed nutrients of the USDA pea (Pisum sativum L.) core collection. *Genes Genom.* **2012**, *34*, 305–320. [CrossRef]

 *agronomy*

Article

QTL Mapping for Drought-Responsive Agronomic Traits Associated with Physiology, Phenology, and Yield in an Andean Intra-Gene Pool Common Bean Population

Aleš Sedlar [1], Mateja Zupin [1], Marko Maras [1], Jaka Razinger [2], Jelka Šuštar-Vozlič [1], Barbara Pipan [1] and Vladimir Meglič [1,*]

1   Crop Science Department, Agricultural Institute of Slovenia, Hacquetova ulica 17, SI-1000 Ljubljana, Slovenia; ales.sedlar@kis.si (A.S.); mateja.zupin@kis.si (M.Z.); marko.maras@kis.si (M.M.); jelka.sustar-vozlic@kis.si (J.Š.-V.); barbara.pipan@kis.si (B.P.)
2   Plant Protection Department, Agricultural Institute of Slovenia, Hacquetova ulica 17, SI-1000 Ljubljana, Slovenia; jaka.razinger@kis.si
*   Correspondence: vladimir.meglic@kis.si; Tel.: +386-(0)1-280-51-80

Received: 18 December 2019; Accepted: 28 January 2020; Published: 4 February 2020

Abstract: Understanding the genetic background of drought tolerance in common bean (*Phaseolus vulgaris* L.) can aid its resilience improvement. However, drought response studies in large seeded genotypes of Andean origin are insufficient. Here, a novel Andean intra-gene pool genetic linkage map was created for quantitative trait locus (QTL) mapping of drought-responsive traits in a recombinant inbred line population from a cross of two cultivars differing in their response to drought. Single environment and QTL × environment analysis revealed 49 QTLs for physiology, phenology, and yield-associated traits under control and/or drought conditions. Notable QTLs for days to flowering (Df1.1 and Df 1.2) were co-localized with a putative QTL for days to pods (Dp1.1) on linkage group 1, suggesting pleiotropy for genes controlling them. QTLs with stable effects for number of seeds per pod (Sp2.1) in both seasons and putative water potential QTLs (Wp1.1, Wp5.1) were detected. Detected QTLs were validated by projection on common bean consensus linkage map. Drought response-associated QTLs identified in the novel Andean recombinant inbred line (RIL) population confirmed the potential of Andean germplasm in improving drought tolerance in common bean. Yield-associated QTLs Syp1.1, Syp1.2, and Sp2.1 in particular could be useful for marker-assisted selection for higher yield of Andean common beans.

Keywords: drought stress; common bean; quantitative trait loci (QTLs); physiology; yield; phenology

1. Introduction

Common bean (*Phaseolus vulgaris* L.) is an important legume crop for human consumption [1]. During the growing season it requires between 300 and 500 mm of rain for optimal development and production of seed according to its genetic potential [1,2]. Water deficit results in a reduction of quality and quantity of yield and is problematic especially in drought-endemic regions in developing countries [2]. Due to ongoing climate changes, traits associated with drought tolerance are being introduced in many common bean breeding programs in Europe and worldwide [3].

Adaption of plants to drought stress can be a result of drought escape or different drought tolerance mechanisms. The plant can escape the drought by completing their life cycle before the onset of severe stress conditions, which is enabled by early maturation and seed development characterized by increased mobilization of carbon to the seed. Resistance mechanisms include avoiding the drought consequences by maintaining high water potential (drought avoidance) or tolerating low water potential (drought tolerance) [4]. Drought avoidance is facilitated by the maximization of water uptake by the root system and optimization of water consumption by shoots resulting in greater biomass production in relation to transpired water [5]. An important characteristic of drought tolerance is osmotic adjustment with an accumulation of compatible solutes [6].

Common bean originates from two major gene pools, Andean and Mesoamerican, that contain a great diversity of genotypes from which traits contributing to increased performance under drought have been identified, including earliness, deep rooting, increased ability to partition dry matter for grain production, and water use efficiency [5]. Common beans of the Mesoamerican gene pool, belonging to races Durango and Mesoamerica, are especially known for their superior drought tolerance and have been used to develop drought-tolerant bean cultivars [7]. Although Andean cultivars with better performance under drought have been reported in Andean races Nueva Granada and Peru [8], the reports of their use in breeding are lacking [9]. On the other hand little progress has been achieved in transferring traits related to drought tolerance across gene pools to large-seeded beans of the Andean origin.

Consequently, the majority of drought responsive quantitative trait loci (QTLs) mapping studies were performed on mapping populations where the resistant parent consisted of one of the more drought-tolerant Mesoamerican cultivars. Inter-genepool crosses enable high coverage of the genetic map with polymorphic markers between Andean and Mesoamerican genepools and were utilized to detect phenological and seed mass QTLs associated with drought tolerance [10]. On the other hand QTLs identified in an intra-genepool population can be easier to transfer to plants within the same gene pool; however, because of a more similar genetic makeup, the development of genetic linkage maps is limited by lower numbers of polymorphic markers [11–14]. Mesoamerican mapping population has been utilized to identify drought tolerance-associated QTL for phenological and yield-related traits [15], as well as QTL for photosynthate acquisition, accumulation, and remobilization traits in drought stress [16]. Although Andean genetic maps have previously been utilized to detect QTL associated with complex traits with polygenic inheritance, such as popping ability [14], only recently has QTL mapping of drought-responsive traits been reported [9]. In this particular study, a population derived from drought-tolerant Ecuadorian cultivar Portillo was used for mapping of phenology, yield component, and partitioning traits in field trials in Uganda [9].

Developing Andean cultivars with enhanced performance under drought conditions and adaptability to local environments remains a challenge [9]. Andean varieties represent the majority of common bean germplasm in Europe (67%), are becoming an important staple food on the international market, and are commonly consumed in parts of Africa [11,17–20]. The potential of Andean genotypes for improving drought tolerance by selection for yield-associated traits has been indicated in phenotyping studies examining genotypes from both gene pools under drought and irrigation conditions [8,21,22]. Perspective cultivars of both gene pools exhibited similar grain yield potential under watered conditions, as well as no significant changes in canopy biomass in drought between the genepools. The sensitivity to drought stress of Andean genotypes was attributed to poor mobilization of photosyntates to pod production resulting in lower yields [22]. However, the same study identified an Andean genotype with the same level of ability to remobilize the photosyntate reserves to pod development and grain filling. Although yield-associated traits remain the most important indicator of genotype performance under drought stress, studies are also directed toward investigation of physiological parameters such as chlorophyll fluorescence with respect to drought susceptibility in common bean [23].

For the improvement of Andean genepool to drought tolerance, it will be crucial to explore drought tolerance alleles and QTLs in additional genotypes of Andean origin in order to identify alternative sources of drought tolerance and more compatible genetic donors for improvement of Andean common bean. Therefore, the objectives of this study were (1) to develop a novel stable Andean common bean recombinant inbred line (RIL) mapping population, (2) to construct an Andean intra-gene pool genetic linkage map of common bean based on simple sequence repeat (SSR) and amplified fragment length polymorphism (AFLP) markers, and (3) to identify QTLs for physiology traits (leaf water potential and effective quantum yield of photosystem II (PSII), termed in this paper as ΦPSII) under drought and control conditions, as well as QTL for phenology traits (days to flowering, days to pods) and yield traits (pods per plant, seed per pod, seed yield per plant, 100 seed mass, and pod harvest index) under drought conditions in two consecutive seasons.

2. Materials and Methods

2.1. Plant Material

A common bean RIL population consisting of 82 lines resulting from Andean intra-gene pool cross of 'Tiber' × 'Starozagorski čern' was used in this study. The population was developed by hand pollination of parental genotypes to develop initial F1 hybrids, and single seed descent method was used to generate F_2 to F_8 generations. 'Tiber' (Clause Semences, France) is a cultivar that exhibits more drought tolerance in comparison to 'Starozagorski čern' (Semenarna Ljubljana, Ljubljana, Slovenia) [24,25].

2.2. Greenhouse Experiments

For the evaluation of drought tolerance-associated traits in parental genotypes and 82 RILs, 15 plants per genotype were planted in two greenhouse environments in two consecutive years. The experiments were carried out in greenhouses at two locations: the Biotechnical faculty, University of Ljubljana (46.05 N, 14.47 E), during the May-July season of 2013; and The Agricultural Institute of Slovenia (46.06 N, 14.52 E), during the April-July season of 2014.

The parental genotypes and the RIL population were planted in 27 cm diameter pots containing a mixture of fertilized peat (Klasmann Potgrond P, Klassman-Deilmann, Geeste, Germany) and vermiculite (1:1, v/v). For each genotype and RIL, five healthy seeds were planted per pot in three pots, after being surface-sterilised with 5% sodium hypochlorite (Kemika, Zagreb, Croatia). When the plants developed the first trifoliate leaf (5 days of growth after plant emergence), three most equally developed plants per pot were kept and the others were discarded.

Equal soil water conditions were maintained for control plants by watering them regularly to 60% of volumetric water content (VWC), close to field capacity. Stressed plants with discontinued irrigation (moderate and severe drought) gradually achieved wilting point, on the basis of measurements of substrate water holding capacity. At each sampling time point, physiological response of plant was assessed by measuring plant water potential. Fresh mass (Fw_s) of the soil mixture sample was weighed in a 10 cm^3 cylinder, and placed in an oven for 24 h at 105 °C to dry completely (Dw_s). VWC was calculated according to the formula:

$$VWC\ [vol\%] = (Fw_s - Dw_s) \times Dw_s^{-1} \times 100 \times \rho_{substrate} \tag{1}$$

where $\rho_{substrate}$ is substrate density. Equal soil water conditions for different pots in the same treatment were maintained by weighing the pots to assess the gradual decrease in VWC between watering.

Drought conditions were induced for parent cultivars and the RIL population by discontinuing watering after 21 days of plant growth after emergence. Measurements of physiology traits were performed at three time-points: 1 day before discontinuation of watering (control), after 10–13 days of induced drought (moderate drought), and after 17–23 days of induced drought (severe drought). After the final measurement, the watering was re-established until the harvest. The differential response between stress and non-stress (continuous watering) conditions was evaluated in parallel blocks for parental genotypes.

Plants were grown under natural light, temperature, and relative humidity conditions. Outside daily temperatures and rainfall were monitored because of their potential effect on humidity and are shown in Figure S1. During the duration of the experiments, the plants were treated three times in 2-week intervals with acaricide Vermitec Pro and insecticide Actara 25 WG (Syngenta, Basel, Switzerland) following the manufacturers' guidelines.

2.3. Trait Evaluation

Plant response to drought stress for parental genotypes and RIL populations was evaluated by recording leaf water potential (Wp) and the effective quantum yield of PSII (ΦPSII) at three time-points: control, moderate drought, severe drought. Water potential was measured on the third trifoliate leaf by using pressure chamber 3005–1223 (Soil Moisture Equipment Corp., Goleta, CA, USA) as described by Scholander et al. [26]. Water potential for each condition was measured on three consecutive days—each day, measurements were performed on all RILs in the time period of 10-13 h, and in total three plants from different pots were measured per each RIL. Chlorophyll *a* fluorescence parameters were measured with a Mini-PAM (pulse-amplitude-modulated) fluorometer (Heinz Walz GmbH, Effeltrich, Germany). The parameter effective quantum yield of PSII:

$$\Phi\text{PSII} = (F_{m'} - F_s) \times (F_{m'})^{-1}, \tag{2}$$

was measured on light-adapted leaves under natural illumination present in the shaded glasshouse (average photosynthetic photon flux density (PPFD), values ranged from 80–90 μmol m^{-2}s^{-1}). $F_{m'}$ and F_s are the maximal and steady-state fluorescence under light conditions, respectively; PPFD is the photosynthetic photon flux density incident on the leaf; 0.5 if the factor that assumes equal distribution of energy between the two photosystems; and 0.84 is the leaf absorbance factor [27,28]. Chlorophyll *a* fluorescence parameters were measured on the same day or on two consecutive days, and each time, three plants from one pot were measured consecutively for all RILs in the time period of 10–13 h or 13–17 h to minimize the changes in light conditions that highly affect these measurements.

Days to flowering (Df) were determined for individual RILs as the number of days from sprouting to number of days when flower opening were observed on more than 50% of the plants. Days to pod-setting (Ds) for individual RILs was determined as the number of days from days to flowering to number of days when pod-setting was observed on more than 50% of the plants. Seed yield-related parameters were recorded when the pods reached maturity (on average after 55 days of plant growth): pods per plant—the number of both full and empty pods (Pp); seeds per pod—the number of seed per full pods (Sp), seed yield per plant (Syp), and 100 seed mass (Hsm). Pod harvest index (Phi) was calculated as

$$\text{Phi} = (\text{dry mass of seed}) \times (\text{dry mass of pod at harvest})^{-1} \times 100. \tag{3}$$

2.4. DNA Extraction and Population Genotyping

Fresh leaf samples of each RIL and parental genotypes were collected and 60-100 mg of tissue was homogenized in 200 μl RLT lysis buffer (Qiagen, Germantown, MA, USA) using TissueLyzer (Qiagen, Stockah, Germany). Total genomic DNA was extracted using MagMAX Express magnetic extractor (Applied Biosystems, Waltham, MA, USA) and BioSprint 15 DNA Plant kit (Qiagen, Germantown, MA, USA) according to the optimized user manual.

Screening for polymorphisms between the parental genotypes was performed using 447 SSR markers and 256 AFLP marker combinations. Primers were obtained from the previously reported studies in common bean. Among them, 103 primer pairs with the prefix 'BM' [29,30], 79 with the prefix 'BMb', 10 with the prefix 'BMc' [31], 50 with the prefix 'BMg', and 15 with the prefix 'PVBR' were tested. Among the gene-based SSR markers, 36 primer pairs with the prefix 'BMd' [32,33], 49 with the prefix 'PvM' [34], and 29 with the prefix 'SSR-IAC' [35] were tested. PCR amplification of SSR markers was performed as previously reported [19]. AFLP amplification was performed following the original protocol [36] with previously reported modifications [37]. Briefly, DNA was cut with *Eco*RI (cuts at G/AATTC) and *Mse*I (cuts at T/TAA) restriction endonuclease enzymes and double-stranded adaptors (*Eco*RI-adaptor: CTC GTA GAC TGC GTA CC CAT CTG ACG CAT GGT TAA; *Mse*I-adaptor: GAC

GAT GAG TCC TGA G TA CTC AGG ACT CAT) were ligated to the fragment ends. Pre-amplification was performed using non-selective primers, followed by a selective amplification using a total of 256 combinations of forward primers (E-AGC, E-AGG, E-ACC, E-ACT, E-ACG, E-AAA, E-AGA, E-AAC, E-ACA, E-AAG, E-AAT, E-ATA, E-ATC, E-ATT, E-ATG, E-AGT; where E is GAC TGC GTA CCA ATT C) and reverse primers (M-CAG, M-CAT, M-CAC, M-CAA, M-CTC, M-CTT, M-CGC, M-CTA, M-CTG, M-CGA, M-CGT, M-CGG, M-CCA, M-CCC, M-CCT, M-CCG; where M is GAT GAG TCC TGA GTA A). The amplified products were genotyped on ABI 3130XL Genetic Analyzer (Applied Biosystems, Waltham, MA, USA), QIAxcel advanced system (Qiagen, Germantown, MA, USA), or separated on agarose gel electrophoresis, depending on the base pair difference between the polymorphic markers. The AFLPs were named by combining the last two selective bases ligated to *Eco*RI and *Mse*I adaptors in the amplification step and by numbering the individual bands in order of decreasing molecular weight [13].

2.5. Genetic Linkage Map Construction and QTL Analysis

The genetic linkage map was constructed using QTL IciMapping integrated software for building linkage maps and mapping quantitative trait genes [38]. Markers were assigned anchor values on the basis of common bean consensus linkage map to inform their linkage group placement and marker order [39]. Grouping of markers was performed using a logarithm of odds (LOD) score of 3.0 as a threshold to declare significant linkage relationship between two markers. Unanchored markers were moved to the appropriate linkage groups on the basis of data from the literature. The final marker order was determined using the algorithms "nnTwoOpt" for the marker ordering and "SARF" for rippling. The mapping function Kosambi [40] was used to convert the recombination fractions to centimorgans (cM).

Single environment QTLs were detected using Windows QTL Cartographer 2.5 [41] with the composite interval mapping method (CIM) set to Model 6: standard model, backward regression, 5 control markers, 10 cM window size, and 0.05 probabilities. The threshold value for QTL detection was calculated using 1000 permutations with a significance level of 0.05. The phenotypic variance associated with each QTL was determined with a determination coefficient (R^2). QTL by environment interaction mapping for each trait was performed using QTL IciMapping using the MET functionality [38,42]. In both types of analysis, the LOD score threshold for significance ($P = 0.05$) was calculated using 1000 permutations. QTLs were named according to the trait and the linkage group they were located on in sequential order. The genetic linkage map was drawn using MapChart v2.3 [43].

2.6. Consensus Map Integration and QTL Validation

A high-density integrated linkage map was constructed using the 'iterative map projection' functionality of BioMercator V4.2 to project the linkage map from our experiment onto a reference common bean consensus linkage map [39,44]. As a validation approach of QTLs detected in our study, BioMercator V4.2 was used to project them on the integrated map together with previously reported QTLs in common bean that were placed on a consensus map by Galeano et al. [39].

3. Results

3.1. A Novel Andean Intra-Gene Pool Genetic Linkage Map of Common Bean

A total of 447 SSR primer pairs developed for common bean and 256 AFLP combinations were screened for polymorphisms in the parental genotypes. Among the successfully amplified markers, 26 polymorphic AFLP combinations and 105 polymorphic SSR markers were detected. In total, 123 markers were placed on the genetic map (Table 1).

Table 1. Distribution of simple sequence repeat (SSR) and amplified fragment length polymorphism (AFLP) markers on the genetic map of common bean recombinant inbred line (RIL) population derived from the 'Tiber' × 'Starozagorski' cross.

LG [1]	SSR	AFLP	Total No. of Markers	Total Linkage Group Length (cM)	Average Distance all Markers (cM)	Anchor Markers	Total Linkage Group Length in Consensus Map (cM)
Pv01	7	1	8	149.9	21.4	8	191.9
Pv02	14	5	19	221.1	12.3	12	192.4
Pv03	3	0	3	65.9	33.0	3	235.8
Pv04	6	6	12	194.4	17.7	4	228.7
Pv05	10	3	13	56.4	4.7	8	132.9
Pv06	7	4	11	40.3	4.0	6	129
Pv07	7	0	7	172.2	28.7	6	180
Pv08	11	3	14	78.4	6.0	7	166.1
Pv09	14	1	15	135.5	9.7	7	128.2
Pv10	13	2	15	100.1	7.1	9	131.9
Pv11	5	1	6	74.2	14.8	5	163.8
Total	97	26	123	1288.4	11.5	73	1880.7

[1] Linkage group ('Pv' signifies *Phaseolus vulgaris*). cM: centimorgans.

Common bean consensus map was used to assign anchor values to 73 markers informing their linkage group placement and marker order [39]. Grouping of markers was on the basis of the LOD value, and 123 markers segregating in the recombinant inbred line population were assigned to all 11 linkage groups of the common bean genome (Table 1). The resulting genetic linkage map was in total 1288.4 cM long, with an average marker density of 11.5 cM. Good coverage of the majority of linkage groups was achieved with 6–19 markers, and with the length in the range of 40.3–221.1 cM. Exceptions were group 3 and 11, with low number of markers covering 30%–50% of the consensus linkage group. In groups 5, 6, and 7, despite higher number of obtained markers, the coverage was below 50% compared to the consensus map. The distribution of markers in these groups was even with the average distance between markers 11.5 cM and ranging from 4.0 to 33.0 cM. The inclusion of AFLP markers contributed to the reduction of the average distance between markers. Linkage groups 2, 9, and 10 were the most populated, containing 15–19 markers. Groups 2 and 4 were the longest, both over 190 cM.

The linkage map with detected QTLs (Figure 1, Figure S2) was projected onto the consensus linkage map [39] on the basis of common markers (Figure 2, Figures S3–S5). The resulting linkage map had a total length of 1880.7 cM and enabled comparison of detected QTLs with previously published QTLs in common bean.

3.2. Phenotypic Variation of Drought Tolerance-Related Traits in the Parents and the Mapping Population

All traits were continuously distributed and displayed transgressive segregation (Figure 3). Descriptive statistics for measured physiology-associated, phenological, and yield traits in the RIL population is presented in Table S3. In watered plants, the coefficient of variation (CV) for physiological traits for a specific year ranged from 18.8%–31.3% for water potential and 3.9%–7.1% for ΦPSII. In drought exposed plants, decreased leaf water potential and reduced ΦPSII were observed and compared to watered plants. In 2014, a temporary reduction in ΦPSII was observed during moderate drought (Figure 3). Evaluation of Wp and ΦPSII during progressive drought conditions in parental genotypes and comparison to simultaneously grown plants that were watered confirmed that reduction of these parameters was due to drought and not due to changes in plant physiology due to ageing. The CV for continuously assessed phenology traits in each season was low, ranging from 5.5%–8.0%, and yield traits assessed in both seasons after re-established watering ranged from 19.4%–26.4% (100 seed mass) to 38.9%–88.0% (seed yield per plant).

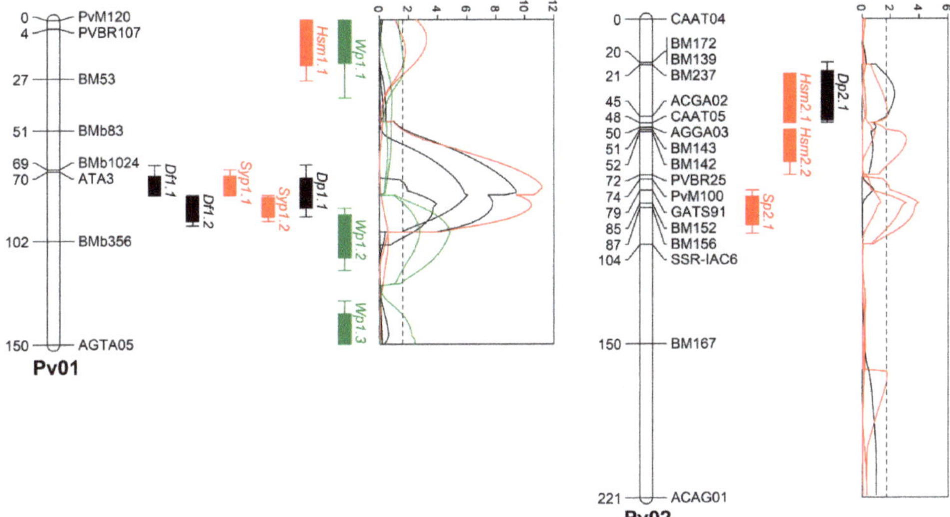

Figure 1. Common bean linkage groups Pv01 and Pv02 showing quantitative trait loci (QTLs) for physiology-associated traits (green), phenological traits (black), and yield-associated traits (red) in 'Tiber' × 'Starozagorski' genetic linkage map detected with single environment analysis. The rest of the linkage groups are presented in Figure S2. For QTL names, see Table 1.

Rainfall distribution and outside average daily temperatures were measured during experiment duration in two consecutive seasons because of their potential effect on the humidity in the greenhouses (Figure S1). In both seasons, the range between average maximum and minimum temperatures was similar (12–24 °C) with the cumulative rainfall being lower in the first season (224.7 mm) compared to the second (281.3 mm). Plants in the second season started to flower before the period of discontinued watering, whereas plants in the first season started to flower after, and had later maturation times and lower yields.

3.3. Drought Tolerance-Related QTL Identification Using the Novel Andean Intra-Gene Pool Genetic Linkage Map

In total, 46 QTLs exceeding the LOD threshold were detected using single environment analysis in control and drought conditions, with two of them identified in both seasons and one of them identified in different treatments during the same season (Table S1). QTL × environment analysis confirmed nine QTLs and identified three additional QTLs (Table 2).

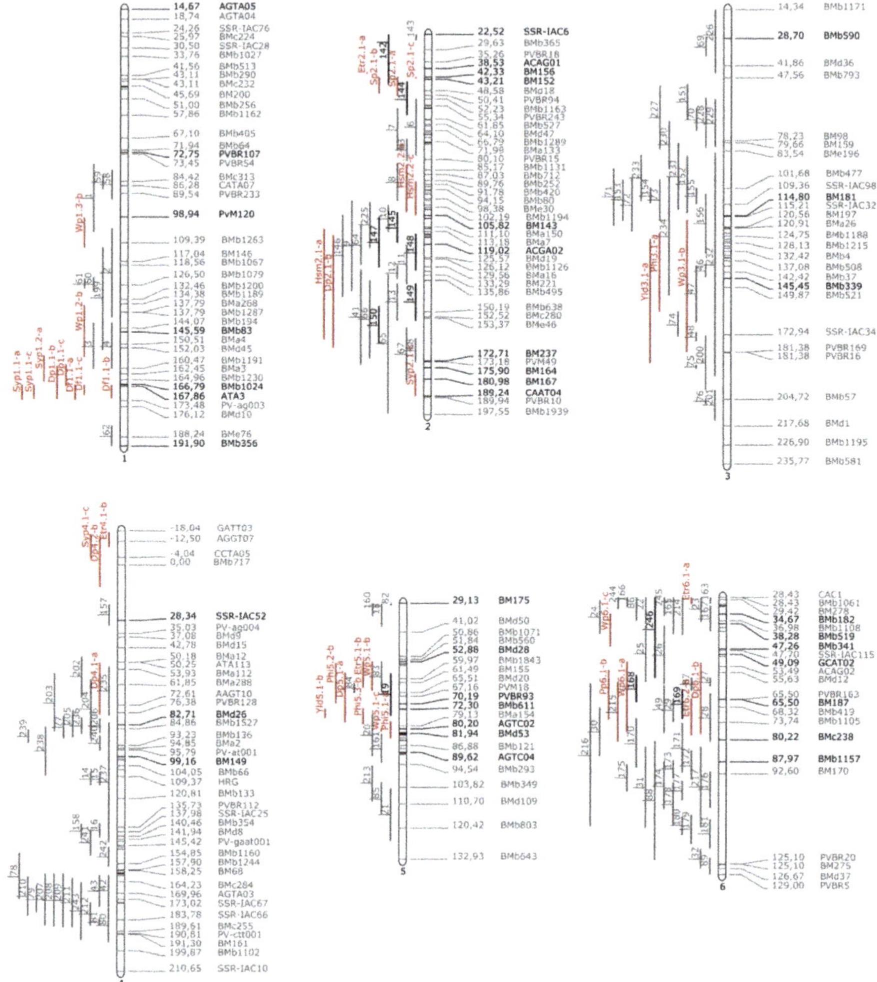

Figure 2. QTLs for physiology-associated traits, phenological traits, and yield-associated traits on the integrated genetic map of the 'Tiber' × 'Starozagorski' recombinant inbred line population and common bean consensus map reported by Galeano et al. [39]. Markers and QTLs from the novel 'Tiber' × 'Starozagorski' genetic map are marked bold and QTLs are colored red (for QTL names, see Table S1). The suffix in QTL names denotes whether they were detected in single environment analysis (-a, -b) or QTL × environment analysis (-c). For clarity reasons, not all loci names are displayed. QTLs from the consensus map are numbered (see Table S2) and colored gray for clarity reasons; only QTLs relevant for discussion are colored black.

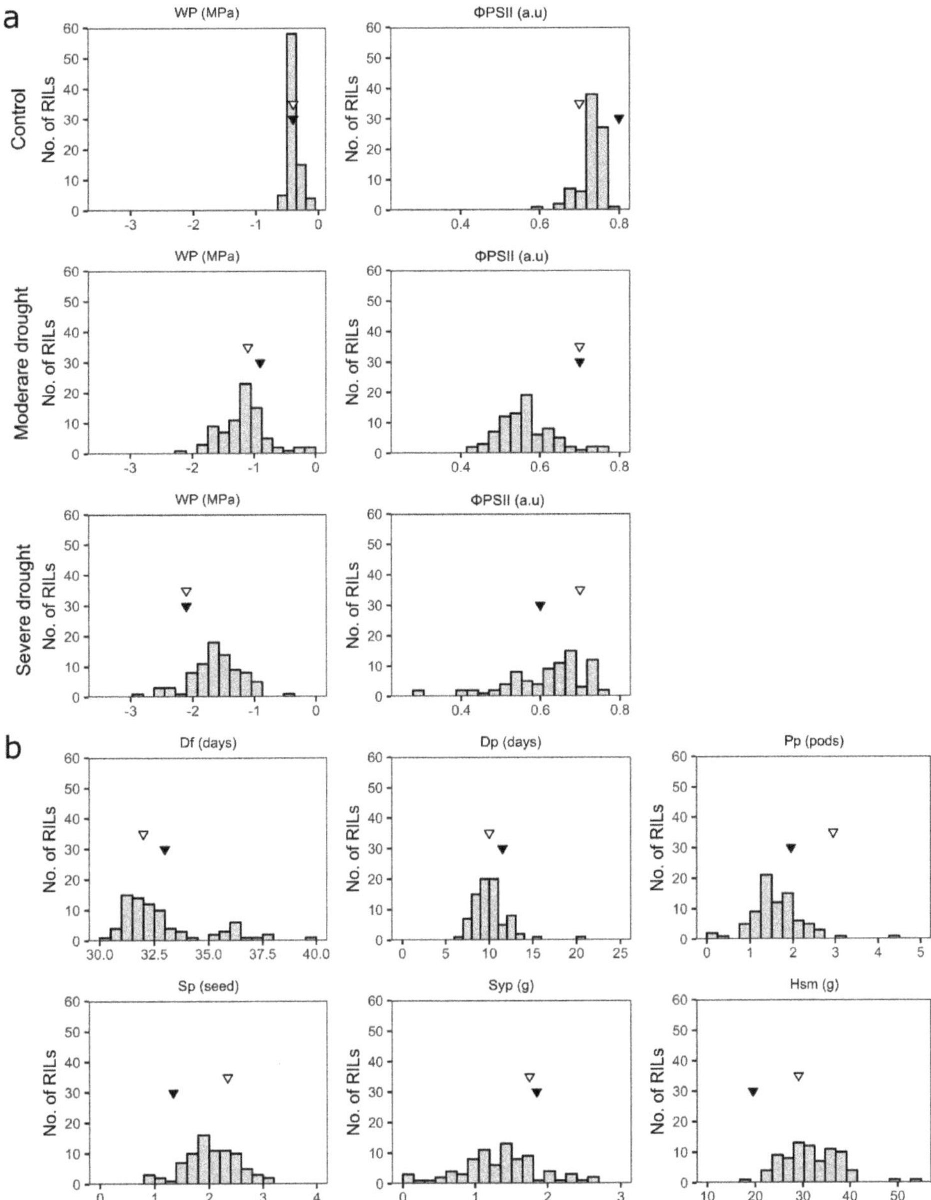

Figure 3. Trait distribution in a recombinant inbred line (RIL) population of 'Tiber' × 'Starozagorski' cross evaluated in greenhouses in Ljubljana in seasons 2013 and 2014. (**a**) Distribution of leaf water potential (Wp) and the effective quantum yield of photosystem II (PSII, ΦPSII) in watered plants and plants exposed to moderate and severe drought. (**b**) Distribution of days to flowering (Df), days to pod-setting (Dp), pods per plant (Pp), seeds per pod (Sp), seed yield per plant (Syp), and 100 seed mass (Hsm) in plants that underwent a period of drought. Values for parental genotypes are indicated by triangle symbols (white—'Tiber', black—'Starozagorski').

3.3.1. Water Potential (Wp)

A total of 12 QTLs for water potential in control plants and drought-exposed plants were identified in both seasons (Table S1), and one of them was identified for different treatments during the same season. QTL Wp1.2 was detected on linkage group 1 and had stable effects at moderate and severe drought of season 2014 and a LOD score in the range from 2.8 to 4.8 (Figure 1). Two different QTL with LOD score in range of 2.4 to 6 LOD were detected on linkage group 5 for either control or moderate drought of season 2014 (Figure S2). On linkage group 6, also, two different QTL with a LOD score in a range of 2.3 to 4 were detected for either control or severe drought of season 2013 (Figure S2). Among additionally detected putative QTLs, two were on linkage group 1 and one each at linkage groups 3, 7, 8, 9, and 10, for either moderate or severe drought in either one season or the other (Figure 1, Figure S2). QTLs for Wp explained 8%–27% of phenotypic variance.

On the projected map, QTLs Wp1.1 and Wp1.2 were surrounded by the QTLs for root hair length (60, 61) and specific root length (62) reported by Yan et al. [45] and Beebe et al. [46] (Figure 2).

3.3.2. The Effective Quantum Yield of PSII (ΦPSII)

Six putative QTLs were detected for ΦPSII in either of the seasons, with LOD scores from 2.4 to 4.6, and explaining up to 20% of phenotypic variance (Figure S2, Table S1).

3.3.3. Days to Flowering (Df)

Two QTLs were detected for days to flowering. QTL Df1.1 with stable effects on days to flowering under environmental conditions of both seasons was detected on linkage group 1 with LOD scores in the range from 7.0 to 9.4 and accounting for 26%–34% of phenotypic variation (Figure 1, Table S1). Df1.1 alleles linked to earlier flowering of the plants originated from 'Tiber'. Additional putative QTL Df1.2 was detected in one of the seasons with LOD 7.8.

The location of detected days to flowering QTL Df1.1 on the projected map was in the vicinity of QTLs for first flower (2) and last flower (4) previously reported by Pérez-Vega et al. [47] (Figure 2).

3.3.4. Days to Pod-Setting (Dp)

Six QTLs were detected for days to pod-setting. QTL on linkage group 1 (Dp1.1) with LOD score 6.1 and accounting for 23% of phenotypic variation was detected on the same locus as a stable QTL for days to flowering Df1.1 (Figure 1, Table S1). Two putative QTLs were detected on different locations on linkage group 2, one in each of the seasons (Figure 1). An additional three putative QTLs were detected on linkage groups 2, 5, and 6 each in only one of the seasons, with LOD scores in the range from 1.8 to 2.5 and explaining 7%–13% of phenotypic variation (Figure 1, Figure S2).

The location of detected days to maturity QTL Dp1.1 on the projected map was in the vicinity of QTL for full maturity (3) previously reported by Pérez-Vega et al. [47] (Figure 2). The location of QTL Dp5.1 on the projected map overlapped with QTL for first flower (19) previously reported by Blair et al. [15] (Figure 2).

3.3.5. Pods per Plant (Pp)

Two putative QTLs for Pp were detected in only one of the seasons on linkage groups 6 and 11, with LOD scores ranging from 2.6 to 2.7 and explaining 11% of phenotypic variation (Figure S2, Table S1).

The location of detected putative QTLs Pp6.1 and Pp11.1 partially overlapped QTLs for seeds per plant (246) and seed mass (168, 169, 198) previously reported by Blair et al. [15,48] (Figure 2 and Figure S5).

3.3.6. Seeds per Pod (Sp)

Four QTLs for Sp were detected (Table S1). QTL with stable effects on number of seeds per pod under environmental conditions of both seasons was detected on linkage group 2 (Sp2.1) with LOD scores in the range from 3.1 to 4.0 and explaining 11%–15% of phenotypic variation (Figure 1). Three additional putative QTLs were detected with LOD in the range of 2.9–3.7 (Figure S2). QTLs on linkage group 7 were detected in both seasons and QTL on linkage group 8 in only one season.

The location of detected QTLs Sp2.1 on the projected map partially overlapped with QTLs for seed mass (142) and seed length (144) previously reported by Blair et al. [48] and Pérez-Vega et al. [47] (Figure 2). Similarly, the location of QTL Sp7.2 overlapped with QTL for seed yield (247) previously reported by Blair et al. [15] (Figure S3).

3.3.7. Seed Yield per Plant (Syp)

Two major putative QTLs for Syp (Syp1.1 and Syp1.2) were found on linkage group 1 and were significant only for environmental conditions in season 2013 with LOD scores 10.6 and 12.6 and a high percentage of phenotypic variation (37%–39%) (Figure 1, Table S1).

On the projected map, QTLs for seed yield per plant were located in the same cluster as QTLs for phenology traits (Figure 2).

3.3.8. One Hundred Seed Mass (Hsm)

Three putative QTLs for Hsm were detected on linkage groups 1 and 2 and were significant only for environmental conditions in season 2013 (Figure 1, Table S1).

On the projected map, QTL for 100 seed mass Hsm2.1 overlapped with a cluster of seed mass QTLs (145, 147-150) reported by Blair et al. [15] and was also adjacent to QTL Hsm2.2 (Figure 2).

3.3.9. Pod Harvest Index (Phi)

Six putative QTLs for Phi were detected. Three of them were on linkage group 5, with QTLs Phi5.1 and Phi5.3 detected in different seasons partially overlapping (Figure S2, Table S1). An additional three putative QTLs were detected on linkage groups 3, 7, and 9 (Figure S2).

3.4. QTL × Environment Interaction

The analysis of QTL × environment interaction was performed by taking into account phenotypic data for a specific trait for both seasons simultaneously, in order to detect QTL with smaller effects in each individual season. The analysis confirmed the presence of nine QTLs already detected in the single QTL analysis (Table 2). Three additional putative QTLs were also detected: ΦPSII10.1, Syp2.1, and Syp4.1 (Table 2).

3.5. Transgressive Segregation

For traits with the greatest additive effects in both directions and observed transgressive segregation of RIL (Wp, Dp, and Pp) graphical genotypes were inspected for genotypes in the top 10% and bottom 10% of the trait values, and allelic composition was determined for significant QTL with high additive effect (Figure S6). For all three traits, higher average positive allele numbers were observed for genotypes in the top 10% of values for individual trait compared to the bottom 10%.

The RILs with the lowest relative water potential between severe drought and control in season 2013 had on average 1.9 positive alleles (1.3 from 'Tiber' and 0.6 from 'Starozagorski') from three contributing loci of parental genotypes. The RILs with lowest number of days to pod-setting averaged from both seasons had on average 4.1 positive alleles (2.1 from 'Tiber' and 2 from 'Starozagorski') from six contributing loci of parental genotypes. The RILs with the lowest number of pods per plant in season 2014 had on average 1.5 positive alleles from parental genotypes (0.9 from 'Tiber' and 0.5 from 'Starozagorski') from two contributing loci of parental genotypes.

Table 2. QTL × environment interaction QTLs for leaf water potential (Wp), effective quantum yield of PSII (ΦPSII), days to flowering (Df), days to pod-setting (Dp), number of seeds per pod (Sp), seed yield per plant (Syp), and 100 seed mass (Hsm) in two seasons for drought and control for the common bean RIL mapping population.

Trait	QTL	Treatment [1]	LG	Position	Left Marker	Right Marker	LOD [2]	LOD Threshold	LOD (A) [2]	LOD (AbyE) [2]	PVE [3]	PVE (A) [3]	PVE (AbyE) [3]	Add [4]
Wp	**Wp5.2**	Control	5	44	AGTC02	BMd53	4.3	3.4	2.0	2.2	62.8	19.2	43.6	0.0
	Wp6.1	Control	6	13	SSR-IAC47	BMb519	3.7	3.4	0.0	3.7	16.7	0.3	16.4	0.0
ΦPSII	**ΦPSII7.1**	Severe	7	95	BMb502	BM150	3.3	3.1	1.9	1.5	9.3	6.3	3.0	0.0
	ΦPSII10.1	Severe	10	23	BM212	BMd42	3.3	3.1	2.0	1.2	9.0	6.6	2.4	0.0
	ΦPSII11.1	Severe	11	0	BMd22	BM239	4.6	3.1	3.2	1.5	13.7	11.1	2.6	0.1
Df	**Df1.1**		1	79	BMb356	ATA3	13.0	3.3	11.0	2.1	30.1	29.1	1.0	−1.1
Dp	**Dp1.1**		1	81	ATA3	BMb1024	5.9	3.2	4.5	1.4	13.6	13.1	0.5	−0.6
Sp	**Sp2.1**		2	86	BM236	BM156	4.9	3.3	4.4	0.5	11.3	11.3	0.0	0.2
Syp	**Syp1.1**		1	78	BMb356	ATA3	10.4	3.4	0.3	10.1	3.3	1.7	1.6	0.1
	Syp2.1		2	4	CAAT04	BM139	3.7	3.4	0.9	2.8	4.8	4.3	0.5	0.1
	Syp4.1		4	187	CCTA05	AGGT07	3.7	3.4	1.9	1.8	10.7	7.8	2.9	−0.2
Hsm	**Hsm2.2**		2	53	BM142	PVBR25	4.1	3.3	3.2	0.9	14.2	10.7	3.5	−2.0

[1] Moderate—moderate drought vs. control, severe—severe drought vs. control. [2] Logarithm of odds (LOD) score is presented for additive and dominance effects (LOD(A)) and additive and dominance by environment effects (LOD(AbyE)). [3] Phenotypic variation explained (PVE) is presented as explained by additive and dominance effect (PVE(A)) and by additive and dominance by environment effect (PVE(AbyE)). [4] The sources of additivity (Add) are 'Tiber' (+value) and 'Starozagorski čern' (−value). QTL also detected in single environment analysis are marked bold.

4. Discussion

An Andean intra-gene pool genetic map integrating two molecular marker types was created in order to explore the potential of common bean genotypes of Andean origin as a source of drought tolerance traits. The employment of AFLP markers in addition to SSR markers proved effective in saturating the linkage map and increasing the map length to 1288.4 cM, which compares favorably to that of other intra-gene pool genetic maps in common bean employing similar marker combination approaches [13,15]. Similarly, as in those studies, the search for polymorphisms in the highly genetically similar genotypes of the same gene pool necessitated a screening of a large number of markers for sufficient coverage of the genetic map [13,15]. In addition, the percentage of polymorphic markers and the map length also compare favorably to studies in common bean and other legumes utilizing SNP markers [9] and recent restriction site-associated DNA sequencing (RADseq) approaches [49,50]. The novel genetic map proved effective in detecting QTLs for drought-responsive agronomic traits associated with physiology, phenology, and yield. However, projection of our linkage map to the consensus map also revealed areas where the coverage of our genetic map is low in cases of lower marker density, which also affected the quality of QTL mapping—these map regions could be further studied by making a denser map before QTL reanalysis.

For water potential, it was notable that more of the QTLs were detected during severe drought than during moderate drought or control. Although QTL Wp1.2 was detected in different treatments of the same season and different QTLs were detected on linkage groups 5 and 6, they were not consistent across seasons. Leaf water potential is an indicator of water saving (isohydric) or water spending (anisohydric) behavior in plants, both when water is available and when it is not. More anisohydric behavior was previously implied for 'Starozagorski' plants on account of high water consumption and seed yield in control conditions, as well as early onset of wilting, decreased water potential, and seed yield in drought [25]. Maintaining high leaf water potential is an important trait that enables plants to avoid the effects of drought. In our study, some of the RIL exhibited better leaf Wp during drought than the parents, which suggests bidirectional transgressive segregation of this trait under drought conditions. This is caused by the allelic combination of multiple loci with additive effects in different directions between parents—RILs with the best relative leaf water potential in severe drought had on average at least two positive alleles coming from both parental genotypes on three loci with additive effects. Interestingly, one of the loci (BM187) also contributed to earlier pod-setting and lower number of pods per plant—four RILs with low relative leaf water potential were observed that were also in the bottom 10% in terms of either day to pod-setting, number of pods per plant, or both.

QTLs for ΦPSII were also detected predominantly during a severe drought. This might have been caused by fluorescence parameters being measured in relatively low light conditions due to shading of the glasshouse, where some of the discriminating power between drought-stressed and irrigated plants was lost [23,51]. However, we were able to discern significant differences in ΦPSII between drought and control treatments of parental cultivars. This is in line with reports that the effectiveness of photosynthesis during drought is more impaired with the increase in the severity of drought conditions. Reduced electron transport rate was reported at the maximum water deficit in common bean, whereas no photo-inhibition was observed in irrigated and moderately water-stressed plants [52]. Reduction of photosynthetic activity and photosynthate accumulation during the drought has been associated with leaf area reduction that benefits the plants by limiting transpiration [53]. Low average R^2 values for ΦPSII (1.6%) suggest low heritability of this trait, which is in line with high dependence of the measured values on the weather conditions, especially the solar radiation.

Stable QTL for days to flowering and putative QTL for days to pod-setting were overlapped and co-localized on linkage group Pv01 (closest markers ATA3 and BMb1024), suggesting pleiotropy for genes controlling these highly correlated traits. In soybean, such a gene involved in both maturation and induction of flowering (*E1* and its homologues) has been reported [54]. The overall numbers of detected QTLs for phenological traits were higher in the season of 2014, which might be associated with observed difference in timing of flowering and maturation with regard to watering conditions.

High R^2 values, however, suggest that phenological traits have high heritability and are less influenced by the interaction of genotype and environment, which is supported by the stability of Df1.1 in both seasons. The alleles seemingly beneficial for earlier flowering were from 'Tiber'. The ability of plants to flower earlier and complete their life cycle before the onset of drought has been described as an important drought escape ability [55]. In plants exposed to the drought, the timing of flowering may, however, impact the final yield production because the severity of drought conditions within the season may change on a day-to-day basis.

For yield-associated traits, the majority of putative QTLs were detected for seed per pod and pod harvest index, with Sp2.1 being the only stable yield-associated QTL detected in both seasons. The additivity scores suggest that the alleles for higher number of seeds per pod were contributed by 'Tiber', whereas most of the alleles for higher pod harvest index were contributed by 'Starozagorski'. Although QTLs for pods per plant were detected in both seasons, their location did not overlap, and QTLs for seed yield per plant and 100 seed mass were detected only in one of the seasons. Some of these differences could have been due to a period of 14 days without rainfall and a daily temperature exceeding 30 °C, resulting in flowers falling off, delayed maturity, and overall lower yields observed in the 2013 season.

Selection for higher yield remains one of the most common approaches of breeding for drought-tolerant plants, with seed mass being an important indicator of drought response. Seed filling is inhibited during drought, and large seed could potentially indicate drought tolerance [56]. In our study, RILs were identified (RIL53 and RIL13) as having both average seed size as well as a pod-harvest index in the top 20% of all RILs averaged between both seasons.

Some of the phenology and yield QTLs in our study were observed on the same linkage groups as previously reported QTLs. QTLs for phenology traits on linkage group 1 were previously reported for inter-gene pool populations (reviewed in Broughton et al. [1]); however, they were not observed in a Mesoamerican population, which was speculated to be because of the genetic differences between the gene pools [15]. In that study QTLs for phenological traits did however map consistently across seasons, with days to flowering and days to maturity QTLs clustered together on linkage group 5 [15], overlapping with QTL for days to pod-setting Dp5.1 from our study.

Positions of some notable QTLs from our study matched previously reported QTL and could be of potential use for marker-assisted selection. Specifically, the position of stable QTL Sp2.1 on linkage group 2 in our study was narrowed down to 2.8 Mb on the basis of the available location for one of the flanking markers, BM156. In the vicinity of this location, a QTL for pod mass per plant PW2.1[PR] was reported in the population 'Portillo' × 'Red Hawk' exposed to drought stress, and was tagged by marker ss715649478 at 0.12 Mb.

Additionally, the QTLs for seed yield per plant Syp1.1 and Syp1.2 observed in our study were located in the region spanning from 41.5 to 51.4 Mb on the genome on the basis of the locations of flanking markers BMb83 and BMb356. The RILs with the highest seed yield per plant had the alleles from 'Tiber' for markers ATA3 and BMb1024 closest to the peaks of Syp1.1 and Syp1.2, respectively. The location suggests one of these QTLs might be the same as those reported in the same region in 'Buster' × 'Roza' and 'Portillo' × 'Red Hawk' populations [9,57]. SY1.1[BR] with an LOD score ranging from 2.4 to 13.9 was detected in 'Buster' × 'Roza' population in multiple stress environments at 47.7 Mb near marker SNP50809 [57]. Recently SY1.1[PR] with LOD score 3.7 was also detected in an Andean 'Portillo' × 'Red Hawk' population exposed to drought with peak region tagged by SNP marker ss715646076 at 45.15 Mb, suggesting it was the same QTL [9]. Additionally, another QTL was detected in the 'DOR 364' × 'BAT 477' population in non-stress conditions; however, it was localized near marker BM200 at 30.8 Mb, which suggests it is a different QTL [16].

To conclude, our study contributes to the evaluation of drought-responsive traits in genotypes belonging to the Andean gene pool. Drought response-associated QTL identified in the novel Andean RIL population, especially seed yield-associated forms Syp1.1, Syp1.2, and Sp2.1, could be useful for marker-assisted selection for higher yield of Andean common beans. Especially valuable are QTLs that are stable across environments (Sp2.1, Df1.1). The presented QTL mapping results confirm the potential of Andean germplasm in improving drought tolerance in common bean by selection for traits associated with phenology, seed yield, and physiology.

Supplementary Materials: The following are available online at http://www.mdpi.com/2073-4395/10/2/225/s1: Table S1. QTLs for leaf water potential (Wp), effective quantum yield of PSII (ΦPSII), days to flowering (Df), days to pod-setting (Dp), number of pods per plant (Pp), number of seeds per pod (Sp), seed yield per plant (Syp), 100 seed mass (Hsm), and pod harvest index (Phi) in two seasons for drought and control for the common bean RIL mapping population. Table S2. Identification numbers used in our study for previously reported Phaseolus vulgaris QTLs with known positions on the consensus linkage map [39]. Table S3. Descriptive statistics for measured physiology-associated, phenological, and yield traits in the RIL population for both seasons and different experiment conditions. Figure S1: Outside temperatures (max, min) and rainfall distribution at greenhouse experiment station location during two successive years. Figure S2: Common bean linkage groups Pv03 to Pv11 showing QTLs for physiology-associated traits, phenological traits, and yield-associated traits. Figure S3: QTLs for physiology-associated traits, phenological traits, and yield-associated traits on the integrated genetic map of the 'Tiber' × 'Starozagorski' recombinant inbred line population and common bean consensus map reported by Galeano et al. [39]. Figure S4: Comparison of the order and distance between the SSR and AFLP loci of 'Tiber' × 'Starozagorski' map, reference map [39], and the projected map for linkage groups 1-6. Figure S5: Comparison of the order and distance between the SSR and AFLP loci of 'Tiber' × 'Starozagorski' map, reference map [39], and the projected map for linkage groups 7–11. Figure S6: Number of positive alleles for QTLs with high additive effects in both directions observed in top and bottom 10% of values for individual trait.

Author Contributions: Conceptualization, V.M., M.M., J.Š.-V., M.Z., and B.P.; methodology, M.Z., M.M., A.S., B.P., and J.R.; software, A.S. and M.Z.; validation, A.S. and M.Z.; formal analysis, A.S., M.Z., and J.R.; investigation, M.Z., M.M., J.R., and A.S.; resources, V.M. and J.Š.-V.; data curation, M.Z., J.R., and A.S.; writing—original draft preparation, A.S.; writing—review and editing, A.S., V.M., B.P., M.Z., M.M., and J.Š.-V.; visualization, A.S.; supervision, V.M.; project administration, V.M. and M.Z.; funding acquisition, V.M. All authors have read and agreed to the published version of the manuscript.

Funding: This research was funded by Slovenian Research Agency (project J4-4126, program P4-0072, young researcher grant for Ms. Zupin) and the FP7 Project CropSustaIn (FP7-REGPOT-CT2012-316205).

Conflicts of Interest: The authors declare no conflict of interest.

References

1. Broughton, W.J.; Hernández, G.; Blair, M.W.; Beebe, S.E.; Gepts, P.L.; Vanderleyden, J. Beans (*Phaseolus* spp.)—model food legumes. *Plant Soil* **2003**, *252*, 55–128. [CrossRef]
2. Beebe, S.E.; Rao, I.M.; Blair, M.W.; Acosta-Gallegos, J.A. Phenotyping common beans for adaptation to drought. *Front. Physiol.* **2013**, *4*, 35. [CrossRef] [PubMed]
3. Beaver, J.S.; Osorno, J.M. Achievements and limitations of contemporary common bean breeding using conventional and molecular approaches. *Euphytica* **2009**, *168*, 145–175. [CrossRef]
4. Levitt, J. *Responses of Plants to Environmental Stresses*; Academic Press: New York, NY, USA, 1972; p. 697.
5. Polania, J.A.; Poschenrieder, C.; Beebe, S.; Rao, I.M. Effective use of water and increased dry matter partitioned to grain contribute to yield of common bean improved for drought resistance. *Front. Plant Sci.* **2016**, *7*, 660. [CrossRef] [PubMed]
6. Yadav, S.; Sharma, K.D. Molecular and morphophysiological analysis of drought stress in plants. In *Plant growth*; Everlon, R., Ed.; IntechOpen: London, UK, 2016; pp. 149–173.
7. Singh, S.P.; Terán, H.; Gutiérrez, J.P. Registration of SEA 5 and SEA 13 drought tolerant dry bean germplasm. *Crop Sci.* **2001**, *41*, 276–277. [CrossRef]
8. Pérez-Vega, J.C.; Blair, M.W.; Monserrate, F.; Ligaretto, M. Evaluation of an Andean common bean reference collection under drought stress. *Agron. Colomb.* **2011**, *29*, 17–26.
9. Dramadri, I.O.; Nkalubo, S.T.; Kelly, J.D. Identification of QTL associated with drought tolerance in Andean common bean. *Crop Sci.* **2019**, *59*, 1007–1020. [CrossRef]
10. Mukeshimana, G.; Butare, L.; Cregan, P.B.; Blair, M.W.; Kelly, J.D. Quantitative trait loci associated with drought tolerance in common bean. *Crop Sci.* **2014**, *54*, 923–938. [CrossRef]
11. Blair, M.W.; Astudillo, C.; Rengifo, J.; Beebe, S.E.; Graham, R. QTL analyses for seed iron and zinc concentrations in an intra-genepool population of Andean common beans (*Phaseolus vulgaris* L.). *Theor. Appl. Genet.* **2010**, *122*, 511–521. [CrossRef]

12. Cichy, K.A.; Blair, M.W.; Galeano Mendoza, C.H.; Snapp, S.S.; Kelly, J.D. QTL analysis of root architecture traits and low phosphorus tolerance in an Andean bean population. *Crop Sci.* **2009**, *49*, 59–68. [CrossRef]

13. Cichy, K.A.; Caldas, G.V.; Snapp, S.S.; Blair, M.W. QTL Analysis of seed iron, zinc, and phosphorus levels in an Andean bean population. *Crop Sci.* **2009**, *49*, 1742–1750. [CrossRef]

14. Yuste-Lisbona, F.J.; Santalla, M.; Capel, C.; García-Alcázar, M.; De La Fuente, M.; Capel, J.; De Ron, A.M.; Lozano, R. Marker-based linkage map of Andean common bean (*Phaseolus vulgaris* L.) and mapping of QTLs underlying popping ability traits. *BMC Plant Biol.* **2012**, *12*, 136. [CrossRef] [PubMed]

15. Blair, M.W.; Galeano, C.H.; Tovar, E.; Muñoz Torres, M.C.; Castrillón, A.V.; Beebe, S.E.; Rao, I.M. Development of a Mesoamerican intra-genepool genetic map for quantitative trait loci detection in a drought tolerant × susceptible common bean (*Phaseolus vulgaris* L.) cross. *Mol. Breed* **2012**, *29*, 71–88. [CrossRef] [PubMed]

16. Asfaw, A.; Blair, M.W.; Struik, P.C. Multienvironment quantitative trait loci analysis for photosynthate acquisition, accumulation, and remobilization traits in common bean under drought stress. *G3 (Bethesda)* **2012**, *2*, 579–595. [CrossRef]

17. Angioi, S.A.; Rau, D.; Nanni, L.; Belluci, E.; Papa, R.; Attene, G. The genetic make-up of the European landraces of the common bean. *Plant Genet. Resour.* **2011**, *9*, 197–201. [CrossRef]

18. Maras, M.; Šuštar-Vozlič, J.; Kainz, W.; Meglič, V. Genetic diversity and dissemination pathways of common bean in Central Europe. *J. Am. Soc. Hortic. Sci.* **2013**, *138*, 297–305. [CrossRef]

19. Maras, M.; Pipan, B.; Šuštar-Vozlič, J.; Todorović, V.; Đurić, G.; Vasić, M.; Kratovalieva, S.; Ibusoska, A.; Agić, R.; Matotan, Z.; et al. Examination of genetic diversity of common bean from the Western Balkans. *J. Am. Soc. Hortic. Sci.* **2015**, *140*, 308–316. [CrossRef]

20. Sedlar, A.; Kidrič, M.; Šuštar-Vozlič, J.; Pipan, B.; Zadražnik, T.; Meglič, V. Drought stress response in agricultural plants: A case study of common bean (*Phaseolus vulgaris* L.). In *Drought—Detection and Solutions*; Ondrasek, G., Ed.; IntechOpen: London, UK, 2019. [CrossRef]

21. Asfaw, A.; Ambachew, D.; Shah, T.; Blair, M.W. Trait associations in diversity panels of the two common bean (*Phaseolus vulgaris* L.) gene pools grown under well-watered and water-stress conditions. *Front. Plant Sci.* **2017**, *8*, 733. [CrossRef]

22. Polania, J.; Rao, I.M.; Cajiao, C.; Rivera, M.; Raatz, B.; Beebe, S. Physiological traits associated with drought resistance in Andean and Mesoamerican genotypes of common bean (*Phaseolus vulgaris* L.). *Euphytica* **2016**, *210*, 17–29. [CrossRef]

23. Sánchez-Reinoso, A.D.; Ligarreto-Moreno, G.A.; Restrepo-Díaz, H. Evaluation of drought indices to identify tolerant genotypes in common bean bush (*Phaseoulus vulgaris* L.). *J. Integr. Agric.* **2019**, *18*, 2–10.

24. Zadražnik, T.; Hollung, K.; Egge-Jacobsen, W.; Meglič, V.; Šuštar-Vozlič, J. Differential proteomic analysis of drought stress response in leaves of common bean (*Phaseolus vulgaris* L.). *J. Proteomics* **2013**, *78*, 254–272. [CrossRef] [PubMed]

25. Zupin, M.; Sedlar, A.; Kidrič, M.; Meglič, V. Drought-induced expression of aquaporin genes in leaves of two common bean cultivars differing in tolerance to drought stress. *J. Plant Res.* **2017**, *130*, 735–745. [CrossRef] [PubMed]

26. Scholander, P.F.; Bradstreet, E.D.; Hemmingsen, E.A.; Hammel, H.T. Sap pressure in vascular plants. *Science* **1965**, *148*, 339–346. [CrossRef] [PubMed]

27. Brooks, M.D.; Niyogi, K.K. Use of a pulse-amplitude modulated chlorophyll fluorometer to study the efficiency of photosynthesis in Arabidopsis plants. In *Chloroplast Research in Arabidopsis*; Jarvis, R., Ed.; Humana Press: Totowa, NJ, USA, 2011; p. 432.

28. Lichtenthaler, H.K.; Buschmann, C.; Knapp, M. How to correctly determine the different chlorophyll fluorescence parameters and the chlorophyll fluorescence decrease ratio R_{Fd} of leaves with the PAM fluorometer. *Photosynthetica* **2005**, *43*, 379–393. [CrossRef]

29. Blair, M.W.; Torres, M.M.; Pedraza, F.; Giraldo, M.C.; Buendía, H.F.; Hurtado, N. Development of microsatellite markers for common bean (*Phaseolus vulgaris* L.) based on screening of non-enriched, small-insert genomic libraries. *Genome.* **2009**, *52*, 772–782. [CrossRef]

30. Gaitán-Solís, E.; Duque, M.C.; Edwards, K.J.; Tohme, J. Microsatellite repeats in common bean (*Phaseolus vulgaris*): Isolation, characterization, and cross-species amplification in *Phaseolus* ssp. *Crop Sci.* **2002**, *42*, 2128–2136. [CrossRef]

31. Blair, M.W.; Torres, M.M.; Giraldo, M.C.; Pedraza, F. Development and diversity of Andean-derived, gene-based microsatellites for common bean (*Phaseolus vulgaris* L.). *BMC Plant Biol.* **2009**, *9*, 100. [CrossRef]

32. Blair, M.W.; Pedraza, F.; Buendia, H.F.; Gaitán-Solís, E.; Beebe, S.E.; Gepts, P.; Tohme, J. Development of a genome-wide anchored microsatellite map for common bean (*Phaseolus vulgaris* L.). *Theor. Appl. Genet.* **2003**, *107*, 1362–1374. [CrossRef]

33. Blair, M.W.; Hurtado, N.; Sharma, P. New gene-derived simple sequence repeat markers for common bean (*Phaseolus vulgaris* L.). *Mol. Ecol. Resour.* **2012**, *12*, 661–668. [CrossRef]

34. Hanai, L.R.; Santini, L.; Camargo, L.E.; Fungaro, M.H.; Gepts, P.; Tsai, S.M.; Vieira, M.L. Extension of the core map of common bean with EST-SSR, RGA, AFLP, and putative functional markers. *Mol. Breed* **2010**, *25*, 25–45. [CrossRef]

35. Benchimol, L.L.; de Campos, T.; Carbonell, S.A.M.; Colombo, C.A.; Chioratto, A.F.; Formighieri, E.F.; Gouvêa, L.R.L.; de Souza, A.P. Structure of genetic diversity among common bean (*Phaseolus vulgaris* L.) varieties of Mesoamerican and Andean origins using new developed microsatellite markers. *Genet. Resour. Crop Evol.* **2007**, *54*, 1747–1762. [CrossRef]

36. Vos, P.; Hogers, R.; Bleeker, M.; Reijans, M.; van de Lee, T.; Hornes, M.; Frijters, A.; Pot, J.; Peleman, J.; Kuiper, M. AFLP: A new technique for DNA fingerprinting. *Nucleic Acids Res.* **1995**, *23*, 4407–4414. [CrossRef] [PubMed]

37. Šuštar-Vozlič, J.; Maras, M.; Javornik, B.; Meglič, V. Genetic diversity and origin of Slovene common bean (*Phaseolus vulgaris* L.) germplasm as revealed by AFLP markers and phaseolin analysis. *J. Am. Soc. Hortic. Sci.* **2006**, *131*, 242–249.

38. Meng, L.; Li, H.; Zhang, L.; Wang, J. QTL IciMapping: Integrated software for genetic linkage map construction and quantitative trait locus mapping in biparental populations. *Crop J.* **2015**, *3*, 269–283. [CrossRef]

39. Galeano, C.H.; Fernandez, A.C.; Franco-Herrera, N.; Cichy, K.A.; McClean, P.E.; Vanderleyden, J.; Blair, M.W. Saturation of an intra-gene pool linkage map: Towards a unified consensus linkage map for fine mapping and synteny analysis in common bean. *PLoS ONE* **2011**, *6*, e28135. [CrossRef]

40. Kosambi, D.D. The estimation of map distances from recombination values. *Ann Eugen* **1943**, *12*, 172–175. [CrossRef]

41. Wang, S.; Basten, C.J.; Zeng, Z.B. Windows QTL Cartographer 2.5. Department of Statistics, North Carolina State University, Raleigh, North Carolina, 2012. Available online: http://statgen.ncsu.edu/qtlcart/WQTLCart.htm (accessed on 3 February 2020).

42. Li, S.; Wang, J.; Zhang, L. Inclusive composite interval mapping of QTL by environment interactions in biparental populations. *PLoS ONE* **2015**, *10*, e0132414. [CrossRef]

43. Voorips, R.E. MapChart: Software for the Graphical Presentation of Linkage Maps and QTLs. *J. Hered.* **2002**, *93*, 77–78. [CrossRef]

44. Sosnowski, O.; Charcosset, A.; Joets, J. BioMercator V3: An upgrade of genetic map compilation and quantitative trait loci meta-analysis algorithms. *Bioinformatics* **2012**, *28*, 2082–2083. [CrossRef]

45. Yan, X.L.; Liao, H.; Beebe, S.E.; Blair, M.W.; Lynch, J.P. QTL mapping of root hair and acid exudation traits and their relationship to phosphorus uptake in common bean. *Plant Soil* **2004**, *265*, 17–29. [CrossRef]

46. Beebe, S.E.; Rojas-Pierce, M.; Yan, X.L.; Blair, M.W.; Pedraza, F.; Muñoz, F.; Tohme, J.; Lynch, J.P. Quantitative trait loci for root architecture traits correlated with phosphorus acquisition in common bean. *Crop Sci.* **2006**, *46*, 413–423. [CrossRef]

47. Pérez-Vega, E.; Pañeda, A.; Rodríguez-Suárez, C.; Campa, A.; Giraldez, R.; Ferreira, J.J. Mapping of QTLs for morpho-agronomic and seed quality traits in a RIL population of common bean (*Phaseolus vulgaris* L.). *Theor. Appl. Genet.* **2010**, *120*, 1367–1380. [CrossRef]

48. Blair, M.W.; Iriarte, G.; Beebe, S. QTL analysis of yield traits in an advanced backcross population derived from a cultivated Andean x wild common bean (*Phaseolus vulgaris* L.) cross. *Theor. Appl. Genet.* **2006**, *112*, 1149–1163. [CrossRef] [PubMed]

49. Pan, L.; Wang, N.; Wu, Z.; Gao, R.; Yu, X.; Zheng, Y.; Xia, Q.; Gui, S.; Chen, C. A high density genetic map derived from RAD sequencing and its application in QTL analysis of yield-related traits in *Vigna unguiculata*. *Front Plant Sci.* **2017**, *8*, 1544. [CrossRef] [PubMed]

50. Valdisser, P.A.M.R.; Pappas, G.J., Jr.; de Menezes, I.P.P.; Müller, B.S.F.; Pereira, W.J.; Narciso, M.G.; Brondani, C.; Souza, T.L.P.O.; Borba, T.C.O.; Vianello, R.P. SNP discovery in common bean by restriction-associated DNA (RAD) sequencing for genetic diversity and population structure analysis. *Mol. Genet. Genomics* **2016**, *291*, 1277–1291. [PubMed]

51. Li, J.; Cang, Z.; Jiao, F.; Bai, X.; Zhang, D.; Zhai, R. Influence of drought stress on photosynthetic characteristics and protective enzymes of potato seedling stage. *J. Saudi Soc. Agric. Sci.* **2017**, *16*, 82–88. [CrossRef]
52. Santos, M.G.; Ribeiro, R.V.; Machado, E.C.; Pimentel, C. Photosynthetic parameters and leaf water potential of five common bean genotypes under mild water deficit. *Biol. Plant* **2009**, *53*, 229–236. [CrossRef]
53. Anyia, A.O.; Herzog, H. Genotypic variability in drought performance and recovery in cowpea under controlled environment. *J. Agron. Crop Sci.* **2004**, *190*, 151–159. [CrossRef]
54. Xu, M.; Yamagishi, N.; Zhao, C.; Takeshima, R.; Kasai, M.; Watanabe, S.; Kanazawa, A.; Yoshikawa, N.; Liu, B.; Yamada, T.; et al. Soybean-specific E1 family of floral repressors controls night-break responses through down-regulation of FLOWERING LOCUS T orthologs. *Plant Physiol.* **2015**, *168*, 1735–1746. [CrossRef]
55. McKay, J.K.; Richards, J.; Nemali, K.S.; Sen, S.; Mitchell-Olds, T.; Boles, S.; Stahl, E.A.; Wayne, T.; Juenger, T.E. Genetics of drought adaptation in Arabidopsis thaliana II. QTL analysis of a new mapping population, KAS-1 x TSU-1. *Evol. Int. J. Org. Evol.* **2008**, *62*, 3014–3026. [CrossRef]
56. Ramirez-Vallejo, P.; Kelly, J.D. Traits related to drought resistance in common bean. *Euphytica* **1998**, *99*, 127–136. [CrossRef]
57. Trapp, J.J.; Urrea, C.A.; Cregan, P.B.; Miklas, P.N. Quantitative trait loci for yield under multiple stress and drought conditions in a dry bean population. *Crop Sci.* **2015**, *55*, 1596–1607. [CrossRef]

Article

Natural Variation in Fatty Acid Composition of Diverse World Soybean Germplasms Grown in China

Ahmed M. Abdelghany [1,2,†], Shengrui Zhang [1,†], Muhammad Azam [1], Abdulwahab S. Shaibu [1], Yue Feng [1], Jie Qi [1], Yanfei Li [1], Yu Tian [1], Huilong Hong [1], Bin Li [1,*] and Junming Sun [1,*]

[1] The National Engineering Laboratory for Crop Molecular Breeding, MARA Key Laboratory of Soybean Biology (Beijing), Institute of Crop Sciences, Chinese Academy of Agricultural Sciences, 12 Zhongguancun South Street, Beijing 100081, China; ahmed.abdelghany@agr.dmu.edu.eg (A.M.A.); zhangshengrui@caas.cn (S.Z.); azaamuaf@gmail.com (M.A.); asshuaibu.agr@buk.edu.ng (A.S.S.); 82101179104@caas.cn (Y.F.); qjycyz@outlook.com (J.Q.); muziyanfei@126.com (Y.L.); 82101181082@caas.cn (Y.T.); 15011290378@163.com (H.H.)

[2] Crop Science Department, Faculty of Agriculture, Damanhour University, Damanhour 22516, Egypt

[*] Correspondence: libin02@caas.cn (B.L.); sunjunming@caas.cn (J.S.); Tel.: +86-10-82105805 (B.L.); +86-10-82105805 (J.S.)

[†] These authors contributed equally to this work.

Received: 16 November 2019; Accepted: 19 December 2019; Published: 23 December 2019

Abstract: Soybean (*Glycine max* L. Merr.) is one of the most important crops in the world. Its major content of vegetable oil made it widely used for human consumption and several food industries. To investigate the variation in seed fatty acid composition of soybeans from different origins, a set of 633 soybean accessions originated from four diverse germplasm collections—including China, United States of America (USA), Japan, and Russia—were grown in three locations, Beijing, Anhui, and Hainan for two years. The results showed significant differences ($P < 0.001$) among the four germplasm origins for all fatty acid contents investigated. Higher levels, on average, of palmitic acid (PA) and linolenic acid (LNA) were observed in Russian germplasm (12.31% and 8.15%, respectively), whereas higher levels of stearic acid (SA) and oleic acid (OA) were observed in Chinese germplasm (3.95% and 21.95%, respectively). The highest level of linoleic acid (LA) was noticed in the USA germplasm accessions (56.34%). The largest variation in fatty acid composition was found in LNA, while a large variation was observed between Chinese and USA germplasms for LA level. Maturity group (MG) significantly ($P < 0.0001$) affected all fatty acids and higher levels of PA, SA, and OA were observed in early maturing accessions, while higher levels of LA and LNA were observed in late maturing accessions. The trends of fatty acids concentrations with different MG in this study further provide an evidence of the importance of MG in breeding for such soybean seed components. Collectively, the unique accessions identified in this study can be used to strengthen the soybean breeding programs for meeting various human nutrition patterns around the globe.

Keywords: fatty acid; germplasm; geographical origin; oil; soybean (*Glycine max* L. Merrill)

---

1. Introduction

Soybean (*Glycine max* L. Merr.) is a leading important oil crop grown worldwide due to its diverse uses of oil and protein for human and livestock. Soybean oil accounts for 60.85% of the world's oil seed production [1]; therefore, it has become the most dominant vegetable oil by far. During 2016 to 2017, the United States of America (USA) and Brazil together accounted for 83% of total world soybean exports [2]. However, China's soybean imports from USA and Brazil accounted for 61% and 77% of their soybean exports, respectively. China accounted for 65% of total world soybean trade value [2]. Soybean is becoming one of the most important oil crops in both China and USA.

Mainly, soybean oil contains saturated fatty acids such as palmitic (16:0) and stearic (18:0) acids. In addition, it contains unsaturated fatty acids such as oleic (18:1), linoleic (18:2), and linolenic (18:3) acids. Fatty acid composition studies are important for improving quality and stability of soybean oil. Typically, soybean has low levels of saturated fatty acids. It was reported that diets rich in saturated fatty acids are associated with an elevated risk of cardiovascular diseases, leading to increased blood serum cholesterol [3,4]. On the other hand, high levels of saturated fatty acids were found to improve the oxidation stability of soybean oil [5]. Particularly, oils with higher stearic acid (SA) levels result in higher melting temperatures, making it suitable for processing of soft margarines with better sensory characteristics, as well as other various baking applications. Higher levels of unsaturated fatty acids improve the quality of the oil for human health [6]. Intrinsically, an oil profile with very high oleic acid (OA) content is influential for enhancing the functionality of soybean oil for food uses free of trans-fatty acids [7]. Linolenic acid (LNA) inhibits triglyceride synthesis and low-density lipoproteins (LDL) in the liver, playing a role in decreasing triglyceride levels in the blood [8]. Some studies pointed out that LNA is involved in maintaining brain nerve functions [9]. Despite these benefits of LNA, polyunsaturated fatty acids, linoleic acid (LA), and LNA, adversely affect the oxidative stability of soybean oil, causing off-flavor problems [10]. Soybean oils with high OA (more than 80%) and low LNA (less than 1%) have been reported in several studies [11–13], and the monounsaturated fatty acid, OA, not only increases oil oxidative stability, but also reduces the production of trans-fat during food processing, making it more desirable.

It is worth noting that wide variations in levels of saturated and unsaturated fatty acids have been detected in several studies on crop germplasm collections [14–16]. Such variations could offer possibilities of developing superior accessions with high quality edible and specialized industrial oils. The variability in fatty acid composition is undoubtedly due to both genetic and weather factors [17,18], which affect their nutritional value and processing property. Basically, much more changes are observed in the preferences for soybean oil owing to the noticeable awareness towards human health care. Consequently, breeding programs designed for altering the soybean oil profile have become a priority for improving both food and industrial uses of soybean oil [19]. Therefore, many studies on soybean seed fatty acid composition have been conducted by soybean breeders in order to develop modified oils that will match the increasing needs of consumers [7,12,14,20–22]. Those several needs of particular uses are divided into nutritional, industrial, or pharmaceutical aspects which generally depend on the vegetable oil quality and its fatty acid composition.

It is generally accepted that introducing new germplasms is indispensable in order to achieve wider genetic diversity and strengthen breeding stock resources. Traditional approaches of breeding methods have been using different sets of germplasm to develop new soybean lines with modified fatty acid composition [10]. Such success in breeding strategies depends largely on available germplasm collections. Germplasm collections largely vary in their origins and may elucidate elite accessions that can be exploited for multiple crop breeding programs. Previous studies on seed fatty acid composition of several germplasm collections from different oil crops—such as sesame [23], *Brassica* species [24,25], and safflower [15,16,26]—have showed a wide variation for fatty acid composition. Likewise, many studies have been conducted on soybean seed fatty acid composition collected from various sources [20,21,27–30]. These studies suggested that soybeans collected from various sources demonstrated key differences in their nutritional composition [31]. Furthermore, classification of soybean accessions into different maturity groups facilitates judgment about the prospects of introducing new varieties, and such an important role in soybean breeding cannot be overemphasized [32].

Despite the extensive research work conducted on variability in soybean seed fatty acids profile, very little is known, to our knowledge, about exploring the variation among diverse germplasms evaluated under same environmental conditions. In an attempt to address this gap, this study exploited four diverse soybean collections which originates from China, USA, Russia, and Japan with varying maturity groups to provide maximal sample heterogeneity. It is expected that these soybean germplasms would have varying nutrient composition and quality profiles. The aims of this study were to comprehensively investigate the variation in seed fatty acid composition among different world soybean germplasms, evaluate the effect of maturity group on seed fatty acid composition, and determine the adaptability of these germplasms in China.

2. Materials and Methods

2.1. Plant Materials

A total of 633 soybean accessions collected from four different regions worldwide—China (451 accessions), USA (138 accessions), Japan (27 accessions), and Russia (17 accessions)—were used in this study. Among the whole collection, 11 maturity groups (MG000-VIII) were identified for only 432 accessions and classified as MG000 (5 accessions), MG00 (8 accessions), MG0 (40 accessions), MGI (63 accessions), MGII (49 accessions), MGIII (104 accessions), MGIV (59 accessions), MGV (44 accessions), MGVI (32 accessions), MGVII (24 accessions), and MGVIII (4 accessions). Accessions with MG000, MG00, and MGVIII were not used in analyzing the effect of MG on seed fatty acid composition due to low number of these accessions, so a total of 415 accessions of MG0–VII were finally used. This panel of soybean germplasm was provided by the germplasm research group of the Institute of Crop Sciences, Chinese Academy of Agricultural Sciences (CAAS). Information of these four germplasm collections is shown in Table S1. Basically, the whole set of accessions in this study is conserved in the Chinese National Soybean Gene Bank (CNSGB).

The Chinese accessions used in this study were collected from the entire Chinese collection of 23,587 soybean accessions conserved in the CNSGB [33,34]. This collection was used to establish a core collection of 2,794 accessions, which represents 11.8% of the entire collection and gains genetic diversity of 73.6% [34]. The 451 Chinese accessions investigated in our recent study account for nearly 1.9% of the total entire Chinese germplasm collection and approximately 16% of the mentioned Chinese core collection. Furthermore, this Chinese collection was originally collected from three major soybean growing regions in China, namely Northern region, Huanghuaihai region, and Southern region, covering an area from 23° N to 51° N [35,36]. The ranges of maturity groups for the Northern, Huanghuaihai, and Southern regions are MG000–II, II–V, and IV–VIII, respectively [37]. With respect to the USA collection, 138 soybean accessions were selected covering various maturity groups from MG000 to VIII, and included many standard varities for maturity group in North America. It is worth noting that information on fatty acid composition of some of these accessions has been reported by the United States Department of Agriculture (USDA) [38] that could have been a basis of systematic selection of soybean accessions. In addition, Russian and Japanese accessions investigated here were selected as representative varieties for their countries of origin and also to provide more diverse panel of accessions along with Chinese and USA accessions.

2.2. Field Experiments

Field trials were conducted at Changping (40°13′ N, 116°12′ E), Beijing, and Sanya (18°24′ N, 109°5′ E), Hainan province, in 2017 and 2018, and at Hefei (33°61′ N 117°0′ E), Anhui province, in 2017. All accessions were planted at Changping, Beijing, on 12 June 2017 and 14 June 2018, respectively; at Sanya, Hainan, on 14 November 2017 and 16 November 2018, respectively; and at Hefei, Anhui, on 5 June 2017.The soil pH, total nitrogen, phosphorus, and potassium levels were 8.22, 80.5 mg kg^{-1}, 68.7 mg kg^{-1}, and 12.31 g kg^{-1} at Changping [39], respectively; 6.6, 35.91 mg kg^{-1}, 56 mg kg^{-1}, and 134.40 g kg^{-1} at Hefei, respectively; and 5.27, 98.59 mg kg^{-1}, 39.68 mg kg^{-1}, and 80.78 g kg^{-1} at Sanya, respectively. The average monthly temperature and rainfall of the three experimental sites are shown in Table S2. In each location, all the soybean accessions were planted following standard local practices. The field experiments were laid out in a randomized incomplete block design, as the different planting locations were used as replications. For each location, soybean seeds of each accession were planted in a 3-m row, spaced 0.5 m apart between rows and 0.1 m between plants within each row. After emergence, the plants were thinned to maintain a uniform and healthy population as well as providing better representative sampling. Plots were fertilized with 15 t ha^{-1} organic fertilizer, 30 kg N ha^{-1}, 40 kg P ha^{-1}, and 60 kg K ha^{-1} during field preparation before sowing. Weeds were controlled by post-emergence application of 2.55 L ha^{-1} of acetochlor (Acetochlor®, 50% EC, Rainbow Chemical, Shandong, China), as well as hand weeding during the growing season. When the plants reached physiological maturity, plots were harvested manually. As the accessions of different maturity groups were grown at different locations and growing times, the harvest dates varied across the three locations. In Changping, the accessions from Northern China were harvested at the same time with the USA, Japanese and Russian accessions in the same maturity groups during the last week of September, while the other Chinese accessions from Huanghuaihai and Southern regions were harvested during the first and second week of October, respectively. At Hefei, all the plant materials were harvested in three batches similar to Changping, but with one week earlier than Changping, while all the plant materials were harvested at the same time at Sanya in mid-February. The total growing durations for the three locations were 102 to 120 days at Changping, 101 to 119 days in Hefei, and 94 to 96 days at Sanya. Some accessions matured very late and could not get mature seeds, therefore they were discarded at the last harvest time. All the plants in each plot were harvested, after which 100 seeds were selected randomly for each accession for fatty acid determination.

2.3. Fatty Acid Extraction and Determination

The five essential fatty acids (palmitic, stearic, oleic, linoleic, and linolenic) were derivatized into their methyl esters and their abundances determined using gas chromatography [40]. In brief, fine powder was obtained from soybean seeds by grinding 20 g of seeds from each accession with a sample preparation mill (Retsch ZM100, Φ = 1.0 mm, Rheinische, Germany). Then, 300 mg of powder from each sample was weighed out using an analytic balance (Sartorius BS124S, Gottingen, Germany) and transferred to a 2 mL centrifuge tube preloaded with 1.0 mL *n*-hexane. This mixture was kept for 20 min at 65 °C and shaken for 10 sec every 5 min. Next, 1.0 mL sodium methoxide solution was added to the mixture, and the mixture was shaken for 10 min on a twist mixer (TM-300, ASONE, Japan) at 65 °C to allow full methyl esterification of the fatty acids, and centrifuged at 12,000× *g* for 2 min. The final supernatant was assayed to determine the concentrations of the methyl esters of the five fatty acids using a GC-2010 gas chromatograph (SHIMADZU Inc., Kyoto, Japan) with flame ionization detector. The chromatographic separation was carried out using an RTX-WAX column (30 m length× 0.25 mm internal diameter × 0.25 mm thickness, Germany) with the following temperature gradient: initially, the temperature was set at 180 °C for 1.5 min, then increased to 210 °C at a rate of 10 °C min^{-1}, kept at 210 °C for 2 min, increased to 220 °C at a rate of 5 °C min^{-1}, and kept at 220 °C for 5 min. The carrier gas was nitrogen, at a flow rate of 54 mL min^{-1}, and 1 µL of each sample was injected. Area was normalized (relative concentrations by mass) to quantify the five fatty acid concentrations using a GC2010 workstation [40]. The content of each fatty acid was expressed as a percentage of total fatty acid content.

2.4. Data Analysis

The analysis of variance (ANOVA) for seed fatty acid composition was conducted using the general linear model (GLM) procedure with a random statement in SAS 9.2 software for Windows [41]. The different planting locations were used as replications and together with year they were considered as random effect. The origin of accessions, accessions and MG were considered as fixed effect. Multiple comparisons among origins of accessions were conducted using Fisher's least significant difference (LSD) test. Boxplots were drawn to show the variation in seed fatty acid composition among the four origins. Scatter plots were used to display the relationship between different MGs and each fatty acid, and also the relationship between mean of each fatty acid with its CV. Pearson's correlation was used to access the associations among the five fatty acids. In R project (version 3.4.5), *ggplot2* package was used to draw boxplot and scatter plot, while the *corrplot* package was used to graphically display correlation matrix.

3. Results and Discussion

3.1. Variation in Seed Fatty Acid Composition in Soybean Accessions

The average fatty acid contents of 633 soybean accessions across three environments for two years are presented in Table 1. Linoleic acid showed predominant average content (54.41%) among all the fatty acids, whereas stearic acid showed the lowest average content (3.92%). The widest range was noticed in oleic acid (13.55–31.88%), followed by linoleic acid (45.64–63.93%), while stearic acid showed the narrowest range (8.23–17.2%). Generally, the lower levels of saturated than unsaturated fatty acids in the current study was in agreement with those reported previously [7,10]. The observed SA level in this study was low in comparison with previous studies [42,43], while levels of palmitic acid (12.12%) and OA (21.63%) in this study were higher than that previously reported by [21,42,43]. The current study also showed a wider range of LNA content when compared with previous studies [42,43]. Such variability between our study and previous works may most likely be due to large collections from various regions exploited in our study. Moreover, LNA content exhibited the highest CV (15.34%) among the five fatty acids, revealing that LNA content had the highest abundant variation which underlies the effects of both diverse germplasms and environments on LNA content. LA content, contrary to LNA, exhibited the lowest CV (5.07%) among all the fatty acids which indicates that growing those accessions with higher LA levels and less changeable effect across different locations is profitable. Theoretically, LA, (omega-6) and LNA (omega-3) are absolutely essential fatty acids which should be obtained from rich diet. Elevated levels of both LA and LNA in soybean oil are necessarily needed for healthier human nutrition [8,9]. Furthermore, serious problems—such as heart diseases, asthma, and other syndromes—could influence human health due to the lack of LA and LNA in diets [44,45]. On the contrary, soybean accessions with high LNA content are not favorable for producing stable oil, as oil with raised concentrations of LNA oxidizes rapidly, inducing off-flavors compounds in cooked foods [10]. Recently, demands for soybean oil as a raw material for producing biodiesel have raised, thus soybean oil higher in OA and lower in LA contents are also preferred for meeting these needs [46].

Table 1. Statistical summary of soybean seed fatty acid composition.

Component	Mean [a]	SD	CV (%)	Min (%)	Max (%)	Skew	Kur
Palmitic acid	12.12	0.78	6.44	8.23	17.20	0.24	4.09
Stearic acid	3.92	0.41	10.36	2.71	5.19	0.23	−0.11
Oleic acid	21.63	2.94	13.60	13.55	31.88	0.48	0.36
Linoleic acid	54.41	2.76	5.07	45.64	63.93	−0.21	0.51
Linolenic acid	7.93	1.22	15.34	3.43	12.76	0.28	0.94

[a] The content of each fatty acid is expressed as a percentage of total fatty acid content.SD, standard deviation; CV, coefficient of variation; Min, minimum; Max, maximum; Skew, skewness; Kur, kurtosis.

Breeding soybean for altered seed fatty acid composition has been extensively studied [7,12,14,20–22] in order to develop modified oils ready for meeting the increasing needs of consumption. Rebetzke et al. [14], used 22 low and 22 normal PA lines derived from two crosses of N87-2122-4 × 'Kenwood' and N87-2122-4 × 'P9273', and reported that OA and LNA contents in those reduced PA lines were significantly higher than that of normal PA lines, while LA content showed no significant changes. They concluded that using soybeans with reduced PA content could improve soybean quality. As mutagenesis is also one of the conventional breeding methods followed to change the oil and its fatty acid composition, Pham et al. [12] studied the incorporation of mutant *FAD3* genes into high OA background to lower the LNA content. Results of their investigation identified lines with less than 2% LNA content. Furthermore, the study of La et al. [21] which aimed to characterize seed composition traits, including fatty acids, in 80 wild soybean plant introductions (PIs) from the USDA soybean collection, observed a lower OA content (122.1 g kg^{-1}) while LNA content was high (163.8 g kg^{-1}). They suggested the possibility of exploiting this collection of wild soybean in improving oil profile constituents in cultivated soybean for human health benefits. genome-wide association studies (GWAS) were recently conducted and resulted in identification of beneficial alleles and candidate genes which are expected to be valuable for generally improving seed quality [47] and particularly improving unsaturated fatty acid of soybean [11].

In the current study, the highest accession in PA content was T309 (17.20%) from USA, while YZY2004-15-W90, from China, had the lowest content of PA (8.23%) (Table 2). For SA, ZDD00294 from China had the highest level (5.19%), whereas WDD00405, from USA, had the lowest level (2.71%) (Table 2). This shows the existence of a wider variability between the Chinese and USA germplasms for PA and SA. The Chinese accession ZDD02925 showed the highest OA content (31.88%), whereas ZDD09581 contained the lowest level of OA (13.55%) (Table 2). The observed result showed a large variation among the Chinese accessions for OA content. The highest level of LA was yielded by accession S01-9391 from USA (63.93%), whereas the Chinese accession ZDD02925, in contrast to its OA content, recorded the lowest LA content (45.64%) (Table 2).

Table 2. Accessions with highest and lowest contents of five fatty acids.

Fatty Acid	ID Number	Name	Mean [a]	Origin
Palmitic acid	-	YZY2004-15-W90	8.23	China
	WDD01709	T309	17.20	USA
Stearic acid	WDD00405	Yellow marvel	2.71	USA
	ZDD00294	Qingdou	5.19	China
Oleic acid	ZDD09581	DLHD	13.55	China
	ZDD02925	DLFB	31.88	China
Linoleic acid	ZDD02925	DLFB	45.64	China
	WDD03084	S01-9391	63.93	USA
Linolenic acid	WDD01482	C1640	3.43	USA
	ZDD03739	PXDZHC	12.76	China

[a] The content of each fatty acid is expressed as a percentage of total fatty acid content.

For LNA, the Chinese accession ZDD03739 had the highest content (12.76%), while C1640 from USA had the lowest LNA content (3.43%) (Table 2). The variability among USA accessions in LA and LNA contents in this study can explicitly provide insights into achieving a goal of dual purpose soybean accessions; accessions with higher LA to produce healthier soybean oil and lower LNA to achieve oil stability. USA accessions with lower level of LNA, such as accession C1640, offers vital source for more advanced research on low-LNA soybean studies with the aim of improving oil stability. Genetically, Fehr et al. [48] reported that the expression of low LNA content is governed by, at least, two recessive alleles and a combination of these alleles will decrease LNA level to less than 3%.

3.2. Effect of Germplasm Origins on Variation in Seed Fatty Acid Composition

We observed highly significant differences ($P < 0.001$) among the germplasm origins for all fatty acids (Table S3a). The significant variation of fatty acids among different germplasm origins was consistent with previous studies [20,21,28]. Grieshop and Fahey [31] reported that soybeans collected from various sources elucidated major differences in their nutritional composition of fatty acids. The contents of SA, LA, and LNA showed highly significant differences ($P < 0.001$) between the two years, whereas PA and OA levels showed no significant differences (Table S3a). A previous study revealed that cultivation year had a significant effect on some soybean seed quality traits, while others were not significantly affected [49]. In addition, the interaction of cultivation year $\times$ origin had a highly significant effect ($P < 0.01$) only on LNA content, while no significant effects were observed for the other fatty acid contents (Table S3a). The significant effect of year × origin interaction on LNA, coupled with the highest CV of 15.34% (Table 1), underline the high sensitivity across various environmental conditions. In contrast, the other fatty acids were not affected by such interaction which was in accordance with the study of Graef et al. [50]. The results also showed that there were highly significant differences ($P < 0.0001$) among accessions for all fatty acids contents (Table S3b). Table S3b further showed that cultivation year × accession interaction did not significantly affect all the fatty acid contents, indicating that genetic factor plays a key role on fatty acid composition between various cultivation years.

The variation in fatty acid composition among the four germplasm origins are shown in Figure 1A–E. Russian accessions had the highest average level of PA (12.31%), followed by Japan (12.26%) and China (12.22%), while USA accessions had the lowest level of PA (11.68%) which significantly differed from the other three germplasms origins (Figure 1A). Chinese accessions had higher content of SA (3.95%), followed by USA (3.89%), then Russian accessions (3.79%), while Japanese germplasm revealed the lowest SA content (3.76%) (Figure 1B). The Chinese germplasm can be used as good parents for the development of soybean varieties with high saturated oils, which could increase the oil shelf life. In respect to unsaturated fatty acids, the Chinese accessions contained the highest OA level (21.95%), followed by Japan (21.54%) and Russian germplasms (21.45 %), while USA germplasm recorded the lowest OA content (20.66%) (Figure 1C). Russian accessions did not differ significantly from USA accessions in OA content, while USA accessions differed significantly from Chinese and Japanese accessions. In contrast with OA content, the USA germplasm had the highest average of LA content (56.34%), while the other germplasms of Japan and Russia recorded 54.55% and 54.30%, respectively (Figure 1D). The Chinese accessions, in contrast with SA level, contained the lowest level of LA (53.81%) (Figure 1D). One important finding in the current study is that the four germplasm sources used varied greatly in LA content, suggesting that LA can be used as a discriminative factor for geographical classification of soybean.

The largest difference in LA was found between Chinese and USA accessions which was in accordance with Song et al. [30], as they reported that USA accessions had higher LA content than Chinese accessions (48.5% and 44.7%, respectively). This result demonstrates that different genetic background between both origins greatly influenced LA content. Russian accessions had the highest LNA content (8.15%), followed by Chinese accessions (8.06%), and Japanese accessions (7.90%), whereas USA accessions had the lowest LNA content and differed significantly from the other origins (7.40%, Figure 1E). The highest LA and lowest LNA levels in observed in USA accessions shows the possibility of decreasing LNA level with an increase in LA level, which suggests that higher content of LA could be acting at the expense of LNA content.

As a result, improving genotypes or/and developing new cultivars with special nutrition pattern largely rely on the noted variations in seed fatty acids. Japanese and Chinese accessions showed no significant differences in PA, OA, and LNA contents, which was also in agreement with Song et al. [29], indicating that the closeness of these two origins may have resulted in the lack of significant effect on the soybean quality components. Generally, the observed variation in saturated and unsaturated fatty acids among different origins could be attributed to environmental and genetic factors, which further confirms the results of Song et al. [29], as the germplasm origins exhibited various oil profiles.

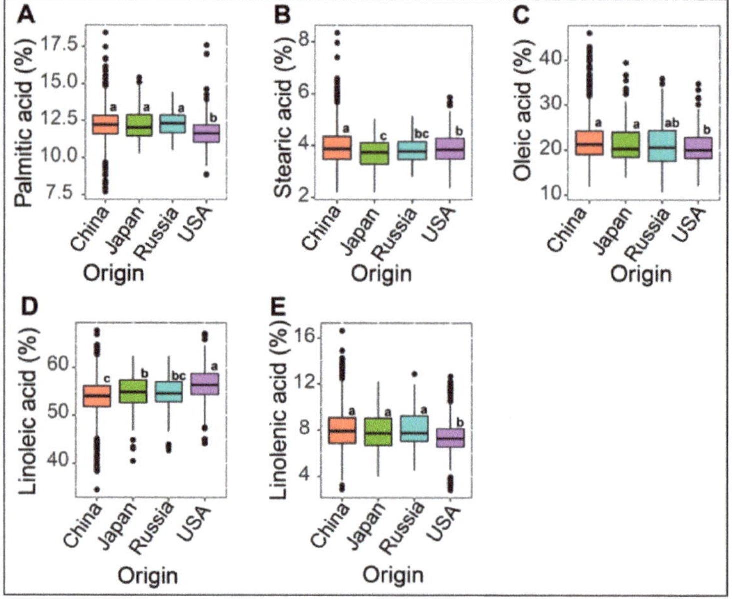

Figure 1. Boxplots of five fatty acid composition of 633 soybean accessions collected from four different germplasm origins (averaged across three locations and two years). (**A**) palmitic acid (PA), (**B**) stearic acid (SA), (**C**) oleic acid (OA), (**D**) linoleic acid (LA), and (**E**) linolenic acid (LNA). Different lowercase letters (a, b, and c) at the top of each boxplot indicate statistically significant differences at $P < 0.05$ level among the four germplasm origins.

3.3. Effect of Maturity Group on Seed Fatty Acid Composition

Highly significant differences ($P < 0.0001$) in fatty acid composition were also observed among different maturity groups (MGs) for all fatty acids (Table S3c). Previous studies have reported that MG significantly affects soybean seed quality [39,51–53]. These findings document the variation among maturity groups in different fatty acids, confirming the fact that MG should be considered as an important factor [39]. Furthermore, the results demonstrated that in the same planting site or even planting date, early MGs could affect soybean seed fatty acid composition in a way that differs from late MGs, which could be attributed to the difference in growth durations of both early and late maturing cultivars [24]. In other words, effects of MG in this study underline that accessions from some MGs had higher content of specific fatty acid than other accessions belonging to other MGs, regardless of the environments. The MG x year interaction had no significant effect on all fatty acids (Table S3c), which indicates the consistency of the MGs in both years of experimentation.

In this study, PA and SA levels ranged from 11.75% to 12.33% in MG0–VII and from 3.69% to 4.16% in MGI–VI, respectively (Table 3). MG0 had significantly higher PA level (12.33%) when compared to MGI-VII which had a decreasing PA level, with the lowest PA level observed in MGVII (11.75%). For SA, MGI yielded the highest level (4.16%) which was significantly higher than all the other maturity groups. Also, MGII-VII had a decreasing SA level which ranged from 3.94% in MGII to the lowest SA level of 3.69% in MGVI (Table 3). Level of OA ranged from 19.73% in MGVII to 22.99% in MGI (Table 3), showing a similar trend to that observed in PA and SA levels as higher contents of PA, SA and OA were presented in early MGs, while lower contents were yielded in late MGs. Level of LA ranged from 53.30% in MGI to 56.20% in MGVII, while LNA level ranged from 7.31% in MG0 to 8.88% in MGVI (Table 3). These findings reflect contrasting trends of fatty acids with different MGs, where higher PA, SA, and OA levels were observed in early, rather than late, maturing soybean accessions, while the higher levels of LA and LNA were observed in late maturing soybean accessions. The MG can be used as an influential factor on soybean seed fatty acids and early maturing soybean accessions had higher levels of PA, SA, and OA components, while late maturing accessions had higher levels of LA and LNA components. Despite the solid knowledge that environment is a major player influencing soybean seed composition since it drives to an influential change in climatic factors [54], the current study further points out the remarkable effect of MG on seed fatty acid composition. In other words, the trend of fatty acid across different MGs observed in this study further highlights that genetic factors play a key role in variation of seed fatty acid constituents among different accessions corresponding to different backgrounds. This assertion that genetic variability is present in soybean seed composition and yield has also been reported by other studies [51,55,56].

Table 3. Fatty acid composition (%) in soybean seeds from different maturity groups (MGs) across two years.

MG		Palmitic Acid	Stearic Acid	Oleic Acid	Linoleic Acid	Linolenic Acid
0	Mean [a]	12.33 a	3.92 b	22.63 a	53.77 ef	7.31 d
	Range	10.27–15.33	2.24–5.5	13.3–46	34.55–63.36	4.3–13.67
I	Mean	12.20 ab	4.16 a	22.99 a	53.30 f	7.35 d
	Range	9.91–14.67	2.80–41.02	10.64–63.93	38.88–63.93	4.08–12.90
II	Mean	12.01 c	3.94 b	21.34 b	55.02 bc	7.51 d
	Range	9.8–15.56	2.21–5.91	12.6–35.49	44.18–66.75	2.94–11.89
III	Mean	12.07 c	4 b	21.89 b	54.21 de	7.89 c
	Range	9.61–17.48	2.37–6.39	13.03–42.48	38.77–62.43	3.94–14.91
IV	Mean	12.15 bc	3.93 b	21.48 d	54.15 de	8.30 b
	Range	9.97–15.00	2.49–6.57	14.08–32.78	43.48–61.71	5.05–12.58
V	Mean	12.09 bc	3.79 c	21.18 d	54.70 cd	8.25 b
	Range	9.77–14.71	2.37–6.48	13.62–37.98	42.93–61.74	4.65–13.37
VI	Mean	12.08 bc	3.69 cd	19.88 c	55.47 ab	8.88 a
	Range	7.93–16.71	2.61–5.49	11.95–33.50	44.67–66.90	3.68–13.91
VII	Mean	11.75 d	3.79 c	19.73 c	56.20 a	8.54 ab
	Range	9.72–18.44	2.62–6.08	13.63–28.85	48.80–62.31	5.04–13.95

[a] The content of each fatty acid is expressed as a percentage of total fatty acid content. Values of means within each column with different letters indicate statistically significant differences at $P < 0.05$.

The relationship between fatty acid composition and MG is shown in Figure 2A–E. All the fatty acids showed a significant relationship with maturity groups. Levels of PA, SA and OA showed negative and significant linear relationship with MG ($r = -0.76$, $P < 0.028$, $r = -0.77$, $P < 0.027$ and $r = -0.92$, $P < 0.001$, respectively) (Figure 2A–C). In contrast, levels of both LA and LNA revealed a positive and significant relationship with MG ($r = -0.8$, $P < 0.017$ and $r = -0.95$, $P < 0.0004$, respectively) (Figure 2D,E). These findings further confirm that there is a decreasing trend in PA, SA, and OA levels from early to late maturity groups, with much stronger relationship in OA level. For LA and LNA levels, there is an increasing trend from early to late maturity groups, with much stronger relationship in LNA level. This relationship pattern was not in agreement with that reported by Bellaloui et al. [57], as they concluded that MG showed no consistent effect on fatty acid composition, except stearic acid which was minimally affected.

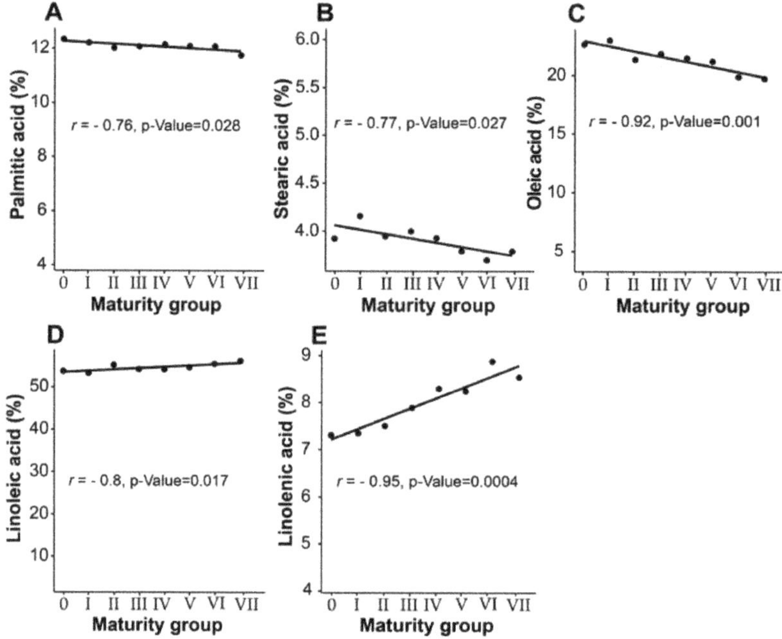

Figure 2. Relationship between means of soybean seed five fatty acid composition (%) with maturity groups (MG0–VII). (**A**) palmitic acid, (**B**) stearic acid, (**C**) oleic acid, (**D**) linoleic acid, and (**E**) linolenic acid.

Despite this conclusion of Bellaloui et al. [57], their study was conducted based on two sets of near-isogenic soybean lines derived from two cultivars Clark and Harosoy with a common genotypic background, which did not enable them to generalize their conclusions. In contrast, the soybean accessions utilized in the current study included diverse soybean accessions in different genetic background. We believe that this contradictory pattern of relationship between fatty acid and MG could be exploited to adopt best management combinations depending on selection purpose and other practices aiming at maximizing specific end use targets.

3.4. Correlation Analysis of Seed Fatty Acids

Pearson's correlation coefficients among the five fatty acids are shown in Figure 3. The highest significant negative correlation was observed between OA and LA contents ($r = -0.85$***), followed by OA with LNA contents ($r = -0.47$, $P < 0.001$, denoted ***). Similar results were previously reported [21,58]. These findings showed that a consistent decrease in LA and LNA contents is linked with a corresponding increase in OA content. This negative correlation may be attributed to the different biochemical pathways of the noted fatty acid biosynthesis [59]. Interestingly, this negative association provides more increasing interest for food industries and consumers to produce oil with high OA and low LA and LNA contents [60]. Non-significant correlation was observed between LA and LNA. The PA content was significantly and negatively correlated with OA and LA ($r = -0.08$* and -0.29***, respectively), but positively correlated with LNA ($r = 0.23$***). The SA content had highly significant negative correlations with LA and LNA ($r = -0.16$*** and -0.23***, respectively), but positive correlation with OA. Previous study of La et al. [21] demonstrated that there is a significant negative association between PA and LA, as well as between SA and LNA. These negative correlations of both PA and SA with unsaturated fatty acid contents could be related to the different pathways in their biosynthesis. Our result is in opposite to other studies [42,61], as they found positive and significant correlations between PA with LA, and SA with both LA and LNA. Such a contradiction between our study with other studies could be due to the difference in materials exploited in both studies.

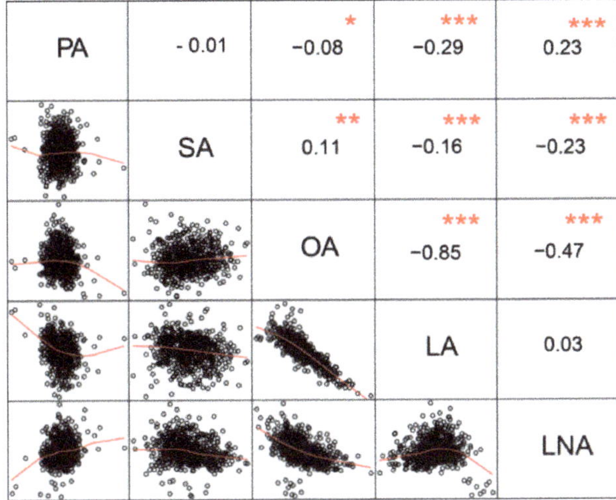

Figure 3. Pearson's correlation coefficients between soybean seed fatty acid composition. * Significant at the $P < 0.05$ probability level, ** significant at the $P < 0.01$ probability level, *** significant at the $P < 0.001$ probability level. Values without asterisks are not significant at $P < 0.05$. PA, palmitic acid; SA, stearic acid; OA, oleic acid; LA, linoleic acid; LNA, linolenic acid.

3.5. Stability of Soybean Fatty Acids across Different Environments

The stability of soybean accessions varies greatly under different environmental conditions [62]. The coefficient of variation (CV) of each seed fatty acid was used to reflect the stability of the fatty acid levels, as a lower CV represents higher stability for a cultivar in different environments [63–65]. The means of the five seed fatty acids were plotted against their CVs to show how much the desirable level of each fatty acid was stable within each germplasm origin (Figure 4A–E). These results indicated that diverse germplasm accessions showed discrepant performance against different environments. The different CV values of accessions from the four origins elucidated the effect of different genetic background for each collection and their performances under unusual climatic conditions [29]. These findings suggest that introducing new germplasms to China will offer promising germplasm resources with specific oil profile depending on their stability across contrasting environments [66]. Importantly, the germplasm accessions which are adapted to local agro-environmental conditions may have beneficial alleles and could be incorporated in soybean breeding for desired fatty acid profiles.

Agronomy **2020**, *10*, 24

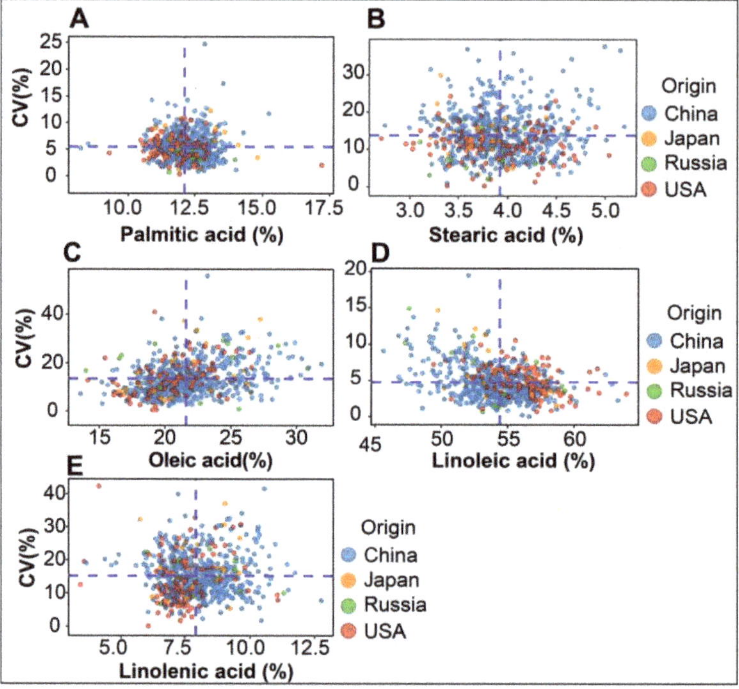

Figure 4. Scatter plots showing the relationship between means of five fatty acids and coefficients of variation (CVs) for 633 soybean accessions grown in five different locations. (**A**) palmitic acid, (**B**) stearic acid, (**C**) oleic acid, (**D**) linoleic acid, (**E**) linolenic acid. Horizontal and vertical dashed lines, in blue, represent average CV and mean of the five fatty acids, respectively.

Accessions with higher level of PA and lower CV values are more preferred form of saturated fatty acids because they showed higher oil stability (Figure 4A). Among these accessions, T309 from USA had higher PA content (17.20%) with lower CV (2.03%), followed by the Japanese accession Saikai 20, as its PA content was 14.81% with lower CV of 3.45%. Other accessions with higher PA showing higher stability are shown in Table 4. A collection of USA and Chinese germplasms with high SA content showed more stability (Figure 4B). The USA accession L72-920 showed higher SA content (5.04%) and lower CV (8.48%), followed by the Chinese accession Laidou24, which had higher SA content of 5.01% and low CV of 12.57% (Table 4). Other stable accessions with higher SA level were shown in Table 4. Figure 4C shows that higher means of OA content were reasonably associated with higher CV values, indicating that majority of the higher OA accessions were not very stable across different environments. Despite that, two Chinese accessions ZDD11235 and ZDD06021 with OA content of 28.47% and 28.14%, respectively, exhibited high stability as their CVs were 9.98 and 4.93%, respectively, in addition to other Chinese accessions with lower CV and higher OA content which are shown in Table 4. In contrast to other fatty acids, higher LA levels showed lower CV values (Figure 4D), revealing that a greater number of accessions with higher LA content were relatively stable across different environments. The current study showed that many stable and favorable LA levels were abundantly found in USA germplasms, whereas few accessions from other origins exhibited such a desired LA scenario (Figure 4D and Table 4). The USA accession S01-9391 had the highest LA content of 63.93% with CV of 3.20%, followed by the Chinese accession ZDD03222 with LA content of 63.30% and CV of 2.32%. Stable accessions with low LA levels were also observed, including ZDD04382 with LA level of 47.36% and CV of 4.55% (Table 4). These results indicate that higher LA levels tend to be less affected by environments compared with other fatty acids. In respect to LNA, the Chinese accession, ZDD03739, had a higher LNA level of 12.76% and maintained relatively higher stability (CV: 7.82%) (Table 4). As Lower LNA levels is required for maintaining higher oil stability, the accession C1640 from USA had a higher stability, CV of 6.8%, in addition to its lowest LNA content (3.43%) (Table 4) which highlights the importance of exploiting some USA soybean accessions for providing oils with higher stability.

Table 4. Accessions showing desired contents of five fatty acids with higher stability (lower CV).

Fatty Acid	ID Number	Name	Mean [a]	CV (%)	Origin
Higher palmitic acid (PA)	WDD01709	T309	17.20	2.03	USA
	WDD01215	Saikai 20	14.81	3.45	Japan
	ZDD24126	gongdou10	14.01	1.18	China
	WDD02708	PSB313	13.97	3.05	Russia
Higher stearic acid (SA)	WDD02225	L72-920	5.04	8.48	USA
	-	Laidou24	5.01	12.57	China
	-	HLT2-Heihe13	4.72	6.39	China
	Z13-633-1	Z13-633-1	4.81	12.82	China
Higher oleic acid (OA)	ZDD11235	SQDDD	28.47	9.98	China
	ZDD06021	Qingpidou	28.14	4.93	China
	ZDD12746	SCQPD	27.31	4.67	China
	WDD02995	PI 468903	27.27	6.58	USA
Higher linoleic acid (LA)	WDD03084	S01-9391	63.93	3.20	USA
	ZDD03222	RNPDS	63.30	2.32	China
	-	YZY2004-15-15W83	62.86	4.51	China
	WDD00713	Dorchsoy	61.88	3.30	USA
	WDD01488	CX1038-14	61.04	3.86	USA
	WDD01618	Camp	60.10	1.55	USA
Lower linoleic acid (LA)	ZDD04382	DYBHSBD	47.36	4.55	China
	ZDD06450	YGZHD	47.71	4.59	China
	ZDD07088	LQDD	48.24	4.40	China
	ZDD06021	Qingpidou	48.29	3.51	China
Higher linolenic acid (LNA)	ZDD03739	PXDZHC	12.76	7.82	China
	ZDD16816	CSHD	11.48	11.02	China
	PI84751	G1593|2017	11.32	10.68	USA
	WDD02708	PSB313	11.31	10.00	Russia
	PI43848913	G1592|2017	11.07	7.90	USA
	WDD01674	T116H	11.05	9.47	USA
Lower linolenic acid (LNA)	WDD01482	C1640	3.43	12.53	USA

[a] The content of each fatty acid is expressed as a percentage of total fatty acid content. CV, Coefficient of variation.

Basically, one of the aims of soybean breeding is to improve oil stability through developing genotypes with reduced LNA and elevated OA levels [21,67,68]. Therefore, taking into consideration these germplasms resources and their adaptability as determined in this study, our findings are expected to contribute greatly in selection of accessions with desirable levels of certain fatty acids which are less affected by contrasting environmental conditions. Furthermore, soybean oil content and fatty acid composition significantly affect soybean flavor attributes [69], that is why great efforts were made to select soybean cultivars with a desired oil profile to improve such flavor attributes [70]. In our study, the accessions which revealed higher stability and higher OA and LNA contents are beneficial for producing soymilk with preferable levels of smoothness and sweetness attributes, achieving the most important quality and preferable soymilk parameters for Chinese consumers [71]. In contrast, the accessions with higher PA and LNA contents are mostly preferable for western consumers due to the significant positive correlation of these two fatty acid levels with color and appearance of soymilk [71].

As a result, these unique accessions reported in this study constitute promising endeavors to improve soybean products with higher quality to meet various consumers' preferences.

4. Conclusions

To conclude, significant variations in seed fatty acid composition were noticed among four diverse soybean collections in this study. Particularly, accessions form USA and China showed significant differences in all the fatty acids and the largest difference between both origins was found in LA. Higher negative correlation observed between OA and LA levels offers a valuable chance for improvement of specific oil patterns with dual purposes. This variability observed in seed fatty acid traits among four worldwide germplasm accessions can provide valuable genetic resources for soybean breeding. In general, contrasting trends of fatty acids with different MGs further emphasized the importance of MG as an influential factor on soybean seed fatty acid composition. Different adaptability patterns of introduced germplasms encourage exploiting diverse genotypic backgrounds in breeding for high quality soybeans. Furthermore, the variation observed in fatty acid composition among different soybean germplasms due to the effect of diverse origins or MGs is of high importance for identification of accessions with desired lower or higher fatty acid contents. These high stable accessions with interested fatty acid profiles identified in this study can be used for enhanced studies and industries, aiming to achieve desired soybean oils to meet various consumption preferences.

Supplementary Materials: The following are available online at http://www.mdpi.com/2073-4395/10/1/24/s1, Table S1: Information of the 633 soybean accessions used in this study, Table S2: The monthly average of temperature ($^{\circ}$C) and precipitation (mm) for the experimental sites of Hainan, and Beijing in 2017 and 2018, and Anhui in 2017, Table S3; Table S3a. Analysis of variance for seed fatty acid composition affected by different origins at three locations for two years, Table S3b. Analysis of variance for seed fatty acid composition affected by 633 soybean accessions at three locations for two years, Table S3c. Analysis of variance for seed fatty acid composition affected by different maturity groups (MGs) at three locations for two years.

Author Contributions: Conceptualization, A.M.A. and J.S.; Formal analysis, A.M.A.; Funding acquisition, B.L. and J.S.; Investigation, J.Q.; Methodology, A.M.A., M.A., S.Z., A.S.S., Y.F., Y.L., Y.T., and H.H.; Project administration, B.L., and J.S.; Resources, S.Z., M.A., and A.S.S.; Supervision, J.S.; Visualization, B.L.; Writing—original draft, A.M.A.; Writing—review and editing, A.M.A., B.L., and J.S. All authors have read and agreed to the published version of the manuscript.

Funding: This study was supported by Ministry of Science and Technology (2017YFD0101401, 2016YFD0100201, and 2016YFD0100504), National Major Science and Technology Project (2016ZX08004-003), National Natural Science Foundation of China (31671716), Beijing Science and Technology Project (Z16110000916005), and the Chinese Academy of Agricultural Sciences (CAAS) Agricultural Science and Technology Innovation Project.

Acknowledgments: We would like to thank Lijuan Qiu from Institute of Crop Sciences, CAAS, for providing the soybean accessions utilized in this study and her helpful suggestions regarding the manuscript. We also would like to thank Xiaobo Wang from Anhui Agricultural University for helping in growing soybean accessions in Anhui province of China.

Conflicts of Interest: The authors declare no competing financial interest in regard to this manuscript.

Abbreviations

PA	palmitic acid
SA	stearic acid
OA	oleic acid
LA	linoleic acid
LNA	linolenic acid
CNSGB	Chinese National Soybean Gene Bank
MG	maturity group
GC	gas chromatography

References

1. *SoyStats a Reference Guide to Important Soybean Facts and Figures. American Soybean Association.* 2018. Available online: http://soystats.com/ (accessed on 22 August 2019).
2. Gale, F.; Valdes, C.; Ash, M. Interdependence of China, United States, and Brazil in Soybean Trade. In *New York: U.S. Department of Agriculture's Economic Research Service (ERS) Report*; USDA, Southwest: Washington, DC, USA, 2019; pp. 1–48.
3. Mensink, R.P.; Zock, P.L.; Kester, A.D.M.; Katan, M.B. Effects of dietary fatty acids and carbohydrates on the ratio of serum total to HDL cholesterol and on serum lipids and apolipoproteins: A meta-analysis of 60 controlled trials. *Am. J. Clin. Nutr.* **2003**, *77*, 1146–1155. [CrossRef] [PubMed]
4. Siri-Tarino, P.W.; Sun, Q.; Hu, F.B.; Krauss, R.M. Saturated fat, carbohydrate, and cardiovascular disease. *Am. J. Clin. Nutr.* **2010**, *91*, 502–509. [CrossRef] [PubMed]
5. Neff, W.E.; List, G.R. Oxidative stability of natural and randomized high-palmitic-and high-stearic-acid oils from genetically modified soybean varieties. *J. Am. Oil Chem. Soc.* **1999**, *76*, 825–831. [CrossRef]
6. Sacks, F.M.; Lichtenstein, A.H.; Wu, J.H.Y.; Appel, L.J.; Creager, M.A.; Kris-Etherton, P.M.; Miller, M.; Rimm, E.B.; Rudel, L.L.; Robinson, J.G. Dietary fats and cardiovascular disease: A presidential advisory from the American Heart Association. *Circulation* **2017**, *136*, e1–e23. [CrossRef]
7. Fehr, W.R. Breeding for modified fatty acid composition in soybean. *Crop Sci.* **2007**, *47*, S-72. [CrossRef]
8. Gebauer, S.K.; Psota, T.L.; Harris, W.S.; Kris-Etherton, P.M. n−3 fatty acid dietary recommendations and food sources to achieve essentiality and cardiovascular benefits. *Am. J. Clin. Nutr.* **2006**, *83*, 1526S–1535S. [CrossRef]
9. Chalon, S. Omega-3 fatty acids and monoamine neurotransmission. *Prostaglandins Leukot. Essent. Fat. Acids* **2006**, *75*, 259–269. [CrossRef]
10. Clemente, T.E.; Cahoon, E.B. Soybean oil: Genetic approaches for modification of functionality and total content. *Plant Physiol.* **2009**, *151*, 1030–1040. [CrossRef]
11. Shi, Z.; Bachleda, N.; Pham, A.T.; Bilyeu, K.; Shannon, G.; Nguyen, H.; Li, Z. High-throughput and functional SNP detection assays for oleic and linolenic acids in soybean. *Mol. Breed.* **2015**, *35*, 176. [CrossRef]
12. Pham, A.-T.; Shannon, J.G.; Bilyeu, K.D. Combinations of mutant *FAD2* and *FAD3* genes to produce high oleic acid and low linolenic acid soybean oil. *Theor. Appl. Genet.* **2012**, *125*, 503–515. [CrossRef]
13. Bachleda, N.; Grey, T.; Li, Z. Effects of high oleic acid soybean on seed yield, protein and oil contents, and seed germination revealed by near-isogeneic lines. *Plant Breed.* **2017**, *136*, 539–547. [CrossRef]
14. Rebetzke, G.J.; Burton, J.W.; Carter, T.E.; Wilson, R.F. Changes in agronomic and seed characteristics with selection for reduced palmitic acid content in soybean. *Crop Sci.* **1998**, *38*, 297–302. [CrossRef]
15. Johnson, R.C.; Bergman, J.W.; Flynn, C.R. Oil and meal characteristics of core and non-core safflower accessions from the USDA collection. *Genet. Resour. Crop Evol.* **1999**, *46*, 611–618. [CrossRef]
16. Fernandez-Martinez, J.; Del Rio, M.; De Haro, A. Survey of safflower (*Carthamus tinctorius* L.) germplasm for variants in fatty acid composition and other seed characters. *Euphytica* **1993**, *69*, 115–122. [CrossRef]
17. Goffman, F.D.; Pinson, S.; Bergman, C. Genetic diversity for lipid content and fatty acid profile in rice bran. *J. Am. Oil Chem. Soc.* **2003**, *80*, 485–490. [CrossRef]
18. Mourtzinis, S.; Marburger, D.; Gaska, J.; Diallo, T.; Lauer, J.; Conley, S. Corn and soybean yield response to tillage, rotation, and nematicide seed treatment. *Crop Sci.* **2017**, *57*, 1704–1712. [CrossRef]
19. Carter, T.E.; Nelson, R.L.; Sneller, C.H.; Cui, Z.; Boerma, H.R.; Specht, J.E.; Carter, T.; Nelson, R.; Sneller, C.; Cui, Z. Genetic Diversity in Soybean. In *Soybeans: Improvement, Production and Use*, 3rd ed.; Boerma, H., Specht, J., Eds.; American Society of Agronomy: Madison, WI, USA, 2004; pp. 303–450.
20. Wee, C.D.; Hashiguchi, M.; Anai, T.; Suzuki, A.; Akashi, R. Fatty acid composition and distribution in wild soybean (*Glycine soja*) seeds collected in Japan. *Asian J. Plant Sci.* **2017**, *16*, 52–64.
21. La, T.; Large, E.; Taliercio, E.; Song, Q.; Gillman, J.D.; Xu, D.; Nguyen, H.T.; Shannon, G.; Scaboo, A. Characterization of select wild soybean accessions in the USDA germplasm collection for seed composition and agronomic traits. *Crop Sci.* **2019**, *59*, 233–251. [CrossRef]
22. Zhao, X.; Jiang, H.; Feng, L.; Qu, Y.; Teng, W.; Qiu, L.; Zheng, H.; Han, Y.; Li, W. Genome-wide association and transcriptional studies reveal novel genes for unsaturated fatty acid synthesis in a panel of soybean accessions. *BMC Genom.* **2019**, *20*, 68. [CrossRef]

23. Uzun, B.; Arslan, Ç.; Furat, Ş. Variation in fatty acid compositions, oil content and oil yield in a germplasm collection of sesame (*Sesamum indicum* L.). *J. Am. Oil Chem. Soc.* **2008**, *85*, 1135–1142. [CrossRef]
24. Sharafi, Y.; Majidi, M.M.; Goli, S.A.H.; Rashidi, F. Oil content and fatty acids composition in *Brassica* species. *Int. J. Food Prop.* **2015**, *18*, 2145–2154. [CrossRef]
25. Kurasiak-Popowska, D.; Ryńska, B.; Stuper-Szablewska, K. Analysis of distribution of selected bioactive compounds in *Camelina sativa* from seeds to pomace and oil. *Agronomy* **2019**, *9*, 168. [CrossRef]
26. La Bella, S.; Tuttolomondo, T.; Lazzeri, L.; Matteo, R.; Leto, C.; Licata, M. An agronomic evaluation of new safflower (*Carthamus tinctorius* L.) germplasm for seed and oil yields under Mediterranean climate conditions. *Agronomy* **2019**, *9*, 468. [CrossRef]
27. Medic, J.; Atkinson, C.; Hurburgh, C.R., Jr. Current knowledge in soybean composition. *J. Am. Oil Chem. Soc.* **2014**, *91*, 363–384. [CrossRef]
28. Park, M.R.; Seo, M.J.; Lee, Y.Y.; Park, C.H. Selection of useful germplasm based on the variation analysis of growth and seed quality of soybean germplasms grown at two different latitudes. *Plant Breed. Biotechnol.* **2016**, *4*, 462–474. [CrossRef]
29. Song, J.Y.; Piao, X.M.; Choi, Y.M.; Lee, G.A.; Chung, J.W.; Lee, J.R.; Jung, Y.; Park, H.J.; Lee, M.C. Evaluation of genetic diversity and comparison of biochemical traits of soybean (*Glycine max* L.) Germplasm collections. *Plant Breed. Biotechnol.* **2014**, *1*, 374–384. [CrossRef]
30. Lee, J.D.; Bilyeu, K.D.; Shannon, J.G. Genetics and breeding for modified fatty acid profile in soybean seed oil. *J. Crop Sci. Biotech.* **2007**, *10*, 201–210.
31. Grieshop, C.M.; Fahey, G.C. Comparison of quality characteristics of soybeans from Brazil, China, and the United States. *J. Agric. Food Chem.* **2001**, *49*, 2669–2673. [CrossRef] [PubMed]
32. Liu, X.; Wu, J.; Ren, H.; Qi, Y.; Li, C.; Cao, J.; Zhang, X.; Zhang, Z.; Cai, Z.; Gai, J. Genetic variation of world soybean maturity date and geographic distribution of maturity groups. *Breed. Sci.* **2017**, *67*, 221–232. [CrossRef]
33. Qiu, L.; Xing, L.; Guo, Y.; Wang, J.; Jackson, S.A. A platform for soybean molecular breeding: The utilization of core collections for food security. *Plant Mol. Biol.* **2013**, *83*, 41–50. [CrossRef] [PubMed]
34. Qiu, L.; Li, Y.; Guan, R.; Liu, Z.; Wang, L.; Chang, R. Establishment, representative testing and research progress of soybean core collection and mini core collection. *Acta Agron. Sin.* **2009**, *35*, 571–579. [CrossRef]
35. Li, Y.; Reif, J.C.; Ma, Y.; Hong, H.; Liu, Z.; Chang, R.; Qiu, L. Targeted association mapping demonstrating the complex molecular genetics of fatty acid formation in soybean. *BMC Genom.* **2015**, *16*, 841. [CrossRef] [PubMed]
36. Yang, R.P.; Song, W.W.; Sun, S.; Wu, C.X.; Wang, H.J.; Han, T.F. Comparison of soybean yield and yield-related traits of agri-technology demonstration counties in different regions of China. *Soybean Sci.* **2012**, *31*, 557–567.
37. Song, W.; Sun, S.; Ibrahim, S.E.; Xu, Z.; Wu, H.; Hu, X.; Jia, H.; Cheng, Y.; Yang, Z.; Jiang, S.; et al. Standard Cultivar Selection and Digital Quantification for Precise Classification of Maturity Groups in Soybean. *Crop Sci.* **2019**, *59*, 1997–2006. [CrossRef]
38. Nelson, R.L.; Amdor, P.J.; Orf, J.H.; Lambert, J.W.; Cavins, J.F.; Kleiman, R.; Laviolette, F.A.; Athow, K.L. *Evaluation of the USDA Soybean Germplasm Collection: Maturity Groups 000 to IV*; Technical Bulletins 157020; United States Department of Agriculture, Economic Research Service: Washington, DC, USA, 1987; 267p.
39. Zhang, J.; Ge, Y.; Han, F.; Li, B.; Yan, S.; Sun, J.; Wang, L. Isoflavone content of soybean cultivars from maturity group 0 to VI grown in northern and southern China. *J. Am. Oil Chem. Soc.* **2014**, *91*, 1019–1028. [CrossRef]
40. Fan, S.; Li, B.; Yu, F.; Han, F.; Yan, S.; Wang, L.; Sun, J. Analysis of additive and epistatic quantitative trait loci underlying fatty acid concentrations in soybean seeds across multiple environments. *Euphytica* **2015**, *206*, 689–700. [CrossRef]
41. SAS Institute. *The SAS system for Windows, Release 9.2*; SAS Inst. Inc.: Cary, NC, USA, 2009.
42. Qin, P.; Song, W.; Yang, X.; Sun, S.; Zhou, X.; Yang, R.; Li, N.; Hou, W.; Wu, C.; Han, T. Regional distribution of protein and oil compositions of soybean cultivars in China. *Crop Sci.* **2014**, *54*, 1139–1146. [CrossRef]
43. Song, W.; Yang, R.; Wu, T.; Wu, C.; Sun, S.; Zhang, S.; Jiang, B.; Tian, S.; Liu, X.; Han, T. Analyzing the effects of climate factors on soybean protein, oil contents, and composition by extensive and high-density sampling in China. *J. Agric. Food Chem.* **2016**, *64*, 4121–4130. [CrossRef]
44. Simopoulos, A.P. The importance of the omega-6/omega-3 fatty acid ratio in cardiovascular disease and other chronic diseases. *Exp. Biol. Med.* **2008**, *233*, 674–688. [CrossRef]

45. Wang, Q.; Liang, X.; Wang, L.; Lu, X.; Huang, J.; Cao, J.; Li, H.; Gu, D. Effect of omega-3 fatty acids supplementation on endothelial function: A meta-analysis of randomized controlled trials. *Atherosclerosis* **2012**, *221*, 536–543. [CrossRef]
46. Santos, E.M.; Piovesan, N.D.; de Barros, E.G.; Moreira, M.A. Low linolenic soybeans for biodiesel: Characteristics, performance and advantages. *Fuel* **2013**, *104*, 861–864. [CrossRef]
47. Zhang, J.; Wang, X.; Lu, Y.; Bhusal, S.J.; Song, Q.; Cregan, P.B.; Yen, Y.; Brown, M.; Jiang, G. Genome-wide scan for seed composition provides insights into soybean quality improvement and the impacts of domestication and breeding. *Mol. Plant* **2018**, *11*, 460–472. [CrossRef] [PubMed]
48. Fehr, W.R.; Welke, G.A.; Hammond, E.G.; Duvick, D.N.; Cianzio, S.R. Inheritance of reduced linolenic acid content in soybean genotypes A16 and A17. *Crop Sci.* **1992**, *32*, 903–906. [CrossRef]
49. Jiang, G.L.; Chen, P.; Zhang, J.; Florez-Palacios, L.; Zeng, A.; Wang, X.; Bowen, R.A.; Miller, A.; Berry, H. Genetic analysis of sugar composition and its relationship with protein, oil, and fiber in soybean. *Crop Sci.* **2018**, *58*, 2413–2421. [CrossRef]
50. Graef, G.; LaVallee, B.J.; Tenopir, P.; Tat, M.; Schweiger, B.; Kinney, A.J.; Van Gerpen, J.H.; Clemente, T.E. A high-oleic-acid and low-palmitic-acid soybean: Agronomic performance and evaluation as a feedstock for biodiesel. *Plant Biotechnol. J.* **2009**, *7*, 411–421. [CrossRef]
51. Mourtzinis, S.; Gaspar, A.P.; Naeve, S.L.; Conley, S.P. Planting date, maturity, and temperature effects on soybean seed yield and composition. *Agron. J.* **2017**, *109*, 2040–2049. [CrossRef]
52. Dardanelli, J.L.; Balzarini, M.; Martínez, M.J.; Cuniberti, M.; Resnik, S.; Ramunda, S.F.; Herrero, R.; Baigorri, H. Soybean maturity groups, environments, and their interaction define mega-environments for seed composition in Argentina. *Crop Sci.* **2006**, *46*, 1939–1947. [CrossRef]
53. Assefa, Y.; Bajjalieh, N.; Archontoulis, S.; Casteel, S.; Davidson, D.; Kovács, P.; Naeve, S.; Ciampitti, I.A. Spatial characterization of soybean yield and quality (amino acids, oil, and protein) for United States. *Sci. Rep.* **2018**, *8*, 14653. [CrossRef]
54. Goldflus, F.; Ceccantini, M.; Santos, W. Amino acid content of soybean samples collected in different Brazilian states: Harvest 2003/2004. *Braz. J. Poult. Sci.* **2006**, *8*, 105–111. [CrossRef]
55. Bajaj, S.; Chen, P.; Longer, D.E.; Hou, A.; Shi, A.; Ishibashi, T.; Zhang, B.; Brye, K.R. Planting date and irrigation effects on seed quality of early-maturing soybean in the Mid-South USA. *J. New Seeds* **2008**, *9*, 212–233. [CrossRef]
56. Assefa, Y.; Purcell, L.C.; Salmeron, M.; Naeve, S.; Casteel, S.N.; Kovács, P.; Archontoulis, S.; Licht, M.; Below, F.; Kandel, H.; et al. Assessing variation in us soybean seed composition (protein and oil). *Front. Plant Sci.* **2019**, *10*, 1–13. [CrossRef] [PubMed]
57. Bellaloui, N.; Smith, J.R.; Ray, J.D.; Gillen, A.M. Effect of maturity on seed composition in the early soybean production system as measured on near-isogenic soybean lines. *Crop Sci.* **2009**, *49*, 608–620. [CrossRef]
58. Rahman, S.M.; Kinoshita, T.; Anai, T.; Takagi, Y. Combining ability in loci for high oleic and low linolenic acids in soybean. *Crop Sci.* **2001**, *41*, 26–29. [CrossRef]
59. Patel, M.; Jung, S.; Moore, K.; Powell, G.; Ainsworth, C.; Abbott, A. High-oleate peanut mutants result from a MITE insertion into the *FAD2* gene. *Theor. Appl. Genet.* **2004**, *108*, 1492–1502. [CrossRef] [PubMed]
60. Patil, A.; Taware, S.P.; Oak, M.D.; Tamhankar, S.A.; Rao, V.S. Improvement of oil quality in soybean [*Glycine max* (L.) Merrill] by mutation breeding. *J. Am. Oil Chem. Soc.* **2007**, *84*, 1117–1124. [CrossRef]
61. Wu, T.; Yang, X.; Sun, S.; Wang, C.; Wang, Y.; Jia, H.; Man, W.; Fu, L.; Song, W.; Wu, C. Temporal–spatial characterization of seed proteins and oil in widely grown soybean cultivars across a century of breeding in China. *Crop Sci.* **2017**, *57*, 748–759. [CrossRef]
62. Wang, X.; Liu, Z.; Yang, C.; Xu, R.; Lu, W.; Zhang, L.; Wang, Q.; Wei, S.; Yang, C.; Wang, H. Stability of growth periods traits for soybean cultivars across multiple locations. *J. Integr. Agric.* **2016**, *15*, 963–972. [CrossRef]
63. Abdulai, M.S.; Sallah, P.Y.K.; Safo-Kantanka, O. Maize grain yield stability analysis in full season lowland maize in Ghana. *Int. J. Agric. Biol* **2007**, *9*, 41–45.
64. Temesgen, T.; Keneni, G.; Sefera, T.; Jarso, M. Yield stability and relationships among stability parameters in faba bean (*Vicia faba* L.) genotypes. *Crop J.* **2015**, *3*, 258–268. [CrossRef]
65. Hemingway, J.; Eskandari, M.; Rajcan, I. Genetic and environmental effects on fatty acid composition in soybeans with potential use in the automotive industry. *Crop Sci.* **2015**, *55*, 658–668. [CrossRef]
66. You, F.M.; Jia, G.; Xiao, J.; Duguid, S.D.; Rashid, K.Y.; Booker, H.M.; Cloutier, S. Genetic variability of 27 traits in a core collection of flax (*Linum usitatissimum* L.). *Front. Plant Sci.* **2017**, *8*, 1636. [CrossRef] [PubMed]

67. Lee, J.; Hwang, Y.S.; Kim, S.T.; Yoon, W.B.; Han, W.Y.; Kang, I.K.; Choung, M.G. Seed coat color and seed weight contribute differential responses of targeted metabolites in soybean seeds. *Food Chem.* **2017**, *214*, 248–258. [CrossRef] [PubMed]
68. Yoon, M.S.; Lee, J.; Kim, C.Y.; Kang, J.H.; Cho, E.G.; Baek, H.J. DNA profiling and genetic diversity of Korean soybean (*Glycine max* (L.) Merrill) landraces by SSR markers. *Euphytica* **2009**, *165*, 69–77. [CrossRef]
69. Terhaag, M.M.; Almeida, M.B.; de Benassi, M.T. Soymilk plain beverages: Correlation between acceptability and physical and chemical characteristics. *Food Sci. Technol.* **2013**, *33*, 387–394. [CrossRef]
70. Suppavorasatit, I.; Lee, S.; Cadwallader, K.R. Effect of enzymatic protein deamidation on protein solubility and flavor binding properties of soymilk. *J. Food Sci.* **2013**, *78*, C1–C7. [CrossRef]
71. Ma, L.; Li, B.; Han, F.; Yan, S.; Wang, L.; Sun, J. Evaluation of the chemical quality traits of soybean seeds, as related to sensory attributes of soymilk. *Food Chem.* **2015**, *173*, 694–701. [CrossRef]

 *agronomy*

Article

Genetic and Genomic Diversity in a Tarwi (*Lupinus mutabilis* Sweet) Germplasm Collection and Adaptability to Mediterranean Climate Conditions

Norberto Guilengue [1], Sofia Alves [1], Pedro Talhinhas [1,2,*] and João Neves-Martins [1]

[1] Instituto Superior de Agronomia, Universidade de Lisboa, 1349-017 Lisbon, Portugal; guilenguen@gmail.com (N.G.); sofiaalves@isa.ulisboa.pt (S.A.); nevesmartins@isa.ulisboa.pt (J.N.-M.)

[2] LEAF, Linking Landscape, Environment, Agriculture and Food, Instituto Superior de Agronomia, Universidade de Lisboa, 1349-017 Lisbon, Portugal

* Correspondence: ptalhinhas@isa.ulisboa.pt; Tel.: +351-2-1365-3249

Received: 7 October 2019; Accepted: 20 December 2019; Published: 22 December 2019

Abstract: *Lupinus mutabilis* (tarwi) is a species of Andean origin with high protein and oil content and regarded as a potential crop in Europe. The success in the introduction of this crop depends in part on in depth knowledge of the intra-specific genetic variability of the collections, enabling the establishment of breeding and conservation programs. In this study, we used morphological traits, Inter-Simple Sequence Repeat markers and genome size to assess genetic and genomic diversity of 23 tarwi accessions under Mediterranean conditions. Phenotypic analyses and yield component studies point out accession LM268 as that achieving the highest seed production, producing large seeds and efficiently using primary branches as an important component of total yield, similar to the *L. albus* cultivars used as controls. By contrast, accession JKI-L295 presents high yield concentrated on the main stem, suggesting a semi-determinate development pattern. Genetic and genomic analyses revealed important levels of diversity, however not relatable to phenotypic diversity, reflecting the recent domestication of this crop. This is the first study of genome size diversity within *L. mutabilis*, revealing an average size of 2.05 pg/2C (2001 Mbp) with 9.2% variation (1897–2003 Mbp), prompting further studies for the exploitation of this diversity.

Keywords: *Lupinus mutabilis*; genetic diversity; morphological traits; ISSR; genome size; Mediterranean climate

---

1. Introduction

The genus *Lupinus* includes more than 280 species [1], approximately 90% of which are native and widely distributed throughout the American continent [2,3], with greater inter- and intra-specific genetic variability than in Euro-African species. *Lupinus mutabilis* Sweet (also known as tarwi, chocho, altramuz and Andean lupin) is native from the Andean region in South America. The species is auto- and allogamic with wide variability of flower, stem and seed colours, and exhibits indeterminate growth [4,5]. It has been domesticated in the Andean region and used for grain production, forage, green manure, fixing atmospheric nitrogen and soil conservation [4,6]. In spite of their high alkaloid content [7,8], tarwi seeds have high nutritional value, containing up to 53% protein and 24% lipids [9]. The nutritional attributes found in tarwi are supposedly better than those in soybeans [10] and for this reason it is called Andean soybean [11]. Tarwi protein is rich in globulins (43%–45%) and albumins (8%–9%) and the oil has high quality and does not require industrial removal of the linolenic acid like in soybean [12,13]. Additionally, low alkaloid (<0.1%) lines have been selected in *L. mutabilis* [14].

Tarwi exhibits key traits of domestication, including indehiscent pods and seeds with permeable tegument, representing a locally important crop in several Andean areas [15]. Recently the species

L. piurensis was considered the wild relative from which tarwi would have evolved until arriving at the domesticated form known nowadays [16]. According to this hypothesis, no wild specimens of *L. mutabilis* exist and the species would have suffered a classic domestication bottleneck no later than 2600 years before present time [16], leading to a recognizably low genetic diversity of tarwi [17]. Nevertheless, the crop conceals important morphological variability, which is related to the high variability of agroecological conditions across its native range [18]. For instance, small plants occur in the Potosi region, where the altitude exceeds 3500 m and of low temperatures and precipitation prevail. Branched and tall plants are found in the Andean valleys of Bolivia and Southern Peru with more than 50% of their production centred on the main stem. Highly branched plants with over 1.8 m in height, with long vegetative period and little production in the main stem occur in Colombia, Ecuador and northern Peru, under frost-free climates [18].

Due to its high plasticity, tarwi has a wide adaptation to varied soils, precipitation and temperature regimes [6]. In the light of this broad adaptation, attempts have been made in order to introduce tarwi to European conditions [12] to reduce local dependence on imported soybeans. As such, seeds harvested in the Andean region have been used during several years to select plants with determined growth in the Mediterranean conditions. As a result, a germplasm collection was created focused on promising accessions. The success in introducing this species in this region will depend in part on the deep knowledge of the genetic variability of this collection. Thus, understanding the genetic variability is extremely important for the establishment of future breeding and conservation programmes [19].

Research on crop genetic variability has been based on morphological descriptors and molecular markers as the main tools [20–23]. The morphological descriptors are used to generate relevant information about the description and classification of the germplasm collections in order to allow efficient use in breeding programmes [24,25]. Morphological analysis and molecular markers can be used together to generate more reliable and consistent information. Contrary to morphological descriptors, molecular markers have the advantage to not depend on the environment, phenotype and stage of development of the plant [26]. Several DNA markers are available and can be used in genetic diversity studies, among which are Inter-Simple Sequence Repeat (ISSR) markers. ISSR markers allow preliminary screening of germplasm collections and have been used to perform genetic mapping, phylogenetic and evolutionary studies because of their good repeatability, high polymorphism, easy handling and low cost [27–30]. ISSR analyses thus enable the selection of contrasting accessions that, together with pertaining morphological traits, can be selected for further characterisation using more informative markers, such as Simple Sequence Repeats. In addition to the use of molecular markers, in recent years the use of nuclear DNA content information to explain intra-specific genetic diversity has been increasing [31–33]. The DNA content is important for understanding molecular, cellular and evolutionary genomic mechanisms [34]. Flow cytometry is widely used for DNA content estimation due to its simplicity and efficacy [35]. This technique has been applied successfully in the estimation of nuclear DNA content in different species. In particular, flow cytometry was employed to differentiate *Lupinus* species based on the genome size [36]. However, there are few studies addressing intraspecific variability in *L. mutabilis* based on morphological traits, ISSR [37,38] and on DNA content. This prompts a need to characterise *L. mutabilis* germplasm collections in depth, both under genotypic and phenotypic perspectives. Instituto Superior de Agronomia (ISA), Portugal, has one of the most important collections of *Lupinus* in the world, containing over 1300 *Lupinus* accessions, including *L. mutabilis*. However, little is known about the genetic variability in this collection. The present study aims to evaluate genetic and genomic diversity in 23 *L. mutabilis* accessions present in ISA collection using 37 morphological traits, six ISSR markers and genome size data, contributing simultaneously to assess its adaptability to Mediterranean climate conditions and to provide genotypic data.

2. Materials and Methods

2.1. Plant Materials

A total of 23 *L. mutabilis* accessions were selected from the ISA *Lupinus* germplasm including five accessions provided by the Julius Kühn-Institut (JKI), Germany (e.g., Table 2). *Lupinus albus* cultivars Misak and Mihai were used as reference in the morphological characterization because of their high adaptation to the Mediterranean conditions and as outgroups/standards in the ISSR marker and genome size analyses.

2.2. Morphological Analysis

Field experiments were conducted at Tapada de Ajuda in Lisbon (coordin: 38.709133, −9.182976, alt: 60 m) on a vertisol in the 2016/17 (sowing date: 29 December) and 2017/18 (sowing date: 18 December) seasons under rain-fed conditions. Meteorological data were collected daily from the weather station located adjacent to the field. Soil water balances were calculated according to Allen et al. [39].

The experimental design adopted was randomized block with three replicates. Each replicate was composed of 26 1.8 m^2-plots with 20 plants in each plot (immediately surrounded by a 60 cm-wide edge of *L. albus* 'Misak' plants to avoid border effects) and the total number of plots in the assay was 78. For morphological characterization, 10 plants of each plot were selected as recommended by Talhinhas et al. [40].

Data of morphological characterization were obtained based on *Lupinus* spp. descriptors [41], as listed in Table 1. Yield components and vegetative traits were analysed considering a two-factor experimental design (genotype and year), with differences being statistically analysed using the Kruskal Wallis test. Characteristics for multivariate analysis were selected based on correlation coefficients and heritability values [40]. Variables with correlation above 0.85 were considered redundant and thus one of them was excluded. Meanwhile, variables with low heritability (<65%) were also excluded, as these were explained by environmental factors.

Univariate analysis (UA) was performed to compare each individual characteristic across the accessions. Before running the UA, normality and homogeneity of variances was tested. Since data did not follow normal distribution and the variance was not homogeneous, an analysis of variance (ANOVA) based on rank transformation for non-parametric analysis was performed [42]. Post-hoc Tukey's honest significant difference (HSD) test of means was performed for all variables at 5% significance. Afterwards, broad sense heritability (H^2), genotypic variance (σ^2_g), phenotypic variance (σ^2_P), phenotypic coefficient of variance (PCV) and genotypic coefficient of variance (GCV) were estimated to understand the genetic variation between accessions and environment, as well as the genetic effects on different traits, following Mazid et al. [43].

Multivariate analysis was performed for all 25 accessions and all characteristics selected and represented in a single graphic, as described by Talhinhas et al. [40]. Standardized morphological data transformation (mean = 0, and standard deviation = 1) was performed before conducting multivariate analysis. Cluster analysis was performed based on Euclidean distance and average method for the 25 accessions. A dendrogram was constructed using an unweighted pair group method of arithmetic mean (UPGMA) algorithm. Principal component analysis (PCA) was performed and eigenvectors and eigenvalues were projected to visualize the components. All analyses were performed in the RStudio program version 1.1.456 (The R consortium, Boston, USA).

Table 1. List of morphological traits evaluated in the experiment, method and unit of measurement.

Acronym	Trait [1]	Method (unit)
DUF	Days from sowing until flowering [3]	Counting (nr [4])
NLMS	Number of leaves on the main stem	Counting (nr)
HUFF	Height up to first flower	Metric meas. [5] (cm)
ADNL	Average distance between leaves	= HUFF/NLMS (cm)
NPMS	Number of pods on the main stem	Counting
PLMS	Pod length on the main stem	Metric meas. (cm)
PWMS	Pod width on the main stem	Metric meas. (cm)
RBLWPMS	Ratio length-width of pods on the main stem	= PLMS/PWMS (dim. [6])
NSMS	Number of seeds on the main stem	Counting (nr)
SLMS	Seed length on the main stem [2]	Metric meas. (cm)
SWMS	Seed width on the main stem	Metric meas. (cm)
RBLWSMS	Ratio length-width of seed on the main stem [2,3]	= SLMS/SWMS (dim.)
NSPMS	Number of seed per pod on the main stem	Counting (nr)
WSMS	Weight of seeds on the main stem	Weighting (g)
TSWMS	Thousand seeds weight on the main stem [2]	= WSMS/NSMS*1000 (g)
NPB	Number of primary branches	Counting (nr)
ADBPB	Average distance between primary branches [2,3]	= HUFF/NPB (cm)
SLPB	Sum of the length of primary branches	Metric meas. (cm)
ALPB	Average length of primary branches	= SLPB/NPB (cm)
PBL	Proportion of leaves with branches [2,3]	= NPB/NLMS (%)
NPPB	Number of pods on primary branches [2]	Counting (nr)
NSPB	Number of seeds on primary branches [2]	Counting (nr)
NSPPB	Number of seeds per pod on primary branches	= NSPB/NPPB (nr)
NPPPB	Number of pods per primary branch	= NPPB/NPB (nr)
WSPB	Weight of seeds on primary branches [2]	Weighting (g)
WSPPB	Thousand seeds weight per primary branches	Weighting (g)
TSWPB	Thousand seeds weight on primary branches	= WSPB/NSPB × 1000 (g)
TBL	Total branch length	= SLPB+ HUFF
TNP	Total number of pods [2]	Counting (nr)
TNS	Total number of seeds	Counting (nr)
TNSPP	Total number of seeds per pod	= TNS/TNP (nr)
TW	Total seed weight	Weighing (g)
PSMS	Percentage of seed weight on the main stem	= WSMS/TW (%)
PSPB	Percentage of seed weight on primary branches [3]	= WSPB/TW (%)
TTSW	Total thousand seeds weight	= TW/TNS × 1000 (g)
SWBLR	Seed weight/total branch length ratio	= TW/TBL × 100 (g/m)

[1] Characteristics related with secondary and tertiary branches were excluded due to insufficiency of data; [2] Redundant or non-independent characteristics excluded of multivariate analysis based on the correlation coefficient ($r > 0.85$); [3] Characteristics excluded of multivariate analysis due to presenting low value of heritability (<0.65); [4] number; [5] Metric measurement; [6] dim.—dimensionless.

2.3. Molecular Analysis

Young but fully expanded leaves of the 23 *L. mutabilis* accessions and of the two *L. albus* reference cultivars were collected and immediately frozen in liquid nitrogen and stored at −80 °C. Freeze-dried vegetal material was used for DNA extraction using the DNeasy® Plant mini kit (Qiagen, Hilden, Germany) according to the manufacturer instructions. The DNA quality and quantity were estimated using spectrophotometry in the Gen5 program, and electrophoresis using a 1% agarose gel. The stock solution of DNA was diluted with sterilized water to make a working solution with a concentration of 10 ng/μL to be used in amplifications.

For molecular characterization, six ISSR primers were selected (Table 6) from those reported by Talhinhas et al. [44] based on preliminary analyses of a limited set of accessions. The Polymerase Chain Reaction (PCR) amplification for all primers was carried out under the following conditions: pre-denaturation 4 min at 94 °C, 40 cycles of 30 s at 94 °C, 45 s at 52 °C and 2 min at 72 °C, and a final extension at 72 °C for 10 min. The PCR reactions were performed in a final volume of 10 μL containing 20 ng of DNA, 0.5 µM of primer and 5 µL of dNTP + *Taq* DNA polymerase (NZYTaq II DNA polymerase, NZYTech, Lisbon, Portugal). After amplification, products were separated by electrophoresis in a 2% agarose gel stained using GreenSafe Premium (NZYTech).

The ISSR bands were scored in a binary matrix as presence (1) or absence (0) for each accession and for each fragment size. Based on the binary matrix, parameters such as percentage of polymorphic and monomorphic bands were determined and discriminatory power of primers was calculated based on the polymorphic information content (PIC), effective multiplex ratio (EMR), resolving power (RP) and marker index (MI). PIC value is the probability for detecting polymorphism by a primer or primers combination between two randomly drawn genotypes and can be calculated using the formula $PIC = 1 - \Sigma pi^2$, where pi is the frequency of occurrence of polymorphic bands in different primers [45]. The effective multiplex ratio was calculated using the formula $EMR = np\beta$; where ß is the fraction of polymorphic markers and is estimated after considering the number of polymorphic loci (np) and non-polymorphic loci (nnp) as ß = np/(np + nnp) [46]. Marker index (MI) is the primer capacity to detect polymorphic loci among different genotypes and was calculated as $EMRxPIC$. Resolving power (RP) is the ability of primers to distinguish between genotypes and was calculated as $RP = \Sigma Ib$, where Ib is the informative fragments and can take values of: $1 - [2|0.5 - p|]$; p is the proportion of total genotypes containing the band [47]. Genetic similarity was obtained according to the Jaccard similarity index. The results were used for the construction of ISSR and morphological traits dendrograms, in order to evaluate the similarity relations between the genotypes. Dendrograms were constructed based on UPGMA grouping and the ISSR results were correlated with morphological traits.

2.4. Flow Cytometry

For each accession, young leaves in healthy conditions were randomly collected and immediately analysed in the laboratory. Nuclear DNA content was measured by flow cytometry. *Solanum lycopersicum* 'Stupické' (2C = 1.96 pg; [48]) was tested as DNA standard but its genome size showed to be too close to that of *L. mutabilis*. Therefore, we tested *L. albus* as DNA standard (2C = 1.20 pg; [49]) and for such *L. albus* 'Misak' was validated as standard by comparison to *S. lycopersicum* 'Stupické' and *Raphanus sativus* 'Saxa' (2C = 1.11 pg; [48]). Each *L. mutabilis* accession, together with the standard, was chopped with a razor blade in the presence of 1 mL of buffer (Woody Plant Buffer; [50]). The nuclear suspension obtained was then separated from plant debris using a 30 µm nylon filter. After filtration, 50 µg/mL of propidium iodide (PI; Sigma-Aldrich) were added to stain DNA and 50 µg/mL of RNase (Sigma-Aldrich) ware added to prevent staining of double stranded RNA. The samples were maintaining at room temperature and analyzed using a CyFlow Space flow cytometer (Sysmex, Norderstedt, Germany) equipped with a 30 mW green solid-state laser emitting at 532 nm for optimal PI excitation. The reproducibility of results were assessed using five independent replicates for each accession. FloMax software v2.4d (Sysmex) was used to measure nuclear DNA content and three graphics were generated from data measurement: fluorescence pulse integral in linear scale (FL); fluorescence pulse integral in linear scale versus time; and fluorescence pulse integral in linear scale versus side light scatter in logarithmic scale (SSC). The absolute DNA amount of a sample was calculated based on the values of the G1 peak means, as suggested by Doležel and Bartoš [51]:

$$Sample\ 2C\ DNA\ content = \frac{Sample\ G1\ peak\ mean}{standard\ G1\ peak\ mean} \times Standard\ 2C\ DNA\ Content \qquad (1)$$

The results generated from 2C DNA (in picogram) were transformed to million base pairs using the following conversion: 1 pg = 978 Mbp [52]. Coefficient of variation (CV, %) of G1 peaks in the FL histograms, and estimates of the CV of the genome size of each accession were used to assess the reliability of the results. Intra-specific genome size comparison was carried out using Kruskal Wallis test ($\alpha = 0.05$) because genome size data did not exhibit normal distribution. Data analysis was done in RStudio Program Version 1.1.456.

3. Results

3.1. Morphological Characterization and Genetic Parameters among Accessions

Studies on the genetic variability are important because they generate relevant data for breeding programmes and can be used as basis for development and selection of superior genotypes. Here, we used morphological characterization and genetic parameters to evaluate the variability of a *L. mutabilis* germplasm collection under Mediterranean conditions.

Meteorological conditions during the trial were typical of the Mediterranean climate (Figure S1), although rainfall was well below average during autumn and winter and above average during spring in 2017/18, while rain was scarce in April 2016. Two-way ANOVA based on rank transformation was performed and revealed that all morphological traits exhibited significant difference at p-value < 0.05. Tables 2 and 3 show the mean values, homogeneous groups and p value obtained in a two-factor experimental design for morphological and reproductive characteristics, respectively. Differences analysis results for morphological traits of each year are given in Supplementary Material (Table S1). The statistical analysis of the results depict those genotypes showing differences that were consistent over the two years.

Figure 1. *Lupinus mutabilis* flowers and stem colours: (**a**) green stem, white wings, white standard with yellow central spot; (**b**) green stem, pale pink wings (more intense as flower matures), pale pink standard with yellow central spot (central spot turning dark pink as the flower matures); (**c**) green stem, blue wings, standard blue in the marginal area, white in the intermediate and yellow in the central spot; (**d**) purple stem, purple wings, purple standard with yellow central spot.

The results presented in Table 2 reveal that the average number of days from sowing to flowering (DUF) ranged between 80.8 (accession JKI-L309) and 103.4 (accession XM1-39) for *L. mutabilis*, spanning 23 days, while for *L. albus* cultivars the average was 98.1 days. The number of leaves on the main stem (NLMS) ranged between 12.8 and 18.0, while the number of primary branches (NPB) ranged from 2.6 to 4.6. The average numbers of leaves (NLMS) and branches (NPB) were 15.9 and 3.4, respectively, showing similar values to *L. albus*. On average, 21% of main stem leaves axillae harboured primary branches (PBL), with a minimum of 16% for accession JKI-L295 and a maximum of 26% for accession LM231. The average height (HUFF) of *L. mutabilis* plants was 55.7 cm, ranging between 40.0 cm (JKI-L309) and 68.9 cm (Inti), while the average for *L. albus* cultivars was 40.4 cm. In both experiments the accession JKI-L309 grew less than other accessions. The total length of primary branches (SLPB) varied between 81.9 cm for accession JKI-L309 and 163.1 cm for accession LM231, with an average of 122.8 cm. The total stem length (main stem and primary branches, TBL) attained a global average of 179.4 cm (164.7 for *L. albus*), varying between 122.0 cm (accession JKI-L309) and 229.4 cm (LM231). The JKI-L309 accession presented low values for TBL, suggesting that this may be a semi-determinate genotype. Stem and flower colors varied among accessions, with no clear correlation to morphologic traits. Figure 1 depicts the four groups of flower and stem colors.

Table 2. Average values (and homogeneous groups [1]) for vegetative traits of 23 *Lupinus mutabilis* accessions and two *L. albus* cultivars ('Mihai' and 'Misak'), obtained upon analysis of variance (ANOVA) based on rank transformation.

Accession	SFC [2]	DUF [3]	NLMS	HUFF	NPB	SLPB	PBL	TBL
JKI-L309	D	80.8 a	12.9 a	40.0 a	2.8 abc	81.9 a	21.6 efgh	122.0 a
JKI-L377	B	92.8 b	12.8 a	47.7 bc	2.7 ab	87.2 a	19.9 cde	137.7 ab
MUTAL	A	93.5 b	14.8 b	60.0 gh	3.3 efg	152.0 h	22.8 efgh	213.1 i
JKI-L210	C	96.4 c	15.8 def	50.0 bcd	3.0 cde	89.1 a	19.1 bc	142.2 bc
LM13	C	97.7 cd	16.5 fg	53.0 cdef	3.9 ijk	132.0 cdefg	23.3 efghi	185.6 efgh
LM81	C	97.8 cd	15.2 bcd	54.3 cdef	3.3 cdef	127.0 bcdef	21.7 defgh	182.4 defgh
JKI-L295	A	98.6 cd	16.2 f	60.7 fgh	2.6 a	114.1 b	16.3 a	175.2 def
LM268	A	98.8 cd	16.3 f	60.9 gh	3.6 hij	133.3 defgh	22.2 efgh	194.2 fghi
PRT79	C	98.8 cd	17.4 hi	65.1 hi	3.5 hij	136.2 efgh	20.4 cde	202.3 ghi
I82	D	99.9 cd	15.9 def	52.4 cde	3.4 fghi	115.8 bc	21.2 def	169.0 de
LM34	D	100.1 cd	15.2 bcd	50.9 bcd	3.7 hijk	124.5 bcde	24.3 i	175.6 de
Potosi-ALE	C	100.3 cd	15.0 b	53.6 cde	3.5 fghi	134.2 bcdef	23.4 fghi	188.7 efgh
LM27	A	100.4 cd	15.9 cdef	55.8 efgh	3.5 fghi	125.3 bcdef	21.9 defgh	181.7 efgh
Potosi-ISA	C	100.5 cd	16.7 fg	58.9 efgh	3.4 fghi	118.5 bcd	20.7 cde	177.5 def
LM18	C	100.5 cd	17.7 ghi	54.6 defg	3.7 hij	127.8 bcdef	20.8 cde	182.5 defg
CM157	C	100.6 cd	15.4 bcde	49.1 bcd	3.0 bcd	111.4 b	18.6 bc	160.7 cd
XM-5	B	101.2 cd	16.4 fg	58.5 efgh	3.5 fghi	125.0 bcdef	21.3 defg	184.1 efgh
LM32	C	101.6 d	16.2 ef	53.7 cdef	3.4 fgh	125.3 bcdef	21.5 def	179.6 defg
LM231	C	102.4 d	18.0 i	64.8 hi	4.6 k	163.2 gh	25.5 ghi	229.4 hi
P20993	B	102.6 d	15.1 bc	55.3 defg	3.6 ghij	125.1 bcdef	23.5 hi	180.4 defg
INTI	C	102.6 cd	17.7 i	68.9 i	3.3 def	147.0 fgh	18.6 b	218.7 i
SBP	C	103.0 d	16.0 f	58.6 efgh	3.2 def	119.7 bcdef	19.7 bcd	179.4 defgh
XM1-39	C	103.4 d	16.2 f	54.6 defg	3.1 def	110.3 bc	18.7 bc	166.4 de
Mihai		97.6 cd	16.6 fgh	42.6 ab	3.7 hijk	122.1 bcdef	22.5 defgh	164.7 cde
Misak		98.6 cd	18.8 j	38.2 a	3.8 jk	120.7 bcdefg	20.4 cde	158.9 cd
p-value [4]		0.000	0.000	0.000	0.000	0.000	0.000	0.000
p-value [5]		0.000	0.000	0.000	0.000	0.000	0.020	0.000
p-value [6]		0.000	0.000	0.000	0.000	0.000	0.000	0.000

[1] Homogeneous groups—accessions sharing the same letter for each trait are not statistically different; [2] SFC—stem and flower colour, according to Figure 1; [3] Full name of acronyms and description of the respective morphologic traits are given in Table 1; [4] *p*-value taking into account the accessions; [5] *p*-value taking into account the experiments; [6] *p*-value taking into account the interaction between accessions and experiments.

The total seed weight (TW) per plant (Table 3) varied 3.7×, ranging between 3.7 g per plant (accession JKI-L210) and 13.8 g per plant (accession LM268), the latter attaining a projected productivity estimated at 1533 kg/ha, although less than half of the total yield of the *L. albus* accessions. Dissecting yield components evidences additional variability among the accessions (Table 3). The total number of pods (TNP) per plant varied nearly 2.1×, with a maximum of 25 pods per plant for accession Potosi-ISA. The total number of seeds (TNS) reached a maximum of 67.9 seeds per plant (accession LM34). The average number of seeds per pod (TNSPP) is 2.7, ranging between 2.2 (accession LM268) and 3.3 (accession JKI-L210). The total thousand seeds weight (TTSW) attained a global average of 187.0 g, varying 2.85× between 101.4 g (accession JKI-L210) and 289.2 g (accession LM268). LM268 was the only *L. mutabilis* accession producing more yield on the primary branches than on the main stem (40% and 60% of total yield on the main stem and on primary branches, respectively), following a similar pattern to that of *L. albus* cultivars. The accessions JKI-L295 and JKI-L210 produced about 80% of seed weight on the main stem. Unlike accession LM268, several *L. mutabilis* accessions reached superior seed yields (over 10 g per plant) while concentrating over 60% of their yield on the main stem: CM157, I82 and LM27. For the comparison of seed yield and vegetative development (Table 3), the seed weight/total branch length ratio (SWBLR) was calculated. SWBLR average was 2.0 g of seeds per meter of branch length in *L. mutabilis* (23.1 g/m in *L. albus*), ranging between 1.1 g/m (accession LM32) and 3.3 g/m (accession Mutal).

Table 3. Average values (and homogeneous groups [1]) for yield components of 23 *Lupinus mutabilis* accessions and two *L. albus* cultivars ('Mihai' and 'Misak'), obtained upon ANOVA based on rank transformation.

Accession	TNP [2]	TNS	TNSPP	TTSW	PSMS	PSPB	TW	SWBLR
JKI-L210	11.5 a	36.6 ab	3.3 fg	101.4 a	87.8	12.2	3.7 a	1.3 a
INTI	13.7 a	37.4 ab	2.7 bcd	154.0 cd	74.6	25.4	5.7 b	2.3 ef
XM1-39	13.9 ab	36.1 ab	2.8 bcd	180.4 jkl	78.1	21.9	6.3 b	1.3 a
JKI-L377	19.8 cdef	61.3 cde	3.0 ef	103.7 a	72.2	27.8	6.3 b	3 def
JKI-L309	16.1 bc	49.2 cd	3.1 efg	159.7 ef	72.2	27.8	7.8 c	3.1 fg
SBP	16.8 cd	43.6 bc	2.6 bc	188.7 lm	77.0	23.0	8.0 cd	1.7 abc
LM32	22.1 efg	56.7 def	2.7 bcd	150.8 cde	67.3	32.7	8.4 cde	1.1 ab
Potosi-ALE	25.0 gh	62.2 efg	2.6 b	182.7 bc	59.1	40.9	8.4 cdef	1.5 abcd
LM34	24.9 gh	67.9 fg	2.7 bcde	131.6 b	55.0	45.0	8.5 cde	1.6 abcd
JKI-L295	18.0 cde	48.0 cd	2.7 bcd	188.4 m	80.3	19.7	9.0 def	2.2 bcdef
LM231	22.9 fgh	59.9 def	2.6 bcd	169.4 fgh	56.2	43.8	9.5 cdef	1.5 abcd
MUTAL	19.2 def	52.2 cd	2.8 bcd	184.7 lm	67.2	32.8	9.6 def	3.3 g
LM81	23.1 gh	63.1 fg	2.9 bcde	157.8 ef	57.7	42.3	9.7 fghi	1.8 abcde
LM13	22 fgh	57.5 def	2.7 bcd	179.0 ijk	70.3	29.7	9.9 efgh	1.8 abcde
XM-5	23.8 gh	59.7 def	2.6 bc	170.1 ghi	61.2	38.8	9.9 defg	1.9 abcde
CM157	23.0 gh	64.9 efg	2.8 cde	157.3 def	61.6	38.4	10.0 fghij	2.1 abcdef
LM27	23.0 efg	62.0 def	2.8 bcde	178.4 ghijk	64.1	35.9	10.5 fghij	1.8 abcde
PRT79	22.5 gh	62.8 fg	2.8 de	177.6 hijk	58.3	41.7	11.1 ghijk	2.1 abcde
P20993	24.8 hi	67.7 g	2.8 bcde	168.9 fg	58.0	42.0	11.1 ijk	2.1 abcdef
I82	21.8 efg	64.5 fg	3.1 efg	175.4 ghij	62.5	37.5	11.2 ghijk	2.2 def
Potosi-ISA	23.7 gh	64.3 fg	3.1 bcde	184.0 klm	55.7	44.3	11.5 jk	2.3 bcdef
LM18	23.2 gh	66.1 efg	2.9 def	190.5 lm	52.1	47.9	12.6 hijk	2.2 cdef
LM268	22.1 gh	50.5 cd	2.2 a	289.2 n	40.9	59.1	13.8 k	2.7 abcdef
Mihai	25.7 hi	101.2 h	4.0 h	370.7 n	28.8	71.1	36.7 l	23.3 abcde
Misak	28.8 i	98.0 h	3.4 g	381.5 n	45.0	54.9	37.2 l	22.9 abcde
p-value [3]	0.000	0.000	0.000	0.000	0.000	0.000	0.000	0.000
p-value [4]	0.040	0.000	0.000	0.000	0.969	0.756	0.004	0.000
p-value [5]	0.000	0.000	0.002	0.000	0.000	0.000	0.000	0.000

[1] Homogeneous groups—accessions sharing the same letter for each trait are not statistically different; [2] Full name of acronyms and description of the respective morphologic traits are given in Table 1; [3] *p*-value taking into account the accessions; [4] *p*-value taking into account the experiments; [5] *p*-value taking into account the interaction between accessions and experiments.

In Table 4 are presented the average, phenotypic and genotypic variance with their respective coefficient of variation and heritability for the 2016/17 and 2017/18 experiments. Higher values of phenotypic and genotypic variances were observed for TSWPB, TSWMS, TBL, TNS, SLPB, NSPB and WSPPB (see Table 1 for definitions). Conversely, low values were observed for ADNL, RBWLPMS, RBWLSMS, NSPMS, NPB, TNSPP, PSMS, PBL, PSPB, PWMS and SLMS. The highest phenotypic and genotypic coefficients of variation were obtained in 2017/18 for NPPB (33.64% and 29.18%) and NSPB (37.68% and 30.89%), while low values were observed in both years for DUF, RBLWSMS and RBLWPMS. In general, values of the phenotypic coefficient of variation were relatively higher than genotypic. For all characteristics, heritability ranged from 0% to 100%. Most of the characteristics exhibited high heritability, while lower and medium values were found for NSPMS (0), RBWLSMS (0.53), RBWLPMS (0.64, 0.45), NSPPB (0.42, 0.39), TNSPP (0.52), DUF (0.49), SWMS (0.57), PSPB and PBL (0.63). These characteristics exhibiting low and medium values of heritability were excluded for the multivariate analysis.

Table 4. Genetic parameters estimated for 38 quantitative traits among 23 *Lupinus mutabilis* accessions.

Traits	Average		Phenotypic Variance		Genotypic Variance		PCV [1]		GCV [2]		H [2,3]	
	2016/17	2017/18	2016/17	2017/18	2016/17	2017/18	2016/17	2017/18	2016/17	2017/18	2016/17	2017/18
TSWMS	184.15	10.94	1368.99	1779.1	1368.82	1599.4	20.09	21.91	20.09	20.77	1	0.9
TSWPB	156	3.04	1423.64	1562.07	1423.39	1466.89	24.19	24.64	24.18	23.88	1	0.94
TTSW	158.29	177.98	1145.83	1561.49	1113.07	1339.14	21.39	22.20	21.08	20.56	0.97	0.86
DUF	87.67	3.04	17.21	23.25	16.59	10.85	4.73	4.52	4.65	3.09	0.96	0.47
WSPPB	52.29	23.32	148.79	382.43	140.77	289.57	23.33	30.14	22.69	26.22	0.95	0.76
HUFF	73.1	31.24	83.85	26.45	78.7	23.64	12.53	12.29	12.14	11.61	0.94	0.89
NLMS	15.21	23.32	2.14	1.36	2	1.22	9.62	7.09	9.29	6.73	0.93	0.9
SLMS	0.95	51.2	0	0.6	0	0.57	6.72	8.19	6.48	7.97	0.93	0.95
PWMS	1.57	7.57	0.01	0.01	0.01	0.01	7.12	8	6.75	7.83	0.9	0.96
SWBLR	4.31	192.52	1.31	0	1.17	0	26.54	20.07	25.09	18.33	0.89	0.83
TBL	220.4	352.93	1183.56	421	1044.3	344.87	15.61	14.55	14.66	13.17	0.88	0.82
ALPB	46.86	10.94	17.46	27.53	15.11	22.04	8.92	14.9	8.29	13.33	0.87	0.8
TW	9.49	31.24	7.7	6391.49	6.73	6049.35	29.24	22.65	27.33	22.04	0.87	0.95
ADNL	4.84	10.48	0.16	0	0.14	0	8.35	9.57	7.72	8.82	0.85	0.85
PSMS	0.66	99	0.01	0.02	0.01	0.01	16.93	19.84	15.5	17.91	0.84	0.81
PSPB	0.34	0.07	0.01	0.01	0.01	0.01	32.29	26.56	29.56	21.14	0.84	0.63
SLPB	147.29	2.64	796.89	291.04	671.56	218.18	19.17	17.23	17.59	14.92	0.84	0.75
NSMS	31.29	0.66	48.96	38.56	40.63	25.37	22.36	19.88	20.37	16.12	0.83	0.66
WSMS	5.69	5.54	1.82	2.76	1.52	2.3	23.74	29.96	21.66	27.39	0.83	0.84
WSPB	4.19	3.6	2.68	2.72	2.19	2.19	39.08	45.87	35.36	41.15	0.82	0.8
ADBPB	25.11	3.56	12.34	14.96	9.59	10.17	13.99	22.22	12.33	18.32	0.78	0.68
NPMS	9.67	7.94	3.85	2.44	2.98	1.88	20.29	14.9	17.84	13.09	0.77	0.77
NPB	3.12	3.03	0.17	0.23	0.13	0.19	13.18	13.41	11.52	12.25	0.76	0.84
NPPPB	3.11	1.39	0.67	0.81	0.51	0.52	26.38	29.55	22.9	23.77	0.75	0.65
PBL	0.21	1.27	0	0	0	0	11.77	17.57	10.18	13.96	0.75	0.63
TNS	62.05	3.56	191.54	140.03	142.42	104.68	22.3	23.11	19.23	19.98	0.74	0.75
NSPB	26.83	0.41	72.78	77.2	52.66	51.89	31.8	37.68	27.05	30.89	0.72	0.67
NPPB	10.09	0.18	8.17	13.53	5.77	10.18	28.32	33.64	23.8	29.18	0.71	0.75
TNP	21.62	64.89	22.19	20.2	15.86	15.97	21.79	22.64	18.42	20.13	0.71	0.79
PLMS	6.96	9.46	0.25	7.88	0.17	7.87	7.17	29.04	5.93	29.65	0.68	1
RBLWSMS	1.28	19.85	0	0	0	0	2.66	4.81	2.13	0	0.64	0.45
SWMS	0.75	160.41	0	0.38	0	0.36	7.29	8.1	5.48	7.88	0.57	0.95
RBLWPMS	4.46	141.02	0.06	0.09	0.03	0.08	5.6	6.61	4.06	6.13	0.53	0.86
TNSPP	2.98	10.48	0.07	0.09	0.04	0.08	9.1	11.59	6.55	10.5	0.52	0.82
NSPPB	2.74	4.64	0.18	6.04	0.08	2.37	15.5	30.97	10.04	19.41	0.42	0.39
NSPMS	3.48	1.52	0.21	0.2	0	0.14	13.28	14.87	0.67	12.17	0	0.67

[1] Phenotypic coefficient of variation; [2] Genotypic coefficient of variation; [3] Broad sense heritability.

Correlation is an important test and is used to assess relationship and associations between variables and is frequently applied in several studies. The results generated after correlation coefficient analysis using the Spearman method for morphological traits in experiments of 2016/17 and 2017/18 are presented in Tables S2-1 and S2-2. Results show positive correlation for most traits. Total weight (TW) was positively correlated ($p \leq 0.05$) with 13 variables in 2016/17: SLPB (0.6**), TBL (0.69**), NPB (0.60**), PBL (0.47**), TNP (0.57**), NPPB (0.65**), WSMS (0.70**), TSWMS (0.59**), NSPB (0.57**), WSPB (0.91**), TWSPB (0.56**), NPPPB (0.59) and TNS (0.65**). In 2017/18 TW was positively correlated with eleven variables: TNS (0.80**), TSWMS (0.68**), NPB (0.56**), SLPB (0.61**), NPPB (0.70**), NSPB (0.72**), NPPPB (0.62**), TSWPB (0.55**), TNP (0.72**), TBL (0.60**), WSMS (0.73**) and WSPB (0.92**). Also, positive correlation was found between TW and NSPPB (0.44**), SLMS (0.76**), SWMS (0.66**), SWBLR (0.84**), NSPPB (0.55**), PSPB (0.61**) and PLMS (0.53**). Significant ($p \leq 0.05$) and positive correlation was also reported between TNS and NSPMS, SLPB, NPPB, NSPB, NPPPB, NSPPB, TNP, ADNL, PBL, WSMS, TSWMS, NPB, SWPB, TW, SWBLR, TBL, PSPB and WSPB in both years of experiments.

From the correlation data, heat maps were constructed (Figures 2 and 3) using euclidean distances and the UPGMA method, where in the vertical columns are the clusters of morphological traits while in the horizontal lines are the clusters of accessions. Dark red colors represent lower values while the dark blue are higher values. Figure 2 corresponds to the heat map obtained from 2016/17 data and Figure 3 was obtained with 2017/18 data. Six groups of morphological traits could be drawn in Figure 2 and five in Figure 3. In both figures, group 1 is related to the reproductive capacity of pods, defined in the Figure 2 by the characteristics NSMS, SWBLR and TW and in Figure 3 by PLMS, WSMS, TNS, TW, SWBLR, NSPPB and NPPB. In Figure 2, group 2 is composed by characteristics related to the distance between leaves and primary branches in the main stem (ADNL and ADBPB), while group 2 in Figure 3 includes vegetative and reproductive traits related to the main stem: average distance between leaves (ADNL), number of seeds per pod on the main stem (NSPMS) and ratio between length and width of pod on the main stem (RBLWPMS). Only one characteristic (PSMS) defines group 3 in both figures. In Figure 2, group 4 includes characteristics related with total number of seeds (TNS), proportion of leaves with branches (PBL) and percentage of seed on the primary branches (PSPB). In Figure 3, group 4 includes vegetative and reproductive traits related to the main stem (NSMS, NPMS, NPB and NLMS). Group 5 (Figure 2), includes traits related to the reproductive capacity on the main stem (TSWMS, TSWPB, PWMS and PLMS). The same group on Figure 3 is related with pod and seed size (thousand seeds weight, pod and seed size parameters) and traits that include primary branches (ALPB, SLPB and WSPPB). Group 6 is mostly defined by vegetative characteristics (TBL, ALPB, HUFF and DUF).

Cluster I represent three JKI accessions in both figures, which is discriminated by morphological groups 5 and 6 (four reproductive and five vegetative characteristics) in Figure 2. In Figure 3 Cluster I is discriminated by morphological groups 1 (seven reproductive characteristics), 4 (with three reproductive and two vegetative characteristics) and 5 (five reproductive and two vegetative characteristics). Cluster II (Figure 2) is composed by 6 accessions defined by groups 1 and 4 (five reproductive and one vegetative characteristics) and, in Figure 3, by 19 accessions and does not exhibit a defined pattern. In both figures, Cluster III with accession LM268 only, is characterized by high values in groups 1 and 5 (SWBLR, TW, PWMS, TSWPB and TSWMS) in Figure 2 and high values in group 5 and low values in groups 2–4 in Figure 3. Cluster IV in Figure 2 does not exhibit a defined pattern. For Figure 3, Cluster IV is represented by the *L. albus* cultivars and is characterized by high values for all heat map in most characteristics. This cluster is defined by three groups of morphological traits: group 1 (seven reproductive variables), group 2 (three reproductive and one vegetative characteristics) and group 5 (five reproductive and two vegetative characteristics).

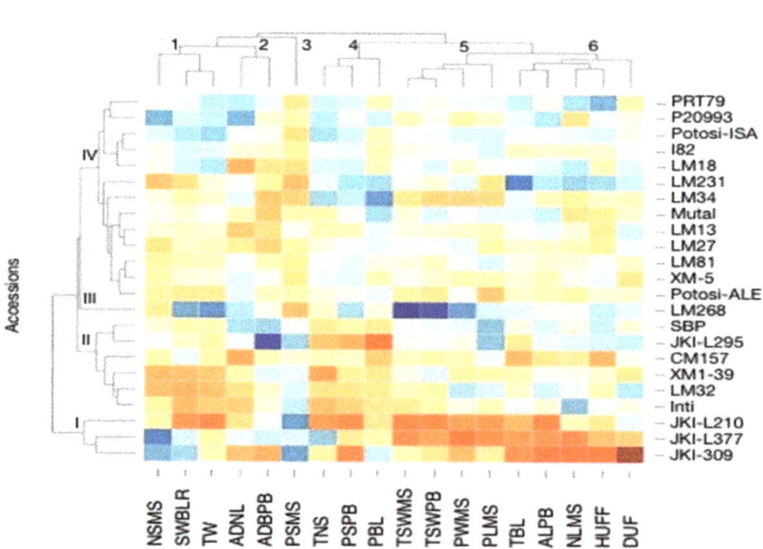

Figure 2. Heat map of the 23 accessions of *L. mutabilis* obtained from morphological characterization data for the 18 traits, where red and blue boxes indicate low values and high values respectively.

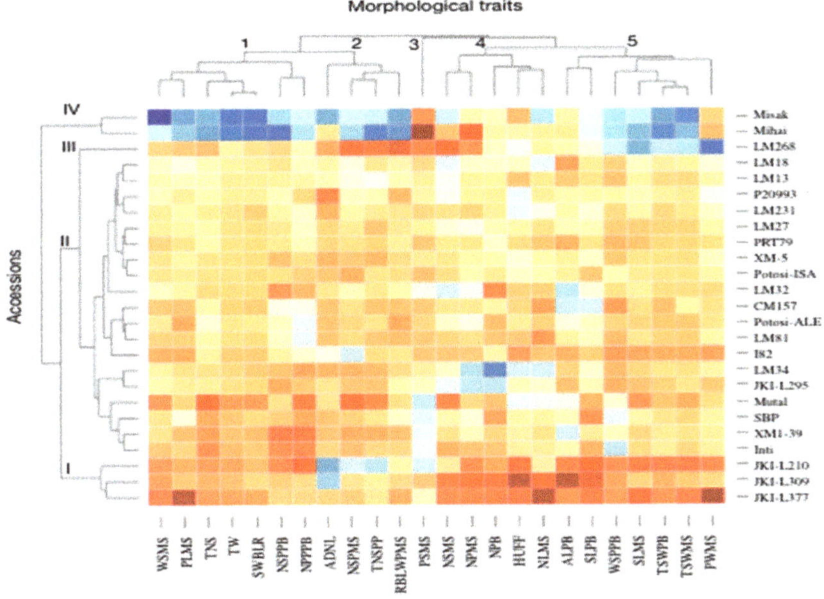

Figure 3. Heat map of the 25 accessions obtained from morphological characterization data for the 24 traits, where red and blue boxes indicate low values and high values respectively.

Principal component analysis (PCA) confirmed the cluster analysis results (Figures 2 and 3). For instance: cluster I is localized oppositely for many vectors of groups 5 and 6 (TSWMS, TSWPB, PWMS PLMS, TBL, ALPB, NLMS, HUFF, and DUF) for Figure 4 and a similar scenario can be observed in Figure 5 were the vectors defined by groups 2 and 4 (HUFF, PWMS, ALPB, NPB, NLMS, SLPB, WSPPB, SLMS, TSWMS, TSWPB, WSMS and PLMS) are in the opposite position, thus justifying the low values of these characteristics in those groups (Figures 4 and 5). In both figures the LM268 accession presents high values for many characteristics among *L. mutabilis* accessions. Cluster IV (Figure 5) is composed of two accessions that present high values for 17 vectors (characteristics marked in cluster analysis with blue color). In this cluster eight vectors (RBLWPMS, TSWMS, TSWPB, NSPPB, SWBLR, TW, WSMS and TNSPP) are highlighted by presenting the highest scores, with TW being the longest vector. The first three PCs projected in the biplot (Figure 4) show a clear separation of the four cluster and all together account for 75.1% of the total variation. The first component accounts for 40.3% of variation, with PWMS, TBL, PSPB, PSMS and ALPB accounting heavily for this variation. The second PC accounts for 21.4% of the variation, with TNS, NSMS and TW being the most important variables. The third PC explains another 13.4% of the variation, with the most important variables being PBL, SWBLR and PLMS. In Figure 5 the first three component explain 76.4% of total variation. For the first principal component, characteristics TW, SWBLR, TSWMS, TSWPB and TNS contribute more, explaining 43.9% of total variation. HUFF, ADNL, PWMS, TNSPP, and NSPMS are most important variables for second component; this component accounts for explanation 20.4% of variation, while NSMS, NPMSM LMS and PSMS account for 12% of variation in the third component.

3.2. Diversity Assessed by Molecular Markers

The six selected ISSR primers used for analysis of 23 accessions resulted in the production of 37 reproducible bands (Table 5 and Figure 6). Of those, 11 (29.7%) bands were polymorphic and the remaining 26 (70.3%) were monomorphic. The total number of bands per primer ranged between four (GT$_8$YC) and eight (HVH(TG)$_7$), while the percentage of polymorphic bands per primer ranged from 0 to 50%. The average for each primer was 6.2 bands. Polymorphism information content (PIC), which is used in genetics as a measure of polymorphism for a marker locus, ranged from 0.23 (HVH(TG)$_7$) to 0.72 (AG$_8$YT). Effective multiplex ratio (EMR) had its minimum value with AG$_8$YC (0) and maximum in GT$_8$YC (2.25). The resolving power (RP) parameter used to detect the differences between a large number of genotypes ranged from 5.48 (AG$_8$YT) to 13.58 (HVH(TG)$_7$). The minimum and maximum values for marker index were registered for AG$_8$YC (0) and GT$_8$YC (0.54) primers, respectively.

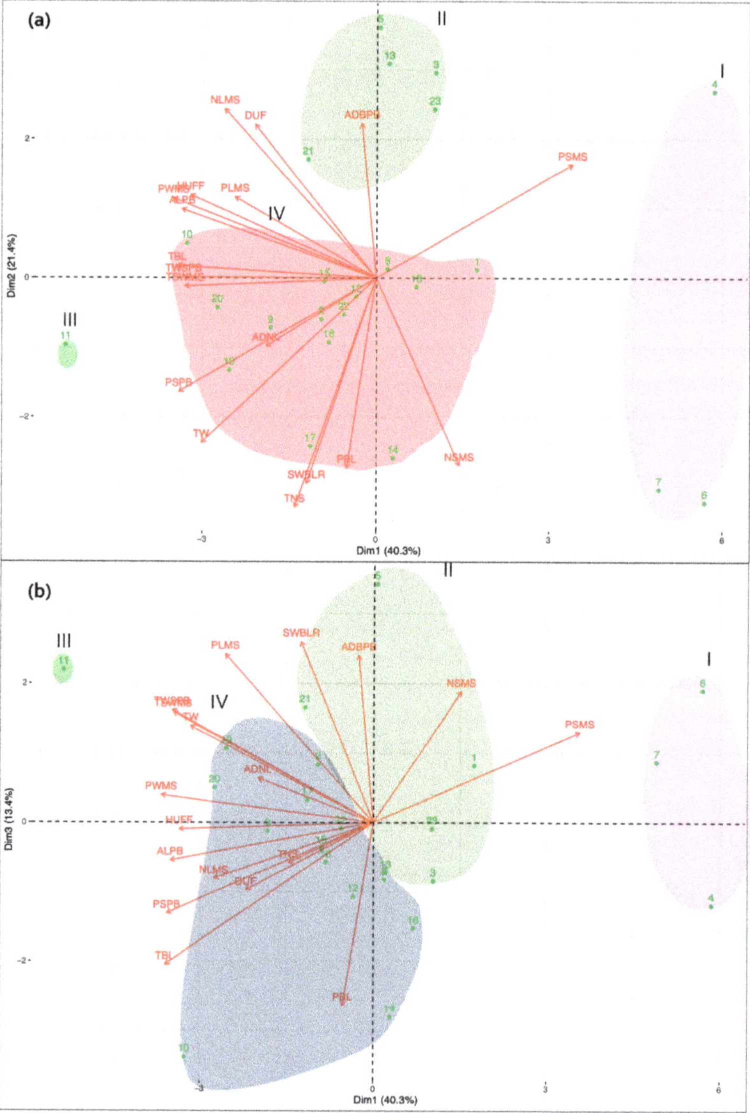

Figure 4. Representation in two dimensions (first and second dimensions—(panel **a**)—of principal component analysis explain 61.7% of the variability, while the inclusion of the third dimension—(panel **b**)—raises the three-dimensional space to explain 75.1% of the variability) of normalized original data of morphological characterization of the 23 *Lupinus mutabilis* accessions in a space defined by the vectors and own values.

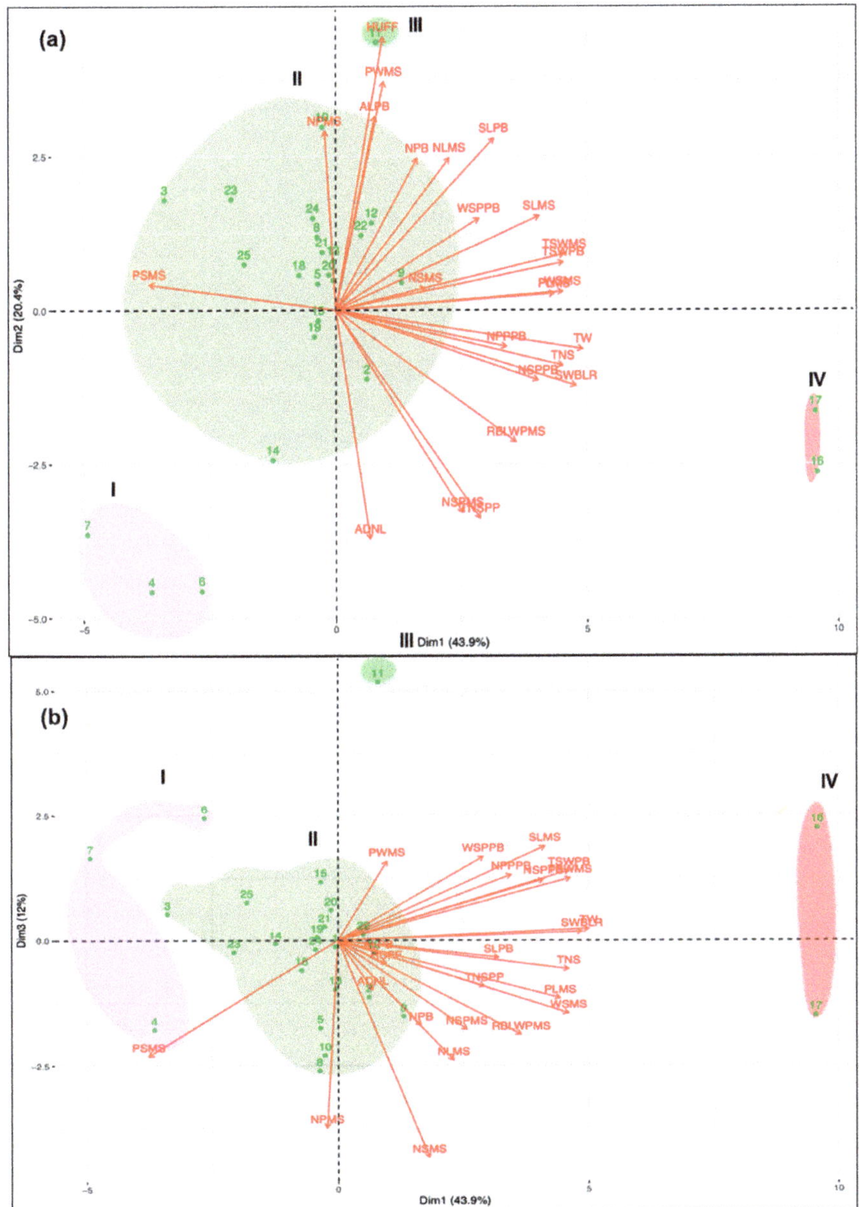

Figure 5. Representation in two dimensions (first and second dimensions—(panel **a**)—of principal component analysis explain 64.3% of the variability, while the inclusion of the third dimension—(panel **b**)—raises the three-dimensional space to explain 76.3% of the variability) of normalized original data of morphological characterization of the 23 *Lupinus mutabilis* accessions in a space defined by the vectors and own values. Numbers 1–25 encode accessions as detailed in Figure 4 (16 and 17 denote *L. albus* 'Mihai' and 'Misak' respectively).

Table 5. List of Inter-Simple Sequence Repeat (ISSR) primers used in this study, their total numbers of band per primer, polymorphic and monomorphic band and polymorphism percentage per primer.

Primer	Bands	PB	MB	PB (%)	MB (%)	PIC	EMR	RP	MI
HVH(TG)$_7$	8	4	4	50	50	0.23	2	13.58	0.46
GA$_8$YT	6	1	5	16.66	83.33	0.71	0.16	5	0.11
AG$_8$YT	6	1	5	16.66	83.33	0.72	0.16	4.58	0.12
GT$_8$YC	4	3	1	75	25	0.24	2.25	6.42	0.54
AG$_8$YC	5	0	5	0	100	0.58	0	6	0.00
AG$_8$YG	8	2	6	25	75	0.48	0.5	10.33	0.24
Total	37	11	26						
Minimum	4	0	1	0	25	0.23	0	4.58	0
Maximum	8	4	6	50	100	0.72	2.25	13.58	0.54
Mean	6.16	1.83	4.33	30.55	69.44	0.49	0.85	7.65	0.24

Notes: PB—polymorphic bands; MB—monomorphic bands; MB (%)—percentage of monomorphic bands; PB (%)—percentage of polymorphic bands; PIC—polymorphism information content; EMR—effective multiplex ratio; RP—resolving power; MI—marker index. The following primers: (CA)$_8$RY, (GA)$_8$YC, (GT)$_8$YC, (TCC)$_5$ and MR were included in the screening test but were rejected during selection. Eight of these primers were previous tested [53].

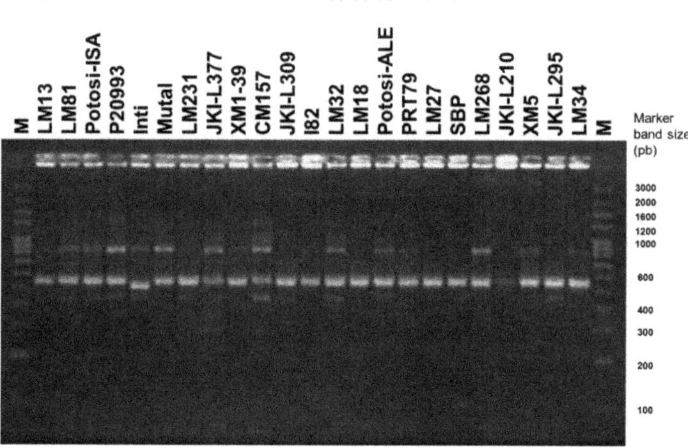

Figure 6. Example of ISSR amplification profiles for 23 *Lupinus mutabilis* accessions using the primer GA$_8$YT separated on a 2% agarose gel. M-NZYDNA Ladder VII marker. Numbers adjacent to accession names refer to coding used in Figure 5.

The similarity matrix was used to construct a dendrogram using the UPGMA method (Figure 7). The cophenetic correlation was 0.9058603, revealing little loss of information with transformation of similarity matrix to dendrogram. The dendrogram reveals five distinct groups. Cluster I is composed by 10 accessions of white, blue and pink flower colors and green and purple stem. Cluster II, containing 9 accessions, can be distinguished from the first group by the absence of purple stem and flower genotypes. Unlike cluster I and II, clusters III, IV and V are composed only by accessions that exhibit green stems and blue flowers. Cluster IV is represented by one accession and cluster III and V by two accessions each.

3.3. Diversity Assessed by Genomic Traits

Lupinus albus 'Misak' was validated as a DNA standard by comparison to *Solanum lycopersicum* 'Stupické' (Figure 8a,b) and *Raphanus sativus* 'Saxa' (data not shown) and estimated at 2C = 1.35 ± 0.0076 pg (1377.6 Mbp), with an average coefficient of variation of 3.47%.

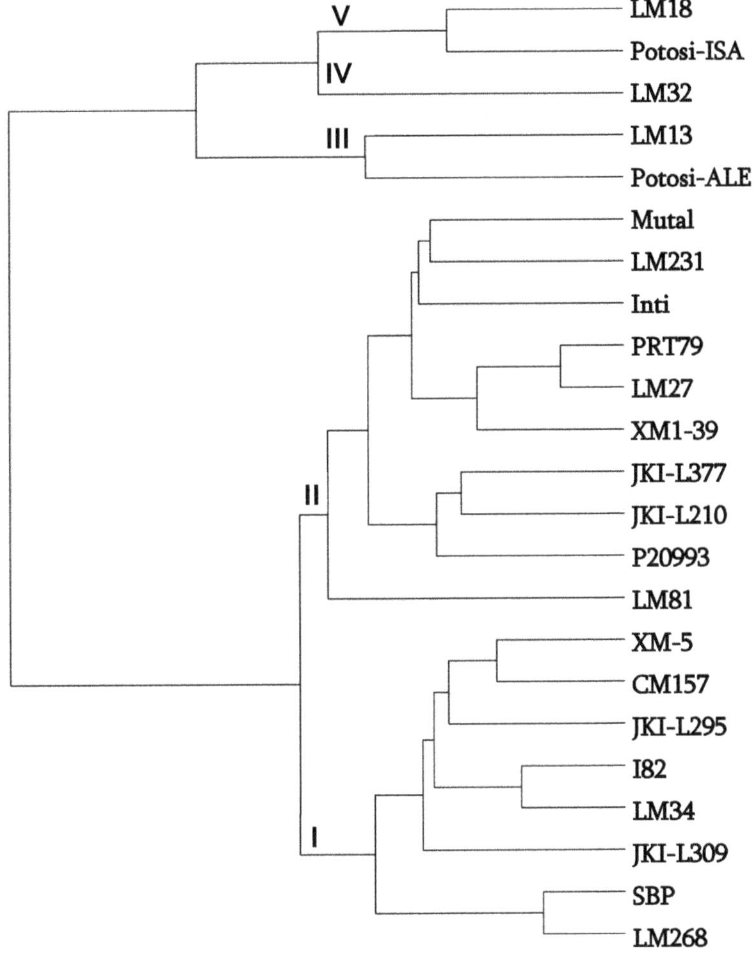

Figure 7. Dendrogram obtained by the unweighted pair group method of arithmetic mean (UPGMA) method from the coefficients of similarity (DICE) between the accessions of *Lupinus mutabilis* from six ISSR markers. $r = 0.9058603$.

The *L. mutabilis* collection was thus analysed by comparison to *L. albus* 'Misak', as exemplified in Figure 8 (panels c,d). The average *L. mutabilis* genome size was estimated at 2C = 2.05 pg (2001.2 Mbp) with a 9.2% coefficient of variation, ranging from 1897.3 Mbp for accession SBP to 2083.2 Mbp for accession LM34 (Table 6). The results from a Kruskal–Wallis test performed for genome size reveal significant difference between accessions ($\chi^2 = 94.845$, $Df = 23$, p value = 0.000). No single accession showed to be statistically different from all the others, rather a continuum of accessions is depicted by the homogeneous groups produced (Table 6).

Genome size is an important criterion to study evolution at the intra-specific level, helping to understand conflicting pattern between morphological traits. In this study we evaluated the associations between genome size and morphological traits using Spearman correlation analysis for all 23 accessions for the two experiments. However, no single morphological trait presented strong correlation with genome size (Figure 9).

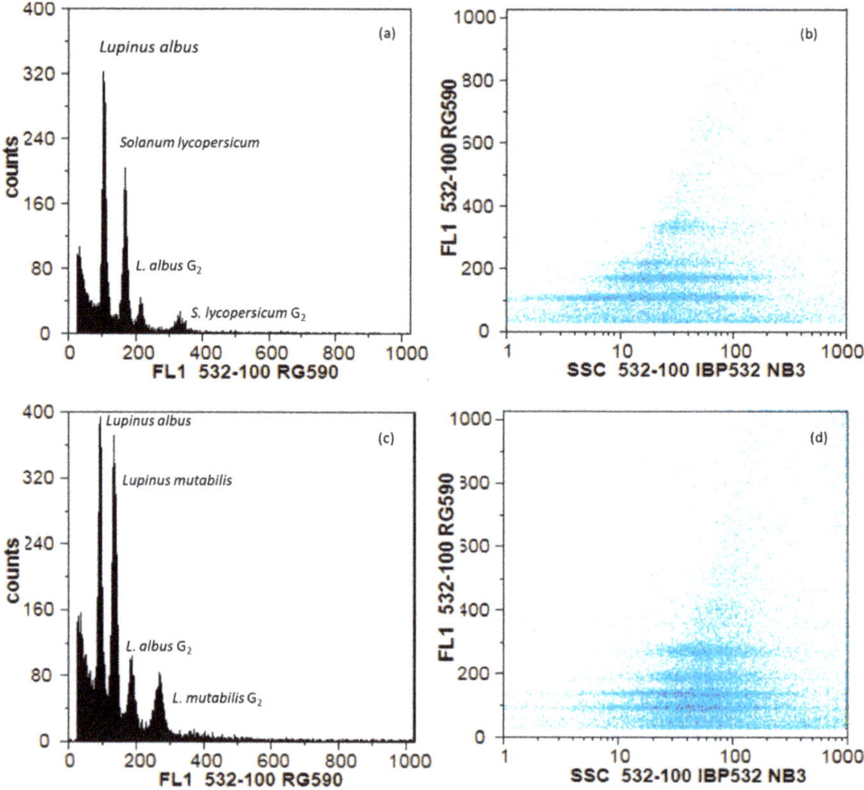

Figure 8. Flow cytometric analysis of relative fluorescence intensities (FL1) of propidium iodide-stained nuclei simultaneously isolated from: (**a**) and (**b**) *Solanum lycopersicum* 'Stupické' and *Lupinus albus*, for the validation of *L. albus* 'Misak' as DNA standard (2C = 1.35 pg); (**c**) and (**d**) *L. albus* 'Misak' and *L. mutabilis* accession LM231. (**a**) and (**c**) Histogram showing relative fluorescence intensities. (**b**) and (**d**) dot plots on side scatter (SSC) versus FL1.

Table 6. Genome size of the *Lupinus mutabilis* accessions estimated by flow cytometry.

Accession	Genome Size (Mbp)		H.G. [2]
	Average	StDev [1]	
SBP	1897.3	18.4	a
XM1-39	1907.3	17.0	a
JKI-L378	1938.0	49.0	ab
Prt-79	1957.4	16.2	ab
JKI-L377	1961.4	16.3	ab
Mutal	1967.6	121.6	abc
JKI-L295	1969.0	11.8	abc
JKI-L210	1973.6	26.9	bcd
LM13	1975.7	118.4	bcd
JKI-L309	1979.7	37.6	bcd
LM231	1984.1	100.4	cd
LM18	1986.4	54.6	cde
XM5	2009.1	20.1	cde
P-20993	2021.5	22.8	cde
Potosi-ISA	2024.3	36.1	cde
Potosi-ALE	2024.8	19.3	cde

Table 6. *Cont.*

Accession	Genome Size (Mbp)		H.G. [2]
	Average	StDev [1]	
LM27	2027.7	23.9	cde
LM32	2040.9	17.9	de
CM157	2040.9	43.8	de
I82	2041.6	10.9	de
Inti	2058.1	23.9	ef
LM268	2078.9	10.1	f
LM81	2080.2	13.1	f
LM34	2083.2	17.3	f

[1] Standard deviation; [2] Homogeneous groups—accessions sharing the same letter are not statistically different.

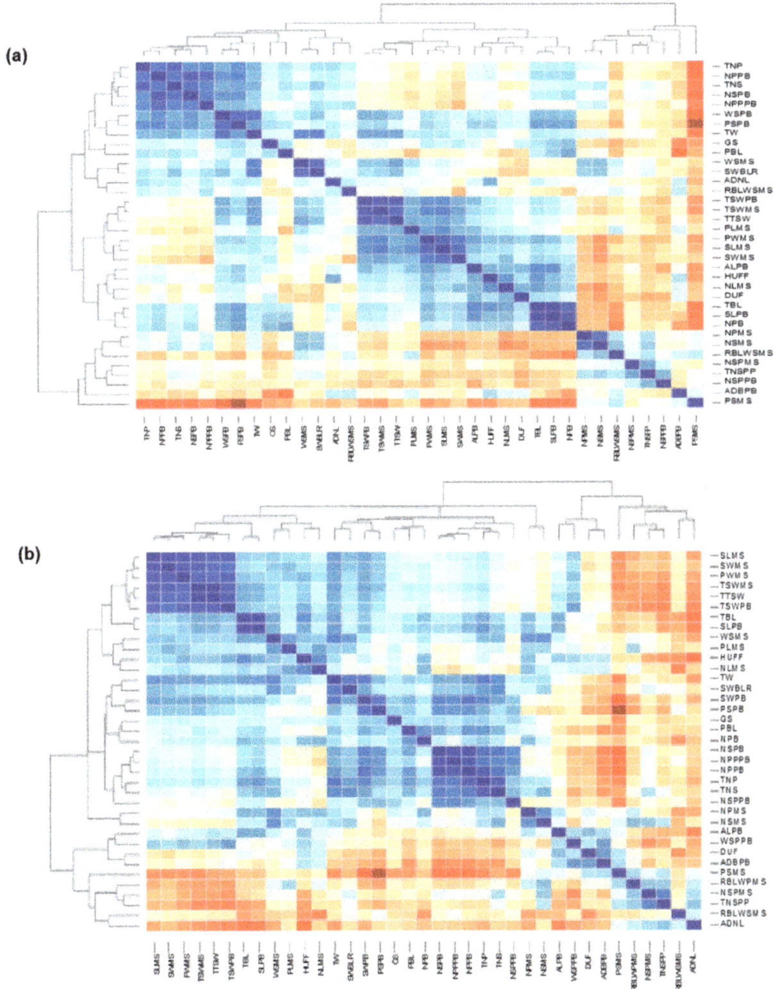

Figure 9. Heat map of the 23 *Lupinus mutabilis* accessions obtained from Spearman correlation among 37 morphological traits and genome size (GS) data for 2016/17 (panel **a**) and 2017/18 data (panel **b**). Dark blue boxes indicate high values and dark red depict low values.

4. Discussion

To assess the diversity in a tarwi germplasm collection under Mediterranean conditions, phenotypic, genetic and genomic analyses were combined, studying morphological traits, ISSR markers and genome size. In general, the morphological traits used to evaluate accessions tested in the trials showed acceptable adaptability to the Mediterranean environment assuming that productivity projected is above 1.5 t/ha, achieved under rain-fed conditions unevenly distributed during the trial periods. Similar yields were previously reported in France and Spain [54]. However, the yields obtained suggest continuing breeding to achieve higher yields. The results showed also significant differences among accessions. While additional years of field trial results would certainly improve the robustness of conclusions, the results obtained are based on traits that presented coherent values between trials.

The knowledge of the correlations between different characteristics is fundamental because it allows the accomplishment of the indirect selection of the complex characteristics that are inherited quantitatively and influenced by genetic effects [43]. In this work we report positive and significant correlation between many variables. Characteristics such as total seed weight (TW) and total number of seeds (TNS) are very important and are directly related to characteristics of reproductive development. Accession LM268 presented higher values for TW and LM34, P20993 and LM18 for TNS. Therefore, these two features can work as criteria of selection in our collection for the breeding programme or to choose the most adapted. Talhinhas et al. [43] verified positive correlation between total weight with plant height, pod width, number of primary branches, proportion of seeds on the primary branches, total number of seeds and number of pods per primary branch in *L. angustifolius*. Georgieva and Kosev [55] also found positive correlation between thousand seed weight and plant height. Clements et al. [56] reported positive correlation between weight and plant height of 1000 seeds in *L. pilosus*. Heritability is a parameter widely used by breeders to genotype selection based on phenotypic expression [57]. Morphological traits exhibiting high values of heritability are chosen for the selection based on this parameter [43,58]. High heritability values enable the identification of important features to be selected for genetic breeding. Concerning our study, tarwi accessions can be selected based on the following traits: TW, SLMS, NPMS, NLMS, TNS, HUFF, PSMS, TSWMS, TSWPB, WSPPB, PWMS, SWBLR, TBL, ALPB, ADNL, SLPB, WSMS, WSPB, NPB, and NPPB. Similar results were noticed by Talhinhas [53] for SLMS, NLMS, HUFF, PSMS, TNS, and TW in *L. albus*, *L. angustifolius* and *L. luteus*. Our results also corroborate those by Georgieva and Kosev [55], who found high values of heritability for pod length and total number of seeds in *L. albus* and *L. luteus*.

In the present study several accessions stood out due to their superior performance in various traits. Along with accession LM268, accessions LM18, LM27, P20993, Potosi-ISA, PRT79 and I82 were the most efficient in converting vegetative growth to seed production, although lagging behind the performance of *L. albus*. Accession LM268 was the only tarwi accession to produce more yield on the primary branches than on the main stem, following the pattern of *L. albus* cultivars.

An important result worth highlighting is that most of the *L. mutabilis* accessions studied concentrate their production on the main stem. This characteristic is very important because it allows adaptability of tarwi to poor growth (soil and/or climatic) conditions. This characteristic may also prove useful to avoid the indeterminate growth habit of tarwi, particularly problematic in areas without summer drought. Breeding programmes should be directed for improving levels of production on the main stem and primary branches for good soil/climate conditions but with summer drought and focus on more determinate growth plants (those concentrating production on the main stem) both for marginal areas and for areas without summer drought. To the latter, accessions such as JKI-L295 and JKI-L210 stand out, as they produced over 80% of their yield on the main stem while attaining relatively high yields (ca. 10 g per plant).

The use of molecular markers in genetic diversity studies at the intra and inter specific levels proved useful in a wide range of species [59,60]. In this study we assessed the efficiency of ISSR markers for the characterisation of genetic diversity of *L. mutabilis* accessions. This technique is important because it allows to make a broad screening of a collection. SSR markers are not optimised yet for

L. mutabilis and the transfer of such markers from other Fabaceae to tarwi did not prove successful [61], leaving ISSRs as a valid tool for preliminary screening of germplasm collections. All six primers used in this investigation revealed a polymorphism of 30.55% for all 23 accessions. Bussell et al. [62] establish 20% as minimum of monomorphic band percentage for genetic diversity study and our study reveals 69.44% monomorphic bands. Similar results were reported by Chirinos-Arias et al. [37] assessing genetic variability among 30 accessions of *L. mutabilis* using eight ISSR markers, finding a total polymorphism of 58.82%. The high level of polymorphism obtained in our study is in accordance with those authors. The parameters PIC, EMR, MI and RP were used to evaluate the efficiency of ISSR primers. However, to the best of our knowledge, there are no studies on *L. mutabilis* assessing the effective multiplex ratio, polymorphic information content, marker index and resolving power. Results show high probability in detecting polymorphism PIC (0.72), for the primer $HVH(TG_7)$. The AG_8YG primer stood out as presenting a high RP value (13.58), being more qualified to distinguish accessions. The highest value of EMR (2.25) was obtained with primer AG_8YC, revealing this to be most efficient. The primer AG_8YC proved to be the most useful because it presented the highest value of MI (0.54). Several studies have been undertaken based on these techniques for selecting efficient ISSR primers in different species [63–67].

The 23 tarwi accessions were divided in five main genetic groups using cluster analysis by the UPGMA method (Figure 5). However, morphological characteristics such as stem and flower colour did not exhibit regular relationships in different clusters. The existence of several distinct groups that aggregate different stem and flower colours probably reflects few differences on the genetic constitution of the accessions. On the other hand, the distinct groups can reflect into distinct morphological characteristics and variations. Talhinhas [53] suggested that low intra-specific diversity in tarwi can be related to the fact that all the accessions originated from a limited number of landraces, reflecting the recent domestication genetic bottleneck effect that is estimated to have occurred no later than 2600 years before the present time in *L. mutabilis* [16]. Similar result was found by Chen et al. [68] in the research done on the 105 genotypes on *Vigna unguiculata*. In this work we verify that the genetic variability is not correlated with phenotypic variability, indicating the need for incorporation of more molecular markers. Similar results have been reported in other species. Previous studies performed by Galek et al. [14] also did not find a relation between genetic and morphological variability in accessions of *L. mutabilis*. In a study aiming to evaluate genetic diversity of *Nelumbo* using analyses of Randomly Amplified Polymorphic DNA (RAPD) and ISSR markers, Li et al. [69] found low correlation between molecular and morphological data. Talhinhas et al. [44] assessing genetic diversity in *Lupinus luteus* using ISSR and Amplified Fragment Length Polymorphism (AFLP) markers did not find any correlation between morphological and molecular data.

In this work we report the existence of significant differences in the intraspecific genome size (GS) variability in 23 accessions of *Lupinus mutabilis*. Our results reveal that the GS ranged from 1.94 pg/2C to 2.13 pg/2C. Naganowska et al. [70], also employing flow cytometry to analyse propidium iodide-stained nuclei, evaluated the nuclear DNA content variation in the genus *Lupinus* and found 1.90 pg/2C for *Lupinus mutabilis*, although a single accession was used in that study. To the best of our knowledge, our study is the first *L. mutabilis* genome size intra-specific analysis, depicting an overall average size of 2.05 pg (2001.2 Mbp). Several studies have reported intraspecific differences in genome size in various species such as *Glycine max*, *Linum austriacum* and *Zea mays* [33,71,72]. The intraspecific variation in genome size can result from repetitive/non-coding regions, hence increasing or decreasing in satellite DNA transposable elements and ribosomal genes [73]. There are studies pointing that transposable elements are largely responsible for notable differences in genome sizes. For instance, in maize, transposable elements are responsible for 85% of differences [74]. According to Petrov [75] these elements have potential of multiplicity of 0.1–1 Mbp in a single generation. The satellite DNA can also contribute greatly to genome size differences [76]. Meanwhile, Garrido-Ramos [74] refer that genomic content variation in plants which are affected by satellite DNA can range from 0.1% to 36%. Small variation of 3.5% in nuclear DNA have been associated with ribosomal genes [77]. The maximum variation of nuclear DNA content obtained in the present research was 9.2%, a value much higher than the 2% maximum genome size variability reported for soybean [71] but smaller than the 36% variation reported for maize [72]. In light of this discussion, one may discard the possibility that differences in *L. mutabilis* genome size are caused by the transposable elements. Only a detailed study could unravel whether this variation is due to repeated sequence differences in satellite DNA or ribosomal genes.

Data on 37 morphological traits and genome size measurement were plotted and no correlation was observed. This is not a surprise, as similar results were also reported from other studies. For instance, Oney and Tabur [31] did not find correlation between genome size and morphological traits on the *Brachypodium distachyon* collected in different locations in Turkey. Realini et al. [72] observed weak association between genome size and morphological traits in maize. Recently Basak et al. [78] assessing the variation of morphological traits with the genome size in turnip found no correlation. This lack of association between morphological traits and genome size suggests that other factors are determinant on the control of such characteristics, reinforcing the view that genome size variations are mainly related to non-coding regions [79].

5. Conclusions

The agronomic performance of *L. mutabilis* in Portuguese conditions was good, assuming that the assay was conducted under rain-fed conditions. Our results highlight the accession LM268 with larger seeds and a total thousand seeds weight similar to *L. albus*, while also achieving the highest yield and being the only tarwi accession producing more on the primary branches than on the main stem. While high yields in lupins depend on the capacity of the plants to produce large amounts of pods and seeds on lateral branches, the indeterminate growth habit of tarwi can be undesirable, either in areas without summer drought or, on the contrary, in areas with limited growing periods where further vegetative growth may impair pod filling. To this end, JKI-L295 accession present high yield concentrated on the main stem, suggesting a semi-determinate development pattern. In either case, this accession is a key point for continued breeding. In fact, the present study has shown that tarwi is still behind white lupin in terms of its adaptability to Mediterranean conditions, namely concerning yield. The genetic diversity revealed in this study, however, prompts further breeding opportunities. Molecular marker and genome size analyses have revealed important levels of genetic/genomic diversity, which could not be related to phenotypic/morphologic diversity. This illustrates a scenario of recent domestication in the absence of a gene flow to wild relatives suggesting, however, that further exploitation of genetic diversity in this tarwi collection is possible and may provide additional sources of useful agronomic traits.

Supplementary Materials: The following are available online at http://www.mdpi.com/2073-4395/10/1/21/s1: Figure S1: Meteorological data collected at Tapada da Ajuda (Lisbon) during 2016/17 and 2017/18: (a) monthly average of daily average temperatures compared to the 30-year climatological normal values for Lisbon; (b) monthly rainfall compared to the 30-year climatological normal values for Lisbon; (c) soil water balance (including daily rainfall values), Table S1: Analysis of morphological traits by Kruskal Wallis test 2016/17 and 2017/18, Tables S2 and S3: Correlation matrix between morphological traits calculated for the 23 *Lupinus mutabilis* accessions under study (values greater than 85% are highlighted) for the experiment carried out in 2016/17 and 2017/18, respectively.

Author Contributions: Conceptualization, N.G. and J.N.-M.; Formal analysis, N.G., S.A. and P.T.; Funding acquisition, J.N.-M.; Investigation, N.G., S.A. and P.T.; Methodology, N.G., P.T. and J.N.-M.; Project administration, J.N.-M.; Supervision, P.T. and J.N.-M.; Writing—original draft, N.G.; Writing—review and editing, P.T. All authors have read and agreed to the published version of the manuscript.

Funding: This research was funded by the European Union (H2020/720726, LIBBIO project) and by Fundação para a Ciência e a Tecnologia, Portugal (UID/AGR/04129/2013, LEAF).

Acknowledgments: The authors acknowledge Julius Kühn-Institut for providing biological material, Instituto Português do Mar e da Atmosfera (IPMA) for meteorological data from the Tapada da Ajuda (Lisboa) weather station and Lusosem for the support provided in field trial conduction.

Conflicts of Interest: The authors declare no conflict of interest. The funders had no role in the design of the study; in the collection, analyses, or interpretation of data; in the writing of the manuscript; or in the decision to publish the results.

References

1. Eastwood, R.J.; Drummond, C.S.; Schifino-Wittmann, M.T.; Hughes, C.E. Diversity and evolutionary history of lupins insights from new phylogenies. In *Lupins for Health and Wealth. Proceedings of the 12th International Lupin Conference, Fremantle, Australia, 14–18 September 2008*; Palta, J.A., Berger, J.B., Eds.; International Lupin Association: Canterbury, New Zealand, 2008; pp. 346–354.
2. Gladstones, J.S. *Lupins of the Mediterranean Region and Africa*; Western Australian Department of Agriculture and Food: Kensington, Australia, 1974; Volume 26, pp. 1–48.
3. Gladstones, J.S. Present situation and potential of Mediterranean/African Lupinus for crop production. In Proceedings of the Third International Lupin Conference, La Rochelle, France, 4–8 June 1984; pp. 18–37.
4. Bellido, L.L.; Garcia, M.F. *El Altramuz*; Ministerio de Agricultura y Alimentación: Córdoba, Spain, 1991.
5. Caligari, P.D.S.; Römer, P.; Rahim, M.A.; Huyghe, C.; Neves-Martins, J.; Sawicka-Sienkiewicz, E.J. *The Potential of Lupinus mutabilis as a crop In Linking Research and Marketing Opportunities for Pulses in the 21st Century*; Kluwer Academic Publishers: Dordrecht, The Netherlands, 2000; Volume 34, pp. 569–574.
6. Jacobsen, S.E.; Mujica, A. El Tarwi (*Lupinus mutabilis* Sweet) y sus parientes silvestres. *Bot. Econ. Andes Cent.* **2006**, *28*, 458–482.
7. Wink, M. Quinolizidine alkaloids. In *Methods in Plant Biochemistry*, 8th ed.; Waterman, P., Ed.; Academic Press: London, UK, 1993; pp. 197–239.
8. Wink, M.; Meißner, C.; Witte, L. Patterns of quinolizidine alkaloids in 56 species of the genus Lupinus. *Phytochemistry* **1995**, *38*, 139–153. [CrossRef]
9. Carvajal-Larenas, F.E.; Linnemann, A.R.; Nout, M.J.R.; Koziol, M.; van Boekel, M.A.J.S. Lupinus mutabilis: Composition, uses, toxicology, and debittering. *Rev. Food Sci. Nutr.* **2016**, *56*, 1454–1487. [CrossRef] [PubMed]
10. Borek, S.; Ratajczak, W.; Ratajczak, L. Regulation of storage lipid metabolism in developing and germinating lupin (*Lupinus* spp.) seeds. *Acta Physiol. Plant.* **2015**, *37*, 119. [CrossRef]
11. Villacrés, E.; Peralta, E.; Álvarez, M. *Recetario Chochos en Su Punto*; INIAP Santa Catalina: Quito, Ecuador, 2003; p. 43.
12. Martins, J.M.N.; Talhinhas, P.; De Sousa, R.B. Yield and seed chemical composition of Lupinus mutabilis in Portugal. *Rev. Ciênc. Agrar.* **2016**, *39*, 518–525. [CrossRef]
13. Santos, C.N.; Ferreira, R.B.; Teixeira, A.R. Seed Proteins ofLupinus mutabilis. *J. Agric. Food Chem.* **1997**, *45*, 3821–3825. [CrossRef]
14. Galek, R.; Sawicka-Sienkiewicz, E.; Zalewski, D.; Stawiński, S.; Spychała, K. Searching for low alkaloid forms in the Andean lupin (*Lupinus mutabilis*) collection. *Czech J. Genet. Plant Breed.* **2017**, *53*, 55–62. [CrossRef]
15. Martínez Flores, L.A.; Ruivemkamp, G.; Jongerden, J. Plant breeding and social rationality: The unintended effects of the release of a lupino seed (*Lupinus mutabilis* Sweet) in Ecuador. *Antípod. Rev. Antropol. Arqueol.* **2016**, *26*, 71–91.
16. Atchison, G.W.; Nevado, B.; Eastwood, R.J.; Contreras-Ortiz, N.; Reynel, C.; Madriñán, S.; Filatov, D.A.; Hughes, C.E. Lost crops of the Incas: Origins of domestication of the Andean pulse crop tarwi, Lupinus mutabilis. *Am. J. Bot.* **2016**, *103*, 1592–1606. [CrossRef]
17. Galek, R.A.; Kozak, B.; Biela, A.; Zalewski, D.; Sawicka- Sienkiewicz, E.; Spychała, K.; Stawiń;ski, S. Seed coat thickness differentiation and genetic polymorphism for Lupinus mutabilis Sweet breeding. *Turk. J. Field Crops* **2016**, *21*, 305–312.
18. Gross, R. *El Cultivo y la Utiliziación del Lupinus mutabilis Sweet*; FAO: Rome, Italy, 1982.
19. Xiao, L.; Ge, X.; Gong, X.; Hao, G.; Zheng, S. ISSR Variation in the Endemic and Endangered Plant *Cycas guizhouensis* (Cycadaceae). *Ann. Bot.* **2004**, *94*, 133–138. [CrossRef] [PubMed]
20. Almajali, D.; Abdel-Ghani, A.H.; Migdadi, H. Evaluation of genetic diversity among Jordanian fig germplasm accessions by morphological traits and ISSR markers. *Sci. Hortic.* **2012**, *147*, 8–19. [CrossRef]
21. Ahmed, B.A.; Ghada, B.; Laila, E.; Hafid, A.; Bouchaib, K.; Amel, S.-H. Use of morphological traits and microsatellite markers to characterize the Tunisian cultivated and wild figs (*Ficus carica* L.). *Biochem. Syst. Ecol.* **2015**, *59*, 209–219. [CrossRef]
22. Ferreira, J.; Garcia-Gonzalez, C.; Tous, J.; Rovira, M. Genetic diversity revealed by morphological traits and ISSR markers in hazelnut germplasm from northern Spain. *Plant Breed.* **2010**, *129*, 435–441. [CrossRef]

23. Yuan, C.-Y.; Wang, P.; Chen, P.-P.; Xiao, W.-J.; Zhang, C.; Hu, S.; Zhou, P.; Chang, H.-P.; He, Z.; Hu, R.; et al. Genetic diversity revealed by morphological traits and ISSR markers in 48 Okras (*Abelmoschus escullentus* L.). *Physiol. Mol. Boil. Plants* **2015**, *21*, 359–364. [CrossRef]

24. Elameen, A.; Larsen, A.; Klemsdal, S.S.; Fjellheim, S.; Sundheim, L.; Msolla, S.; Masumba, E.; Rognli, O.A. Phenotypic diversity of plant morphological and root descriptor traits within a sweet potato, *Ipomoea batatas* (L.) Lam., germplasm collection from Tanzania. *Genet. Res. Crop Evol.* **2011**, *58*, 397–407. [CrossRef]

25. Pagnotta, M.A.; Mondini, L.; Codianni, P.; Fares, C. Agronomical, quality, and molecular characterization of twenty Italian emmer wheat (*Triticum dicoccon*) accessions. *Genet. Res. Crop Evol.* **2009**, *56*, 299–310. [CrossRef]

26. Miller, J.C.; Tanksley, S.D. RFLP analysis of phylogenetic relationships and genetic variation in the genus *Lycopersicon*. *Theor. Appl. Genet.* **1990**, *80*, 437–448. [CrossRef]

27. Heidari, E.F.; Rahimmalek, M.; Mohammadi, S.; Ehtemam, M.H. Genetic structure and diversity of ajowan (*Trachyspermum ammi*) populations based on molecular, morphological markers, and volatile oil content. *Ind. Crops Prod.* **2016**, *92*, 186–196. [CrossRef]

28. Kalpana, D.; Choi, S.H.; Choi, T.K.; Senthil, K.; Lee, Y.S. Assessment of genetic diversity among varieties of mulberry using RAPD and ISSR fingerprinting. *Sci. Hortic.* **2012**, *134*, 79–87. [CrossRef]

29. Vanijajiva, O. Genetic variability among durian (*Durio zibethinus* Murr.) cultivars in the Nonthaburi province, Thailand detected by RAPD analysis. *J. Agric. Technol.* **2011**, *7*, 1105–1114.

30. Zietkiewicz, E.; Rafalski, A.; Labuda, D. Genome Fingerprinting by Simple Sequence Repeat (SSR)-Anchored Polymerase Chain Reaction Amplification. *Genomics* **1994**, *20*, 176–183. [CrossRef]

31. Oney, S.; Tabur, S. Genome size and morphological variation in *Brachypodium distachyon* along altitudinal levels. *Pak. J. Bot.* **2018**, *50*, 1923–1933.

32. Petrova, G.; Dzhambazova, T.; Moyankova, D.; Georgieva, D.; Michova, A.; Djilianov, D.; Möller, M. Morphological variation, genetic diversity and genome size of critically endangered *Haberlea* (Gesneriaceae) populations in Bulgaria do not support the recognition of two different species. *Plant Syst. Evol.* **2014**, *300*, 29–41. [CrossRef]

33. Sheidai, M.; Afshar, F.; Keshavarzi, M.; Talebi, S.-M.; Noormohammadi, Z.; Shafaf, T. Genetic diversity and genome size variability in *Linum austriacum* (Lineaceae) populations. *Biochem. Syst. Ecol.* **2014**, *57*, 20–26. [CrossRef]

34. Yan, J.; Zhang, J.; Sun, K.; Chang, D.; Bai, S.; Shen, Y.; Huang, L.; Zhang, J.; Zhang, Y.; Dong, Y. Ploidy Level and DNA Content of *Erianthus arundinaceus* as Determined by Flow Cytometry and the Association with Biological Characteristics. *PLoS ONE* **2016**, *11*, e0151948. [CrossRef]

35. Wu, Y.Q.; Taliaferro, C.M.; Bai, G.H.; Martin, D.L.; Anderson, J.A.; Anderson, M.P.; Edwards, R.M. Genetic Analyses of Chinese *Accessions* by Flow Cytometry and AFLP Markers. *Crops Sci.* **2006**, *46*, 917. [CrossRef]

36. Obermayer, R.; Świecicki, W.K.; Greilhuber, J. Flow Cytometric Determination of Genome Size in some Old World *Lupinus* Species (Fabaceae). *Plant Biol.* **1999**, *1*, 403–407. [CrossRef]

37. Chirinos-Arias, M.C.; Jiménez, J.E.; Vilca-Machaca, L.S. Analysis of Genetic Variability among thirty accessions of Andean Lupin (*Lupinus mutabilis* Sweet) using ISSR molecular markers. *Sci. Agropecu.* **2015**, *6*, 17–30. [CrossRef]

38. Galek, R.; Kozak, B.; Sawicka-Sienkiewicz, E.; Zalewski, D.; Nowosad, K. Searching for the most useful genotypes of *Lupinus mutabilis* sweet for breeding purpose. *Electron. J. Pol. Agric. Univ.* **2017**, *20*. [CrossRef]

39. Allen, R.G.; Pereira, L.S.; Raes, D.; Smithm, M. *Crop Evapotranspiration—Guidelines for Computing Crop Water Requirements—FAO Irrigation and Drainage Paper 56*; Food and Agriculture Organization of the United Nations: Rome, Italy, 1998.

40. Talhinhas, P.; Leitão, J.; Martins, J. Collection of *Lupinus angustifolius* L. germplasm and characterisation of morphological and molecular diversity. *Genet. Res. Crop Evol.* **2006**, *53*, 563–578. [CrossRef]

41. IBPGR. *Lupin Descriptors*; IBPGR: Rome, Italy, 1981.

42. Conover, W.J.; Iman, R.L. Rank Transformations as a Bridge between Parametric and Nonparametric Statistics. *Am. Stat.* **1981**, *35*, 124.

43. Mazid, M.; Rafii, M.; Hanafi, M.M.; Rahim, H.; Shabanimofrad, M.; Latif, M.A. Agro-morphological characterization and assessment of variability, heritability, genetic advance and divergence in bacterial blight resistant rice genotypes. *S. Afr. J. Bot.* **2013**, *86*, 15–22. [CrossRef]

44. Talhinhas, P.; Neves-Martins, J.; Leitão, J. AFLP, ISSR and RAPD markers reveal high levels of genetic diversity among Lupinus spp. *Plant Breed.* **2003**, *122*, 507–510. [CrossRef]

45. Sehgal, D.; Rajpal, V.R.; Raina, S.N.; Sasanuma, T.; Sasakuma, T. Assaying polymorphism at DNA level for genetic diversity diagnostics of the safflower (*Carthamus tinctorius* L.) world germplasm resources. *Genetica* **2009**, *135*, 457–470. [CrossRef]

46. Varshney, R.K.; Chabane, K.; Hendre, P.S.; Aggarwal, R.K.; Graner, A. Comparative assessment of EST-SSR, EST-SNP and AFLP markers for evaluation of genetic diversity and conservation of genetic resources using wild, cultivated and elite barleys. *Plant Sci.* **2007**, *173*, 638–649. [CrossRef]

47. Kumar, J.; Agrawal, V. Assessment of genetic diversity, population structure and sex identification in dioecious crop, *Trichosanthes dioica* employing ISSR, SCoT and SRAP markers. *Heliyon* **2019**, *5*, e01346. [CrossRef]

48. Dolezel, J.; Sgorbati, S.; Lucretti, S. Comparison of three DNA fluorochromes for flow cytometric estimation of nuclear DNA content in plants. *Physiol. Plant.* **1992**, *85*, 625–631. [CrossRef]

49. Bennett, M.D.; Smith, J.B. Nuclear DNA Amounts in Angiosperms. *Philos. Trans. R. Soc. B Boil. Sci.* **1976**, *274*, 227–274. [CrossRef]

50. Loureiro, J.; Rodriguez, E.; Doležel, J.; Santos, C. Two New Nuclear Isolation Buffers for Plant DNA Flow Cytometry: A Test with 37 Species. *Ann. Bot.* **2007**, *100*, 875–888. [CrossRef]

51. Doležel, J.; Bartoš, J. Plant DNA Flow Cytometry and Estimation of Nuclear Genome Size. *Ann. Bot.* **2005**, *95*, 99–110. [CrossRef] [PubMed]

52. Doležel, J.; Bartoš, J.; Voglmayr, H.; Greilhuber, J. Nuclear DNA content and genome size of trout and human. *Cytometry* **2003**, *51*, 127–128. [CrossRef] [PubMed]

53. Talhinhas, P. Caracterização de Germoplasma do Género *Lupinus*, Avaliação da Resistência à Antracnose e Estudos da Diversidade e Taxinomia do Agente Causal (*Colletotrichum Acutatum* Simmonds ex Simmonds). Ph.D. Thesis, Universidade Técnica de Lisboa, Lisboa, Portugal, 2002.

54. Hardy, A.; Huyghe, C.; Rahim, M.A.; Roemer, P.; Neves-Martins, J.M.; Sawicka-Sienkiewicz, E.; Caligari, P.D.S. Effects of genotype and environment on architecture and flowering time of indeterminate Andean lupins (*Lupinus mutabilis* Sweet). *Aust. J. Agric. Res.* **1998**, *49*, 1241. [CrossRef]

55. Georgieva, N.A.; Kosev, V.I. Analysis of Character Association of Quantitative Traits in *Lupinus* Species. *J. Agric. Sci.* **2016**, *8*, 23. [CrossRef]

56. Clements, J.; Buirchell, B.; Cowling, W. Relationship between morphological variation and geographical origin or selection history in *Lupinus pilosus*. *Plant Breed.* **2006**, *115*, 16–22. [CrossRef]

57. Johnson, H.W.; Robinson, H.F.; Comstock, R.E. Estimates of Genetic and Environmental Variability in Soybeans. *Agron. J.* **1955**, *47*, 314. [CrossRef]

58. Hussain, Q.A.; Tariq, K.; Iftikhar, A.; Nazir, A.; Muhammad, A.; Muhammad, F.; Muhammad, U. Estimation of heritability and selection response for some yield traits in F3 populations of wheat. *Int. J. Agric. Appl. Sci.* **2017**, *9*, 6–13.

59. Girma, G.; Tesfaye, K.; Bekele, E. Inter-simple sequence repeat (ISSR) analysis of wild and cultivated rice species from Ethiopia. *Afr. J. Biotechnol.* **2010**, *9*, 5048–5059.

60. Hashimoto, Z.; Mori, N.; Kawamura, M.; Ishii, T.; Yoshida, S.; Ikegami, M.; Takumi, S.; Nakamura, C. Genetic diversity and phylogeny of Japanese sake-brewing rice as revealed by AFLP and nuclear and chloroplast SSR markers. *Theor. Appl. Genet.* **2004**, *109*, 1586–1596. [CrossRef]

61. Chirinos-Aria, M.C.; Jiménez, J.E. Transference of some microsatellite molecular markers from Fabaceae family to Andean Lupin (*Lupinus mutabilis* Sweet). *Sci. Agropecu.* **2015**, *6*, 51–58. [CrossRef]

62. Bussell, J.D.; Waycott, M.; Chappill, J.A. Arbitrarily amplified DNA markers as characters for phylogenetic inference. *Perspect. Plant Ecol. Evol. Syst.* **2005**, *7*, 3–26. [CrossRef]

63. Adhikari, S.; Saha, S.; Bandyopadhyay, T.K.; Ghosh, P. Efficiency of ISSR marker for characterization of Cymbopogon germplasms and their suitability in molecular barcoding. *Plant Syst. Evol.* **2014**, *301*, 439–450. [CrossRef]

64. Aykut Tonk, F.; Tosun, M.; Ilker, E.; Istipliler, D.; Tatar, O. Evaluation and comparison of ISSR and RAPD markers for assessment of genetic diversity in triticale genotypes. *Bulg. J. Agric. Sci.* **2014**, *20*, 1413–1420.

65. Grativol, C.; Lira-Medeiros, C.D.F.; Hemerly, A.S.; Ferreira, P.C.G. High efficiency and reliability of inter-simple sequence repeats (ISSR) markers for evaluation of genetic diversity in Brazilian cultivated *Jatropha curcas* L. accessions. *Mol. Boil. Rep.* **2010**, *38*, 4245–4256. [CrossRef]

66. Hajiyeva, S.V.; Akparov, Z.I.; Hasanov, N.A.; Mustafayeva, Z.P.; Hajiyev, E.S.; Mammadov, A.T.; Izzatullayeva, V.I.; Babayeva, S.M.; Sharifova, S.S.; Mammadov, A.M.; et al. ISSR Analysis of Variability of Cultivated Form and Varieties of Pomegranate (*Punica granatum* L.) from Azerbaijan. *Russ. J. Genet.* **2018**, *54*, 188–197. [CrossRef]

67. Lamare, A.; Rao, S.R. Efficacy of RAPD, ISSR and DAMD markers in assessment of genetic variability and population structure of wild Musa acuminata colla. *Physiol. Mol. Boil. Plants* **2015**, *21*, 349–358. [CrossRef]

68. Chen, H.; Chen, H.; Hu, L.; Wang, L.; Wang, S.; Wang, M.L.; Cheng, X. Genetic diversity and a population structure analysis of accessions in the Chinese cowpea [*Vigna unguiculata* (L.) Walp.] germplasm collection. *Crop J.* **2017**, *5*, 363–372. [CrossRef]

69. Li, Z.; Liu, X.; Gituru, R.W.; Juntawong, N.; Zhou, M.; Chen, L. Genetic diversity and classification of Nelumbo germplasm of different origins by RAPD and ISSR analysis. *Sci. Hortic.* **2010**, *125*, 724–732. [CrossRef]

70. Naganowska, B.; Wolko, B.; Śliwińska, E.; Kaczmarek, Z. Nuclear DNA Content Variation and Species Relationships in the Genus *Lupinus* (Fabaceae). *Ann. Bot.* **2003**, *92*, 349–355. [CrossRef]

71. Rayburn, A.L.; Biradar, D.P.; Nelson, R.L.; McCloskey, R.; Yeater, K.M. Documenting Intraspecfic Genome Size Variation in Soybean. *Crops Sci.* **2004**, *44*, 261–264. [CrossRef]

72. Realini, M.F.; Poggio, L.; Cámara-Hernández, J.; González, G.E. Intra-specific variation in genome size in maize: Cytological and phenotypic correlates. *AoB Plants* **2015**, *8*, 138. [CrossRef]

73. Wang, W.; Kerstetter, R.A.; Michael, T.P. Evolution of Genome Size in Duckweeds (Lemnaceae). *J. Bot.* **2011**, *2011*, 1–9. [CrossRef]

74. Garrido-Ramos, M.A. Satellite DNA in Plants: More than Just Rubbish. *Cytogenet. Genome Res.* **2015**, *146*, 153–170. [CrossRef]

75. Petrov, D.A. Evolution of genome size: New approaches to an old problem. *Trends Genet.* **2001**, *17*, 23–28. [CrossRef]

76. Garrido-Ramos, M.A. Satellite DNA: An Evolving Topic. *Genes* **2017**, *8*, 230. [CrossRef]

77. Cullis, C.A. Mechanisms and Control of Rapid Genomic Changes in Flax. *Ann. Bot.* **2005**, *95*, 201–206. [CrossRef]

78. Basak, S.; Sun, X.; Wang, G.; Yang, Y. Genome Size Unaffected by Variation in Morphological Traits, Temperature, and Precipitation in Turnip. *Appl. Sci.* **2019**, *9*, 253. [CrossRef]

79. Beaulieu, J.M.; Moles, A.T.; Leitch, I.J.; Bennett, M.D.; Dickie, J.B.; Knight, C.A. Correlated evolution of genome size and seed mass. *New Phytol.* **2007**, *173*, 422–437. [CrossRef]

 *agronomy*

Article

Quantile Regression Applied to Genome-Enabled Prediction of Traits Related to Flowering Time in the Common Bean

Ana Carolina Nascimento [1], Moyses Nascimento [1,*], Camila Azevedo [1], Fabyano Silva [2], Leiri Barili [3], Naine Vale [3], José Eustáquio Carneiro [3], Cosme Cruz [4], Pedro Crescencio Carneiro [4] and Nick Serão [5]

[1] Department of Statistics, Federal University of Viçosa, Viçosa 36570-977, Brazil; ana.campana@ufv.br (A.C.N.); camila.azevedo@ufv.br (C.A.)

[2] Department of Animal Science, Federal University of Viçosa, Viçosa 36570-977, Brazil; fabyano.fonseca@ufv.br

[3] Department of Plant Sciences, Federal University of Viçosa, Viçosa 36570-977, Brazil; leyridaiana@hotmail.com (L.B.); nainemartinsdovale@hotmail.com (N.V.); eustaquiocarneiro@yahoo.com.br (J.E.C.)

[4] Department of General Biology, Federal University of Viçosa, Viçosa 36570-977, Brazil; cruzcd@ufv.br (C.C.); pcscarneiro@gmail.com (P.C.C.)

[5] Department of Animal Science, Iowa State University, Ames, IA 50011, USA; serao@iastate.edu

* Correspondence: moysesnascim@ufv.br; Tel.: +55-31-3612-2408

Received: 1 November 2019; Accepted: 21 November 2019; Published: 23 November 2019

Abstract: Genomic selection (GS) aims to incorporate molecular information directly into the prediction of individual genetic merit. Regularized quantile regression (RQR) can be used to fit models for all portions of a probability distribution of the trait, enabling the conditional quantile that "best" represents the functional relationship between dependent and independent variables to be chosen. The objective of this study was to predict the individual genetic merits of the traits associated with flowering time (DFF—days to first flower; DTF—days to flower) in the common bean using RQR and to compare the predictive abilities obtained from Random Regression Best Linear Unbiased Predictor (RR-BLUP), Bayesian LASSO (BLASSO), BayesB, and RQR for predicting the genetic merit. GS was performed using 80 genotypes of common beans genotyped for 380 single nucleotide polymorphism (SNP) markers. Considering the "best" RQR fit models (RQR0.3 for DFF, and RQR0.2 for DTF), the gains in predictive ability in relation to BLASSO, BayesB, and RR-BLUP were 18.75%, 22.58%, and 15.15% for DFF, respectively, and 15.20%, 24.65%, and 12.55% for DTF, respectively. The potential cultivars selected, considering the RQR "best" models, were among the 5% of cultivars with the lowest genomic estimated breeding value (GEBV) for the DFF and DTF traits—the IAC Imperador, IPR Colibri, Capixaba Precoce, and IPR Andorinha were included in the list of early cycle cultivars.

Keywords: *Phaseolus vulgaris* L.; linear model; conditional quantile; genome-enabled prediction

---

1. Introduction

Meuwissen et al. [1] introduced genomic selection (GS) as a means of incorporating molecular information directly into the prediction of individual genetic merit. GS has been successfully used in breeding to increase genetic gain per generation through early selection [2] and to improve prediction accuracy [3].

However, some statistical issues, for example, longitudinal [4] and non-normal [5] traits, still pose a challenge for GS. Although several statistical methods have been proposed for GS as solutions for multicollinearity and dimensionality, we believe there are limited reports in the literature that generalize these methods to non-normal traits. Non-normal distributions can be found for some traits in the fields of plant and animal breeding, for example, traits that measure the time until the occurrence of specific events (such as flowering and parity [6,7]) and hormone concentrations [8].

Another issue is related to the residual heteroscedastic variance. This tends to be neglected by existing methods, which focus on the mean of the conditional distribution, $E(X|Y)$. Generally, in the presence of heteroscedastic variance, which is frequently observed in high dimensional data sets such as those found in GS studies, the sets of relevant covariates may differ when the different segments of conditional distribution are considered [9].

Quantile regression (QR) is a method that can be used to address these issues [10]. QR allows models to be fitted to all portions of the probability distribution of the trait, enabling a more complete picture of the conditional distribution than a single estimate of the center [11]. This feature allows QR to examine all of the conditional distributions, in order to investigate skewness and heteroscedasticity. From a GS viewpoint, we can choose the "best" conditional quantile to represent the relationship between the dependent and independent variables, thus increasing the accuracy of the genomic estimated breeding value (GEBV) prediction of individual genetic merits [12]. However, because of the high dimensionality commonly found in GS studies, a variation of QR, denoted by regularized quantile regression (RQR) [13] should be considered. RQR uses an L1-norm penalty for simultaneously controlling the variance of the fitted coefficients and performing automatic variable selection.

Among several breeding programs, the common bean (*Phaseolus vulgaris* L.) fits well in low-input agricultural systems, which are commonly practiced in African and Latin American countries. Moreover, the common bean is a key commodity for improving food security [14]. Besides productivity, traits associated with the flowering time such as days to flowering (DTF) and days to first flower (DFF) are important in the selection of the common bean. The identification of cultivars with an early cycle (shorter time from planting to harvest) allows for the planning of harvests in periods of less rain, the reduction of water consumption by irrigated crops, and earlier freeing of the area for crop succession [15]. In addition, cultivars with an early cycle are exposed to the risk of plague and disease for a shorter period of time.

In this context, we aimed to (1) predict the individual genetic merits of the traits associated with the flowering time (DFF and DTF) in the common bean using RQR, and (2) to compare the predictive abilities obtained for RQR, Random Regression Best Linear Unbiased Predictor (RR-BLUP), BayesB [1], and Bayesian LASSO (BLASSO) [16] for predicting genetic merit.

2. Materials and Methods

2.1. Phenotypic and Genotypic Data

The phenotypic data were obtained from two experimental stations at the Federal University of Viçosa, Minas Gerais State (MG), Brazil. One is located in Viçosa, MG (latitude 20° 45′ 14″ S, longitude 42° 52′ 55″ W, altitude 648m asl), and the other is in Coimbra, MG (latitude 20° 51′ 24″ S, longitude 42° 48′ 10″ W, altitude 720 m asl). Two phenological traits, DFF and DTF, were measured in eighty common bean cultivars, which were studied between 1970 and 2013 by research institutions in Brazil (Brazilian agricultural research corporation—Embrapa, Campinas Agronomic Institute—IAC, Federal University of Viçosa—UFV, Paraná Agronomic Institute—IAPAR, Agricultural Research Company of Minas Gerais—Epamig, Federal University of Lavras—UFLA, State Foundation for Agricultural Research—Fepagro, Rural Extension and Agricultural Research Enterprise—Epagri, and FT seeds).

The cultivars were planted at each location in the dry-summer (February) and winter (July) seasons of 2013 following a randomized complete block design with three replicates. The experimental plots consisted of four 3 m long rows, spaced 0.5 m apart with 15 seeds sown per meter. Fertilization was applied according to the results of the soil analyses in order to ensure ideal conditions for development and production. Insect pests, invasive plants, and weeds were controlled as needed, according to the official recommendations for the common bean [17]. DFF was measured as the number of days from planting until at least one plant presented a flower. DTF was measured as the number of days from planting until at least 50% of the plants in a plot (replicate) had at least one open flower. These experiments were developed by the bean breeding program at the Federal University of Viçosa, Brazil. More details, such as the list of cultivars and research institutions are described in the literature [17,18].

The DNA samples were genotyped using the Vera Code1 BeadXpress platform (Illumina) at the Biotechnology Laboratory of Embrapa (Goiania, GO, Brazil). A set of 384 single nucleotide polymorphism (SNP) markers was selected as the oligo pool assay (OPA) SNP marker panel. The genotype call was performed using GenomeStudio software version 2011.1 (Illumina, San Diego, CA, USA), with call rate values ranging from 0.96 to 0.99, and GenTrain $\geq$ 0.46 for SNP clustering.

2.2. Phenotypic Data Analyses

The following statistical model of the phenological traits (DFF and DTF) was fit to the phenotypic data set as follows:

$$y_{ijk} = \mu + g_i + a_j + ga_{ij} + b_{k(j)} + \varepsilon_{ijk} \tag{1}$$

where y_{ijk} is the observed phenotype (DFF and DTF); μ is the overall mean; g_i is the random effect of the genotype (cultivar), $i = 1$ to 80, assumed to follow a normal distribution, with a mean of 0 and a variance of σ_g^2, with $i = 1$ to 80; a_j is the fixed effect of environment j, with $j = 1$ to 4; ga_{ij} is the random effect of the interaction of genotype i with environment j, assumed to follow a normal distribution with a mean of 0 and variance of σ_{ga}^2; $b_{k(j)}$ is the random effect of block (replicate) k ($k = 1, 2$, and 3) within environment j, assumed to follow a normal distribution with a mean of 0 and variance of σ_b^2; and ε_{ijk} is the experimental error for genotype i in block k of environment j, assumed to follow a normal distribution with a mean of 0 and variance of σ_ε^2. Nascimento et al. [19] used the same data set to evaluate models with and without the interaction between genotypes and environments through a likelihood ratio test (LRT), and showed that the full model had better fit.

The corresponding bivariate model is as follows:

$$
\begin{bmatrix} y_{DFF} \\ y_{DTF} \end{bmatrix} = \begin{bmatrix} X_{DFF} & 0 \\ 0 & X_{DTF} \end{bmatrix} \begin{bmatrix} b_{DFF} \\ b_{DTF} \end{bmatrix}
$$
$$
+ \begin{bmatrix} Zg_{DFF} & 0 \\ 0 & Zg_{DTF} \end{bmatrix} \begin{bmatrix} g_{DFF} \\ g_{DTF} \end{bmatrix} + \begin{bmatrix} Zga_{DFF} & 0 \\ 0 & Zga_{DTF} \end{bmatrix} \begin{bmatrix} ga_{DFF} \\ ga_{DTF} \end{bmatrix}
$$
$$
+ \begin{bmatrix} Zr_{DFF} & 0 \\ 0 & Zr_{DTF} \end{bmatrix} \begin{bmatrix} r_{DFF} \\ r_{DTF} \end{bmatrix} + \begin{bmatrix} \varepsilon_{DFF} \\ \varepsilon_{DTF} \end{bmatrix} \tag{2}
$$

where y_{DFF} and y_{DTF} denote the vectors of observed DFF and DTF, respectively; X_{DFF} and X_{DTF} denote the design matrices of the fixed effects for DFF and DTF, respectively; b_{DFF} and b_{DTF} denote the vectors of the fixed effects associated with X_{DFF} and X_{DTF}, respectively; Zg_{DFF} and Zg_{DTF} denote the design matrices of the random effect of the genotype for DFF and DTF, respectively; g_{DFF} and g_{DTF} denote the vectors of the random effects with Zg_{DFF} and Zg_{DTF}; respectively; Zga_{DFF} and Zga_{DTF} denote the design matrices of the random effect of the interaction of the genotype with the environment for DFF and DTF, respectively; ga_{DFF} and ga_{DTF} denote the vectors of the random effects with Zga_{DFF} and Zga_{DTF}, respectively; Zr_{DFF} and Zr_{DTF} denote the design matrices of the random effect of the block (replicates) for DFF and DTF, respectively; r_{DFF} and r_{DTF} denote the vectors of the random effects with Zr_{DFF} and Zr_{DTF}; respectively; and ε_{DFF} and ε_{DTF} denote the vectors of the random errors associated with y_{DFF} and y_{DTF}, respectively. Assuming random effects distributed as a multivariate normal with the mean equal to zero and a covariance matrix, is as follows:

$$
\begin{bmatrix} g_{DFF} \\ g_{DTF} \\ ga_{DFF} \\ ga_{DTF} \\ r_{DFF} \\ r_{DTF} \\ \varepsilon_{DFF} \\ \varepsilon_{DTF} \end{bmatrix} =
$$

$$
\begin{bmatrix}
I\sigma^2_{g,DFF} & I\sigma_{g,DFF;DTF} & I\sigma_{g,}00000 & I\sigma_{g,}00000 & I\sigma_{g,}00000 & I\sigma_{g,}00000 & I\sigma_{g,}00000 & I\sigma_{g,}00000 \\
I\sigma_{g,DFF;DTF} & I\sigma^2_{g,DTF} & I\sigma_{g,}00000 & I\sigma_{g,}00000 & I\sigma_{g,}00000 & I\sigma_{g,}00000 & I\sigma_{g,}00000 & I\sigma_{g,}00000 \\
I\sigma_{g,}00000 & I\sigma_{g,}00000 & I\sigma^2_{ga,DFF} & I\sigma_{g,}00000 & I\sigma_{g,}00000 & I\sigma_{g,}00000 & I\sigma_{g,}00000 & I\sigma_{g,}00000 \\
I\sigma_{g,}00000 & I\sigma_{g,}00000 & I\sigma_{g,}00000 & I\sigma^2_{ga,DTF} & I\sigma_{g,}00000 & I\sigma_{g,}00000 & I\sigma_{g,}00000 & I\sigma_{g,}00000 \\
I\sigma_{g,}00000 & I\sigma_{g,}00000 & I\sigma_{g,}00000 & I\sigma_{g,}00000 & I\sigma^2_{r,DFF} & I\sigma_{g,}00000 & I\sigma_{g,}00000 & I\sigma_{g,}00000 \\
I\sigma_{g,}00000 & I\sigma_{g,}00000 & I\sigma_{g,}00000 & I\sigma_{g,}00000 & I\sigma_{g,}00000 & I\sigma^2_{r,DTF} & I\sigma_{g,}00000 & I\sigma_{g,}00000 \\
I\sigma_{g,}00000 & I\sigma_{g,}00000 & I\sigma_{g,}00000 & I\sigma_{g,}00000 & I\sigma_{g,}00000 & I\sigma_{g,}00000 & I\sigma^2_{\varepsilon,DFF} & I\sigma_{\varepsilon,DFF;DTF} \\
I\sigma_{g,}00000 & I\sigma_{g,}00000 & I\sigma_{g,}00000 & I\sigma_{g,}00000 & I\sigma_{g,}00000 & I\sigma_{g,}00000 & I\sigma_{\varepsilon,DFF;DTF} & I\sigma^2_{\varepsilon,DTF}
\end{bmatrix} \tag{3}
$$

where $\sigma^2_{g,DFF}$ and $\sigma^2_{g,DFF}$ denote the random genotypes variance for DFF and DTF, respectively; $I\sigma_{g,DFF;DTF}$ denote the random genotype covariance between DFF and DTF; $\sigma^2_{ga,DFF}$ and $I\sigma^2_{ga,DTF}$ denote the random genotype by the environment variance for DFF and DFF, respectively; $I\sigma^2_{r,DFF}$ and $I\sigma^2_{r,DTF}$ denote the effect of the block (replicates) variance for DFF and DFF, respectively; $I\sigma^2_{\varepsilon,DFF}$ and $I\sigma^2_{\varepsilon,DTF}$ denote the random error variance for DFF and DTF, respectively; $I\sigma_{\varepsilon,DFF;DTF}$ denotes the random error covariance between DFF and DTF; and I denotes the identity matrix. The genetic parameters (broad sense heritability and correlations) were also estimated for the flowering traits using components of variance estimated by the bivariate model. The analyses were carried out in ASReml 3.0 [19].

2.3. Genomic Prediction Models

Prior to performing the genomic predictions, the adjusted phenotypes (Y_i^*) were obtained as the sum of the random effects (genotypes and error) from the selected model for analyzing the phenotypic data. Specifically, the adjusted phenotypes were obtained to correct for non-genetic sources of variation, for example, blocks, environments, and the interaction between the genotypes and environments (i.e., the combination of two locations (Viçosa and Coimbra, MG) and two seasons (dry and winter)).

The basic genomic model is represented by the following:

$$
Y_i^* = \mu + \sum_{m=1}^{384} X_{im}\beta_m + e_i \tag{4}
$$

where Y_i^* is the adjusted phenotype of the genotype, which is obtained as the sum of the random effects (genotypes and error); μ is the population mean; X_{im} is the incidence of the mth SNP in the ith adjusted phenotype of the genotype; and e_i is the random error term associated with Y_i^*.

Genomic prediction was performed using regularized quantile regression (RQR) [13]. RQR allows the GEBV at different portions of a probability distribution of the traits to be obtained. This method consists of obtaining parameter effects at level τ (θ_τ) that solve the following optimization problem:

$$
\hat{\theta}_\tau = \mathrm{argmin}_{\hat{\theta}_\tau} \left[\sum_{i=1}^{80} \rho_\tau(e_i) + \lambda \sum_{m=1}^{384} |\beta_{m\tau}| \right] \tag{5}
$$

where τ indicates the quantile of interest, $\sum_{m=1}^{384} |\beta_{m\tau}|$ is the sum of the absolute values of the regression coefficients, and λ is the parameter that controls the strength of regularization. In the current study, we evaluated five quantiles—(τ): 0.2, 0.3, 0.5, 0.7, and 0.8. The parameter $\rho_\tau(\cdot)$ is denoted as a check function [10] and is defined by the following:

$$\rho_\tau(e_i) = \begin{cases} \tau \cdot e_i, & \text{if } e_i \geq 0, \\ -(1-\tau)\cdot e_i, & \text{if } e_i < 0. \end{cases} \tag{6}$$

where $\tau \in \,]0,1[$ indicates the quantile of interest. The constrained optimization problem was solved using an algorithm that computes the exact solution for the model parameters [10].

GEBVs were obtained by $\text{GEBV}_\tau = \hat{u}_\tau = X\hat{\beta}_\tau$, where τ represents the τth quantile of interest. To define the "best" RQR fit, a grid of pair values given by combinations of τ ($\tau = 0.2$, 0.3, 0.5, 0.7, 0.8) and λ (from 0 until λ estimated by BLASSO, varying by 1), was considered. The predictive ability was used as a criterion to define the optimal pair. The GEBVs were also obtained using RR-BLUP, BLASSO, and BayesB.

In order to assess the predictive ability of all of the fitted models, the Pearson's correlation between the predicted values and the phenotypes were calculated using a four-fold cross-validation (CV) random process. This process was repeated randomly 10 times. The folds were the same for each fitted model.

The predictive abilities (PAs) and standard errors (SEs) were estimated from 10 estimates of the predictive ability.

After obtaining the GEBVs through the fit models, the Cohen's kappa [20] coefficient, and the Spearman's and Kendall's τ correlations were calculated to assess the agreement between the methods. The Cohen's kappa coefficient was used to calculate the percentage of individuals in common between the better 10% of individuals ranked according to the GEBVs. The Cohen's kappa coefficient is given by $C = \text{NC} - C_{\text{Random}} / 1 - C_{\text{Random}}$, where NC is the relative observed agreement among raters, and C_{Random} is the hypothetical probability of the random agreement. The Spearman's correlation was calculated between the GEBV obtained by the different genomic selection models.

The RQR, RR-BLUP, BLASSO, and BayesB model fittings were carried out using the rq (for RQR) and BGLR (RR-BLUP, BLASSO, and BayesB) functions in the quantreg [21] and BGLR [22] packages, respectively, of R software [23]. The Bayesian methods were implemented using 200,000 Markov chain Monte Carlo iterations, with burn-in and thin values at 10,000 and five iterations, respectively. The convergence of the Markov chains was checked with the Geweke's Diagnostic [24].

3. Results

A summary of the descriptive statistics including the means, standard deviations (SD), ranges, and skewness for the phenological traits is presented in Table 1. The average days from planting until at least one plant presents one flower (DFF) was 34.74. After an average of 42 days, more than 50% of the plants in a plot presented at least one open flower (DTF; Table 1). The skewness coefficient shows that the phenotypes have left-skewed distributions.

Table 1. Means, standard deviations (SD), ranges, and skewness for days to first flower (DFF) and days to flowering (DTF), measured in 80 common bean genotypes.

Trait	Mean (SD)	Minimum	Maximum	Skewness
DFF (days)	34.74 (5.52)	20.00	49.00	−1.08
DTF (days)	42.30 (5.56)	27.00	55.00	−1.32

3.1. Model Selection and Genetic Parameters

The full model (model with the interaction effect) presented lower AIC (AIC_{DFF}: 2206, and AIC_{DTF}: 2318) and BIC ($\text{BIC}_{\text{DFF}} = 2225$, and $\text{BIC}_{\text{DTF}} = 2337$) values compared with those obtained from the reduced model (model without the interaction effect; $\text{AIC}_{\text{DFF}} = 2233$, and $\text{AIC}_{\text{DTF}} = 2334$;

BIC_{DFF} = 2248, and BIC_{DTF} = 2348). The LRTs (LRT_{DFF} = 29.8, and LRT_{DTF} = 17.26) between the full and reduced models were significantly different ($p < 0.01$) for both traits.

The estimates of heritability for DFF and DTF were moderate to high, with 0.58 ± 0.06 and 0.49 ± 0.08, respectively. For the estimates of the correlations between DFF with DTF, the genetic correlations were positive and strong, with a correlation of 0.98 ± 0.01 and a phenotypic correlation of 0.68 ± 0.04.

3.2. Prediction Accuracy of Traits

The estimated predictive abilities for two traits (DFF and DTF) ranged from −0.06 (0.04) to 0.38 (0.02) and are presented in Table 2. For DFF and DTF, the highest accuracy values were 0.38 and 0.34, obtained from $RQR_{0.3}$ and $RQR_{0.2}$, respectively (Table 2).

Table 2. Predictive ability for days to first flower (DFF) and to flowering (DTF) measured in 80 common bean genotypes, using a four-fold cross validation repeated 10 times for all of the fitted models (Bayesian LASSO (BLASSO), BayesB, Random Regression Best Linear Unbiased Predictor (RR-BLUP), and regularized quantile regression (RQR)).

Trait		BLASSO	BayesB	RR-BLUP	$RQR_{0.2}$	$RQR_{0.3}$	$RQR_{0.5}$	$RQR_{0.7}$	$RQR_{0.8}$
DFF	PA [1]	0.32	0.31	0.33	0.37	0.38	0.25	−0.02	−0.05
	SE [2]	0.03	0.02	0.03	0.03	0.02	0.04	0.03	0.06
DTF	PA	0.29	0.27	0.30	0.34	0.32	0.10	0.01	−0.06
	SE	0.03	0.02	0.03	0.02	0.01	0.04	0.03	0.04

[1] Predictive ability. [2] Standard error of parameters estimates.

Considering the "best" RQR fit models ($RQR_{0.3}$ for DFF, and $RQR_{0.2}$ for DTF), the gains in predictive ability in relation to BLASSO, BayesB, and RR-BLUP were 18.75% (the ratio between the predictive ability obtained from "best" RQR model fit and the other methods), 22.58%, and 15.15% for DFF, respectively, and 15.20%, 24.65%, and 12.55% for DTF, respectively.

Estimates of the Spearman's correlation (lower triangle) and Cohen's kappa concordance coefficient (upper triangle) between the GEBVs obtained by the "best" quantile fit models ($RQR_{0.3}$ for DFF, and $RQR_{0.2}$ for DTF), and the BLASSO, BayesB, and RR-BLUP for the two traits are shown in Table 3.

Table 3. Estimates of Spearman's correlation (lower triangle), Kendall's τ rank correlation coefficient (lower triangle—in parenthesis), and Cohen's Kappa concordance coefficient[2] (upper triangle) between the genomic estimated breeding value (GEBV) values, obtained considering four different genomic selection models (BLASSO, RR-BLUP, BayesB, and RQR[1]) for days to first flower (DFF) and to flowering (DTF), measured in 80 common bean genotypes.

Traits	Method	BLASSO	BayesB	RR-BLUP	RQR [1]
DFF	BLASSO		1.00	1.00	0.63
	BayesB	0.99 (0.90)		1.00	0.63
	RR-BLUP	0.99 (0.99)	0.98 (0.89)		0.63
	RQR [1]	0.63 (0.36)	0.69 (0.41)	0.62 (0.32)	
DTF	BLASSO		1.00	1.00	0.63
	BayesB	0.99 (0.91)		1.00	0.63
	RR-BLUP	1.00 (0.99)	0.99 (0.91)		0.63
	RQR [1]	0.67 (0.38)	0.73 (0.43)	0.65 (0.38)	

[1] The "best" RQR fit models ($RQR_{0.3}$ for DFF and $RQR_{0.2}$ for DTF) considering the predictive ability. [2] Based on the lowest 10% GEBVs.

Overall, the Spearman's correlations presented high positive values. The lowest Spearman's correlation was estimated between the RR-BLUP and $RQR_{0.3}$ (0.62) models for DFF. The highest Spearman's correlations were estimated for BLASSO with BayesB (0.99) and RR-BLUP (0.99) for the DFF trait, and between the BLASSO and RR-BLUP (1.00) models for the DTF trait (Table 3). The Kendall's τ rank correlation coefficient estimates vary from moderate to high positive values. For instance, the Kendall's τ rank correlations between the RQR and BayesB, RR-BLUP, and BLASSO present moderate values (0.32–0.43) for both traits.

After ranking the individuals according to the GEBVs, the percentage of selected individuals in common was calculated using Cohen's kappa coefficient and based on the lowest 10% GEBVs. The lowest GEBVs represent those cultivars with an early cycle (shorter time from planting to harvest).

The lowest Cohen's kappa coefficient value was observed between RR-BLUP and $RQR_{0.2}$ (0.25) for the DTF trait. The highest values were observed between BayesB with RR-BLUP (1.00) and BLASSO (1.00) for DFF and DTF, respectively (Table 3, upper triangle).

The computational time to fit a model considering all of the four-fold processes varied from 0.28–264.08 s (Table 4). The Bayesian methods (BLASSO and BayesB) required a higher computational time compared with the RR-BLUP (0.28) and RQR (178.38).

Table 4. Average computational cost (seconds).

Method	Computational Time
BLASSO	264.08
BayesB	258.90
RR-BLUP	0.28
RQR [1]	178.38 [1]

[1] Time related to the fit of several models considering different quantile and shrinkage values.

4. Discussion

In this study, we predicted the individual genetic merits of the traits associated with the flowering time (DFF and DTF) in the common bean using RQR models. Using 80 genotypes of common beans genotyped for 384 markers, we compared the predictive ability of RQR to that obtained by BLASSO, BayesB, and RR-BLUP. The predictive ability of these methods was assessed using a four-fold cross-validation (CV) repeated 10 times. The Spearman's correlation and Cohen's kappa concordance (based on the lowest 10% GEBVs) coefficients between the GEBV values estimated by the four different genomic selection models used in this work were also estimated. However, first, the genetic parameters were estimated for DFF and DTF.

The heritability estimates for DFF (0.58) and DTF (0.49) were consistent with those reported in the literature. Specifically, the heritability estimate for DFF was within the range of estimates for the different crosses in beans (0.29–0.75 [25]; 0.59 [26]). In addition, for DTF, the estimate was close to that reported by the authors of [27], namely, 0.54. The estimates of the genetic and phenotypic correlations between the flowering traits were positive and high, which are similar to those reported in [28] for the phenological traits.

Overall, RQR outperformed the traditional methods for evaluating the two traits of DFF and DTF. Indeed, the better results are reasonable since the RQR allows for estimating the functional relationships between the variables for all portions of the probability distribution of the trait. The ability to choose the "best" relationship between the phenotype and markers increases the predictive performance of the model. According to the authors of [11] and [12], when the conditional distributions of Y are non-normal (for instance, skewed), the mean might not be the best way to describe the functional relationship between the variables.

The skewness coefficient for both traits (DFF and DTF) indicates a negative-skewed phenotypic distribution, and the "best" quantile fit models were $RQR_{0.3}$ for DFF, and $RQR_{0.2}$ for DTF. Therefore, our results indicate that to improve predictive ability, an effective strategy is to evaluate all of the phenotypic distributions so as to choose the "best" quantile fit model. In addition, the heterogeneous variance that is frequently observed in high dimensional data sets suggests that a single slope is not able to characterize changes over the probability distribution, therefore indicating that RQR is a good tool to deal with those situations [29].

Overall, the models presented moderate to high values of Spearman's correlations (Table 3). On the other hand, the Kendall's τ rank correlation, which is a statistic that measures the ordinal association between two measured quantities, presented moderate correlations between RQR and the traditional genomic selection methodologies. Additionally, based on Cohen's kappa coefficient, the classification agreement between the RQR and non-quantile regression models (BLASSO, BayesB, and RR-BLUP) also showed moderate values (0.63) [30], suggesting differences in the classifications obtained by these models.

Among the 5% of cultivars with the lowest GEBVs for the DFF and DTF traits, the IAC Imperador, IPR Colibri, Capixaba Precoce, and IPR Andorinha were included in the list obtained by RQR "best" models. These are considered as early cycle cultivars [31–33], indicating the model ability to select. On the other hand, considering the GEBV obtained by BLASSO, BayesB, and RR-BLUP, the cultivar IPR Andorinha was replaced by BRSMG Madrepérola, which is not characterized as an early cycle cultivar [34].

Altogether, these results show that the use of RQR to predict the individual genetic merits of flowering time-related traits in the common bean (DTF and DFF) is worthy of interest. RQR showed similar or higher estimates of predictive ability compared with traditional methods (RR-BLUP, BLASSO, and BayesB), and was able to find cultivars with an early cycle. Moreover, RQR allows for the fitting of the regression models to other parts of the distribution of the trait, enabling a more informative study of the relationship between variables. This approach can be useful for representing different selection strategies (individuals with higher or lower values of the trait), or to give more information about potential cultivars for selection.

Overall, the computational time does not present any problems as the higher value was less than five minutes. The higher computational cost observed in the Bayesian methods is related to the estimation process, which is based on Markov Chain Monte Carlo (MCMC) algorithms. The RQR requires more computational time compared with RR-BLUP. The higher time observed in the RQR fitting is related to the necessity of evaluating several quantile and shrinkage values to choose the "best" model.

The potential of quantile regression (QR) has been confirmed by many studies. Briollais and Durrieu [11] pointed out some aspects of the use of QR in genome-wide association studies (GWAS). According to these authors, QR allows for direct estimation at the extremes, and specific sampling is not needed. Extreme sampling is used to enrich the genetic signal, where the main idea is to sample individuals with extreme phenotypes in the hope that rare causal variants will be enriched [35]. Recently, Nascimento et al. [18] used QR to identify genomic regions for phenological traits in the common bean. Unlike the traditional single-SNP GWAS model, the QR methodology was able to find SNP-trait associations considering one extreme quantile ($\tau = 0.1$). Barroso et al. [29] successfully used RQR for the SNP marker effect estimation of pig growth curves, as well as to identify the chromosome regions of the most relevant markers and to estimate the genetic individual weight trajectory over time under different quantiles. However, to the best of our knowledge, reports in the literature about the use of QR for GS are limited. Therefore, here, we introduce the QR for plant breeders, bringing new insights for GS studies.

Finally, QR uses all of the data set in the estimation process, which is different to using a subsample of data, which can result in smaller sample sizes for each regression [36] and introduces sample selection bias [37].

5. Conclusions

The regularized quantile regression (RQR) method was able to predict the individual genetic merits of the traits associated with the flowering time (DTF and DFF) in the common bean. In addition, considering the estimates of predictive ability, RQR presented similar or better results compared with those obtained for RR-BLUP, BLASSO, and BayesB. Moreover, RQR was able to find early cycle cultivars.

Author Contributions: Conceptualization, A.C.N., M.N., and N.S.; formal analysis, A.C.N. and M.N.; investigation, L.B.; methodology, A.C.N., M.N., and N.S.; software, A.C.N., M.N., C.A., and F.S.; writing (original draft), A.C.N., M.N., and N.S.; writing (review and editing), A.C.N., M.N., C.A., F.S., L.B., N.V., J.E.C., C.C., P.C.C., and N.S.

Funding: This research was funded by CAPES, CNPq, FAPEMIG, and FUNARBE.

Conflicts of Interest: The authors declare no conflict of interest.

References

1. Meuwissen, T.H.E.; Hayes, B.J.; Goddard, M.E. Prediction of total genetic value using genome-wide dense marker maps. *Genetics* **2001**, *157*, 1819–1829. [PubMed]
2. Resende, M.F.R., Jr.; Muñoz, P.; Resende, M.D.V.; Garrick, D.J.; Fernando, R.L.; Davis, J.M.; Jokela, E.J.; Martin, T.A.; Peter, G.F.; Kirst, M. Accuracy of Genomic Selection Methods in a Standard Data Set of Loblolly Pine (*Pinus taeda* L.). *Genetics* **2012**, *190*, 1503–1510. [CrossRef] [PubMed]
3. Crossa, J.; Beyene, Y.; Kassa, S.; Pérez, P.; Hickey, J.M.; Chen, C.; de los Campos, G.; Burgueño, J.; Windhausen, V.S.; Buckler, E.; et al. Genomic prediction in maize breeding populations with genotyping-by-sequencing. *G3-Genes Genomes Genet.* **2013**, *3*, 1903–1926. [CrossRef]
4. Crispim, A.C.; Kelly, M.J.; Guimarães, S.E.; Silva, F.F.; Fortes, M.R.S.; Wenceslau, R.R.; Moore, S. Multi-trait GWAS and new candidate genes annotation for growth curve parameters in Brahman cattle. *PLoS ONE* **2015**, *10*, e0139906. [CrossRef] [PubMed]
5. Campos, C.F.; Soares, M.L.; Silva, F.F.; Veroneze, R.; Knol, E.F.; Lopes, P.S.; Guimarães, S.E.F. Genomic selection for boar taint compounds and carcass traits in a commercial pig population. *Livest. Sci.* **2015**, *174*, 10–17. [CrossRef]
6. Maurer, A.; Draba, V.; Jiang, Y.; Schnaithmann, F.; Sharma, R.; Schumann, R.; Kilian, B.; Reif, J.C.; Pillen, K. Modelling the genetic architecture of flowering time control in barley through nested association mapping. *BMC Genom.* **2015**, *16*, 290. [CrossRef]
7. Varona, L.; Ibañez-Escriche, N.; Quintanilla, R.; Noguera, J.L.; Casellas, J. Bayesian analysis of quantitative traits using skewed distributions. *Genet. Res.* **2008**, *90*, 179–190. [CrossRef]
8. Mathur, P.K.; Ten Napel, J.; Bloemhof, S.; Heres, L.; Knol, E.F.; Mulder, H.A. A human nose scoring system for boar taint and its relationship with androstenone and skatole. *Meat Sci.* **2012**, *91*, 414–422. [CrossRef]
9. Wang, L.; Wu, Y.; Li, R. Quantile regression for analyzing heterogeneity un Ultra-high dimension. *J. Am. Stat. Assoc.* **2012**, *107*, 214–222. [CrossRef]
10. Koenker, R.; Bassett, G.W. Regression quantiles. *Econometrica* **1978**, *46*, 33–50. [CrossRef]
11. Briollais, L.; Durrieu, G. Application of quantile regression to recent genetic and omic studies. *Hum. Genet.* **2014**, *133*, 951–966. [CrossRef] [PubMed]
12. Nascimento, M.; Silva, F.F.; Resende, M.D.V.; Cruz, C.D.; Nascimento, A.C.C.; Viana, J.M.S.; Azevedo, C.F.; Barroso, L.M.A. Regularized quantile regression applied to genome-enabled prediction of quantitative traits. *Genet. Mol. Res.* **2017**, *16*, gmr16019538. [CrossRef] [PubMed]
13. Li, Y.; Zhu, J. L_1-norm quantile regression. *J. Comp. Graph. Stat.* **2008**, *17*, 1–23. [CrossRef]
14. Kamfwa, K.; Cichy, K.A.; Kelly, J.D. Genome-Wide Association Study of Agronomic Traits in Common Bean. *Plant Genome* **2015**, *8*, 1–12. [CrossRef]
15. Buratto, J.S.; Cirino, V.M.; Fonseca Junior, N.S.; Prete, C.E.C.; Faria, R.T. Agronomic performance and grain yield in early common bean genotypes in Paraná state. *Semina Ciênc Agrár* **2007**, *28*, 373–380. [CrossRef]
16. De los Campos, G.; Naya, H.; Gianola, D.; Crossa, J.; Legarra, A.; Manfredi, E.; Weigel, K.; Cotes, J.M. Predicting Quantitative Traits with Regression Models for Dense Molecular Markers and Pedigree. *Genetics* **2009**, *182*, 375–385. [CrossRef]

17. Barili, L.D.; Vale, N.M.; Prado, A.L.; Carneiro, J.E.S.; Silva, F.F.; Nascimento, M. Genotype-environment interaction in common bean cultivars with carioca grain cultivated in Brazil in the last 40 years. *Crop Breed. Appl. Biotechnol.* **2015**, *15*, 244–250. [CrossRef]

18. Nascimento, M.; Nascimento, A.C.C.; Silva, F.F.; Barili, L.D.; Vale, N.M.; Carneiro, J.E.S.; Carneiro, P.C.; Cruz, C.D.; Serao, N.V.L. Quantile regression for genome-wide association study of flowering time-related traits in common bean. *PLoS ONE* **2018**, *3*, e0190303. [CrossRef]

19. Gilmour, A.R.; Gogel, B.J.; Cullis, B.R.; Thompson, R. *ASReml User Guide Release, 3.0*; VSN International Ltd.: Hemel Hempstead, UK, 2009.

20. Cohen, J. A coefficient of agreement for nominal scales. *Educ. Psychol. Meas.* **1960**, *20*, 37–46. [CrossRef]

21. Koenker, R. Quantreg: Quantile Regression. Available online: https://CRAN.R-project.org/package=quantreg (accessed on 20 April 2018).

22. De los Campos, G.; Rodriguez, P.P. BGLR: Bayesian Generalized Linear Regression. Available online: https://cran.r-project.org/web/packages/BGLR/index.html (accessed on 1 March 2018).

23. R Core Team. *A Language and Environment for Statistical Computing*; Foundation for Statistical Computing: Vienna, Austria, 2017.

24. Geweke, J. Evaluating the accuracy of sampling-based approaches to calculating posterior moments. In *Bayesian Statistics*; Bernardo, L.M., Berger, J.O., Dawid, A.P., Smith, A.F.M., Eds.; Oxford University: New York, NY, USA, 1992; pp. 625–631.

25. Cerna, J.; Beaver, J.S. Inheritance of early maturity of indeterminate dry bean. *Crop Sci.* **1990**, *30*, 1215–1218. [CrossRef]

26. Msolla, S.N.; Mduruma, Z.O. Estimate of Heritability for Maturity Characteristics of an Early x Late Common Bean (*Phaseolus Vulgaris* L.) Cross (TMO 216 x CIAT 16-1) and Relationships Among Maturity Traits with Yield and Components of Yield. *J. Agric. Sci.* **2007**, *8*, 11–18.

27. Moghaddam, S.F.; Mamidi, S.; Osorno, J.M.; Lee, R.; Brick, M.; Kelly, J.; Miklas, P.; Urrea, C.; Song, Q.; Cregan, P.; et al. Genome-Wide Association Study Identifies Candidate Loci Underlying Agronomic Traits in a Middle American Diversity Panel of Common Bean. *Plant Genome* **2016**, *9*, 1–21. [CrossRef] [PubMed]

28. Scully, B.T.; Wallace, D.H.; Viands, D.R. Heritability and correlation of biomass, growth rates, harvest index and phenology to the yield of common beans. *J. Am. Soc. Hortic. Sci.* **1991**, *116*, 127–130. [CrossRef]

29. Barroso, L.M.A.; Nascimento, M.; Nascimento, A.C.C.; Silva, F.F.; Serão, N.V.L.; Cruz, C.D.; Resende, M.D.V.; Silva, F.L.; Azevedo, C.F.; Lopes, P.S.; et al. Regularized quantile regression for SNP marker estimation of pig growth curves. *J. Anim. Sci. Biotechnol.* **2017**, *8*, 1–9. [CrossRef] [PubMed]

30. McHugh, M.L. Interrater reliability: The kappa statistic. *Biochem. Med.* **2012**, *22*, 276–282. [CrossRef]

31. Infoteca-e: Repositório de Informação Tecnológica da Embrapa. Available online: https://www.infoteca. cnptia.embrapa.br/handle/doc/217045 (accessed on 30 May 2018).

32. Chiorato, A.F.; Carbonell, S.A.M.; Carvalho, C.R.L.; Barros, V.L.N.P.; Borges, W.L.B.; Ticelli, M.; Gallo, P.B.; Finoto, E.L.; Santos, N.C.B. 'IAC IMPERADOR': Early maturity "carioca" bean cultivar. *Crop Breed. Appl. Biotechnol.* **2012**, *12*, 297–300. [CrossRef]

33. IAPAR. Instituto Agronômico Do Paraná—IAPAR. Available online: http://www.iapar.br/modules/conteudo/ conteudo.php?conteudo=1960 (accessed on 30 May 2018).

34. Carneiro, J.E.S.; Abreu, A.F.B.; Ramalho, M.A.P.; de Paula, T.J.; Del Peloso, M.J.; Melo, L.C.; Pereira, H.S.; Pereira Filho, I.A.; Martins, M.; Vieira, R.F.; et al. BRSMG Madrepérola: Common bean cultivar with late-darkening carioca grain. *Crop Breed. Appl. Biotechnol.* **2012**, *12*, 281–284. [CrossRef]

35. Wang, K.; Li, W.D.; Zhang, C.K.; Wang, Z.; Glessner, J.T.; Grant, S.F.A.; Zhao, H.; Hakonarson, H.; Price, R.A. A genome-wide association study on obesity and obesity-related traits. *PLoS ONE* **2011**, *7*, e18939. [CrossRef]

36. Cook, B.L.; Manning, W.G. Thinking beyond the mean: A practical guide for using quantile regression methods for health services research. *Shanghai Arch. Psychiatry* **2013**, *25*, 55–59.

37. Heckman, J.J. Sample selection bias as a specification error. *Econometrica* **1979**, *47*, 153–161. [CrossRef]

Article

Genetic Analysis and Gene Mapping for a Short-Petiole Mutant in Soybean (*Glycine max* (L.) Merr.)

Meifeng Liu, Yaqi Wang, Junyi Gai, Javaid Akhter Bhat, Yawei Li, Jiejie Kong and Tuanjie Zhao *

National Center for Soybean Improvement, Key Laboratory of Biology and Genetics and Breeding for Soybean, Ministry of Agriculture, State Key Laboratory for Crop Genetics and Germplasm Enhancement, Nanjing Agricultural University, Nanjing 210095, China; 2014201075@njau.edu.cn (M.L.); 2015201005@njau.edu.cn (Y.W.); sri@njau.edu.cn (J.G.); javid.akhter69@gmail.com (J.A.B.); 2018201083@njau.edu.cn (Y.L.); 2012094@njau.edu.cn (J.K.)
* Correspondence: tjzhao@njau.edu.cn; Tel.: +86-025-8439-9531

Received: 15 October 2019; Accepted: 28 October 2019; Published: 1 November 2019

Abstract: Short petiole is a valuable trait for the improvement of plant canopy of ideotypes with high yield. Here, we identified a soybean mutant line *derived short petiole* (*dsp*) with extremely short petiole in the field, which is obviously different from most short-petiole lines identified previously. Genetic analysis on 941 F_2 individuals and subsequent segregation analysis of 184 $F_{2:3}$ and 172 $F_{3:4}$ families revealed that the *dsp* mutant was controlled by two recessive genes, named as *dsp1* and *dsp2*. Map-based cloning showed that these two recessive genes were located on two nonhomologous regions of chromosome 07 and chromosome 11, of which the *dsp1* locus was mapped at a physical interval of 550.5-Kb on chromosome 07 near to centromere with flanking markers as BARCSOYSSR_07_0787 and BARCSOYSSR_07_0808; whereas, the *dsp2* locus was mapped to a 263.3-Kb region on chromosome 11 with BARCSOYSSR_11_0037 and BARCSOYSSR_11_0043 as flanking markers. A total of 36 and 33 gene models were located within the physical genomic interval of *dsp1* and *dsp2* loci, respectively. In conclusion, the present study identified markers linked with genomic regions responsible for short-petiole phenotype of soybean, which can be effectively used to develop ideal soybean cultivars through marker-assisted breeding.

Keywords: *derived short-petiole* (*dsp*) mutant; soybean (*Glycine max*); map-based cloning; simple sequence repeats (SSR) marker; bulked segregant analysis (BSA)

---

1. Introduction

Plant canopy architecture is an important agronomic trait for improving the yield potential in soybean and other legumes crops [1,2]. Ideal canopy structure with suitable leaf area index values has advantages of increasing light interception efficiency, that leads to increased photosynthesis as well as accumulation of photosynthetic assimilates, and eventually results in a higher yield [3,4]. It has been suggested that the desirable leaf area index of a population can be achieved by developing new cultivars for dense planting [1]. The interception capacity of crops on both direct and diffuse solar radiation is expected to increase under a horizontal canopy of dense seeding [5]. However, closed canopy profile also results in large within-canopy shading, which subsequently induces shade avoidance response and promotes lodging by excessive vegetative growth [6]. Kokubun (1988) investigated the characteristics of high-yielding soybean cultivars and proposed a high-yield ideotype model with the upper leaves vertically closed. The fraction of light absorption by the lower surface will be increased in this model [7]. Hence, it is an immediate prerequisite to modify plant canopy in a desirable direction through genetic manipulation for increasing crop yield.

Petiole length is an important trait that influences canopy architecture besides leaf petiole angle and branching capacity [8,9]. The short petiole changes geometric architecture of soybean plants and makes it possible for group canopy closing in a vertical plane when applied in dense seeding. Although the petiole length in soybean varies depending on photoperiod and light quality [9], environmental factors do not have much effect on this trait. Therefore, investigating short-petiole genetic resources is an efficient way to understand petiole development for further plant canopy improvement. Until now, various soybean mutants with short petiole have been identified as well as characterized. For example, D76-1609 and SS98206SP are two short-petiole lines which are controlled by two different single recessive genes, *lps1* and *lps3*, respectively [10,11]. You (1998) described a short-petiole line NJ90L-1SP, which is controlled by two duplicated loci of a recessive gene, *lps1*, that also controls short-petiole trait in D76-1609, and *lps2*, which controls the abnormal pulvinus [12]. These findings suggest that *lps1* and *lps2* might control different stages of petiole development. In addition, Cary and Nickell (1999) described a short petiole LN89-3502TP that is controlled by a single gene with incomplete dominance (*lc*), fitting a 1:2:1 segregation ratio in F_2 population [13]. So far, only *lps3* gene has been mapped on chromosome 13 within the flanking markers Sat_234 and Sct_033 [14]. Hence, identifying short-petiole mutant in soybean as well as the underlying gene will be helpful in promoting ideotype breeding in soybean for increasing yield.

By keeping the above in view, the present study identified a short-petiole mutant named as *dsp*. The objectives of our study were to elucidate the inheritance of the genes controlling short-petiole trait in *dsp* and to map underlying genes by using bulked segregant analysis (BSA) method.

2. Materials and Methods

2.1. Plant Material

The *dsp* is a derived soybean mutant with extremely short petiole, which is identified during soybean breeding in the field. Two crosses were made by crossing two soybean cultivars, viz., HDS-1 and BW-2, with the common *dsp* short-petiole mutant for genetic analysis and gene mapping of *dsp*. All F_1 seeds were planted and single-plant threshed. The 703 and 238 F_2 seeds were planted for HDS-1 × *dsp* and BW-2 × *dsp* F_2 populations, respectively (Table 1). All the 941 F_2 lines together with three parents were evaluated by visual inspection in the summer of 2014. In the summer of 2015, 50 seeds of 184 random $F_{2:3}$ progenies of HDS-1 × *dsp* were sown and used to determine the genotype of each F_2 plant.

Table 1. Genetic analysis of the short-petiole trait in the F_2 and $F_{3:4}$ populations.

Population	Year	No. of Wild-type Plants	No. of Mutant-type Plants	Total Number	Expected Ratio	x^2	*p-Value*
F_2(HDS-1 × *dsp*)	2014	681	22	703	15:1	11.16	0.00
F_2(BW-2 × *dsp*)	2014	230	8	238	15:1	2.91	0.09
Population	Year	No. of Segregating Lines	No. of Non-segregating Lines	Total Number	Expected Ratio	x^2	*p-Value*
$F_{2:3}$(HDS-1 × *dsp*)	2015	87	97	184	8:7	2.47	0.12
$F_{3:4}$(HDS-1 × *dsp*)-Dsp1_dsp2dsp2	2016	54	27	81	2:1	0.01	0.91
$F_{3:4}$(HDS-1 × *dsp*)-dsp1dsp1 Dsp2_	2016	59	32	91	2:1	0.07	0.80

To further validate the inheritance of *dsp* controlled by two loci, F_2 plants of HDS-1 × *dsp* with one locus as recessive homozygous and the other locus as heterozygotic were specifically selected by simple sequence repeats (SSR) markers according to gene mapping results. In this process, the flanking markers BARCSOYSSR_07_0787 and BARCSOYSSR_07_0808 for *dsp1* locus and BARCSOYSSR_11_0037 and BARCSOYSSR_11_0049 for *dsp2* locus were used to screen out those F_2 plants of which genotypes as *Dsp1dsp1dsp2dsp2* or *dsp1dsp1Dsp2dsp2*. Then 81 and 91 $F_{2:3}$ dominant individuals derived from the above selected F_2 plants were randomly harvested, and 50 seeds for each $F_{3:4}$ progenies were grown to evaluate the genotype of above $F_{2:3}$ individuals in the summer of 2016 (Table 1). A total of 30 mutant individuals from F_2 populations of HDS-1 × *dsp* and BW-2 × *dsp* were collected for gene mapping. There were 132 $F_{2:3}$ recessive individuals derived from *Dsp1dsp1dsp2dsp2* or *dsp1dsp1Dsp2dsp2* F_2 plants used to verify the mapping results of *dsp1* and *dsp2* loci.

The *dsp* mutant has extremely short petiole, however, a very small, nonsignificant variation exists for petiole length at different leaf positions, and thus were ignored. Therefore, short petiole of *dsp* was regarded as a qualitative trait in our study, and the petiole length was phenotypically evaluated as mutant-type (MT) and wild-type (WT) by visual inspection.

All materials were planted at the research field in Jiangpu Experimental Station of Nanjing Agricultural University (Nanjing, China).

2.2. Genetic Analysis of dsp

A Chi-square (x^2) test (1) was used to analyze the segregation ratio of alleles with the expected ratio at a significance threshold of *p*-value>0.05 ($x^2 < 3.84$) [15,16]. The formulas used are shown as below

$$x2 = \sum \frac{(|O - E| - 0.5)^2}{E} \tag{1}$$

where O and E represent observed and expected value, respectively, under the expected ratio.

2.3. DNA Extraction and SSR Markers Analysis

A plant tissue kit from Tiangen Biotech (Beijing, China) was used to extract DNA from the young and healthy fresh leaves of three parents, F_2 generations, and 132 $F_{2:3}$ individuals derived from the crosses HDS-1 × *dsp* and BW-2 × *dsp*. PCR amplifications were performed in 10 µL reactions containing 50–100 ng of template DNA, 1 × PCR buffer, 2.0 mM $MgCl_2$, 75 µM of each dNTP, 0.2 µM each of the forward and reverse primers, and 0.1 U of Taq DNA polymerase. DNA polymerase, deoxy-ribonucleoside triphosphate (dNTP) mix, and DNA Ladder (50 base pairs) were purchased from Tiangen Biotech (Beijing, China). Primer sequences were obtained from SoyBase website (http://www.soybase.org) and were synthesized by Invitrogen Biological Technology (Shanghai, China). The PCR reaction was performed under the following condition: Initial denaturation at 95 °C for 5 min, followed by 29 cycles with 30 seconds of denaturation at 94 °C, 30 seconds of annealing at 43–56 °C, depending on the optimum annealing temperature for each primer pair, and 30 seconds of extension at 72 °C, with a final 10 min extension at 72 °C on a Peltier thermal cycler (PTC-225, MJ Research, Quebec, QC, Canada). The PCR products were separated by electrophoresis through 8% non-denaturing polyacrylamide gels, and then the gels were stained with 1 g L^{-1} $AgNO_3$ for 15 min, followed by a 1% NaOH and 1% CH_3OH solution for 10 min before visualizing under LED light box.

2.4. Bulked Segregant Analysis (BSA) and Target Gene Mapping of the dsp Mutant

Bulked segregant analysis (BSA) was performed to identify SSR markers potentially linked to the genes responsible for short-petiole trait [17]. A total of 1015 pairs of SSR (simple sequence repeat) primers covering all the 20 chromosomes were included in this process. The normal bulk was formed by pooling DNA of six individuals with normal petiole from the F_2 populations. Similarly, the mutant bulk was created with the DNA of six individuals with short petiole. The normal and mutant DNA bulks as well as the DNA of two parents for the F_2 population of HDS-1 × *dsp* cross were screened with 1015

SSR markers to identify polymorphic markers that are potentially linked to the short-petiole trait of *dsp*. Linkage relationship of the locus and SSR markers were calculated with the program Mapmaker 3.0 [18], using a minimum LOD (logarithm of the odds) score of 3.0 and a maximum recombination value of 0.4 as a threshold. Linkage calculations were completed using the Kosambi mapping function [19]. SSR markers identified to be linked were consequently screened against the entire mapping population. Then subsequent mapping processes were conducted according to Song (2004) [20]. Searching of the physical position of the primer sequence was performed via BLASTN engine on National Center for Biotechnology Information database (NCBI) (http//www.ncbi.nlm.nih.gov/) and converting the genetic map to a physical map based on the physical position of SSR markers.

2.5. Synteny Information and Homologous Protein Information Retrieval

For every locus, the synteny information on genome was retrieved from the SoyBase website (https://soybase.org/) through "Genome Browser" [21]. The duplicated region of this locus (if it exists) would be displayed under the precondition of checking "old duplicated blocks" or "recent duplicated blocks" options in select track. Phytozome website (http://www.phytozome.net) [22] was used to retrieve the corresponding homologous proteins information of every candidate protein.

3. Results and Discussion

3.1. Characteristics of Short-Petiole Trait in the dsp Mutant

Soybean short-petiole mutant *dsp* has a compact stature in the natural field conditions (Figure 1), and also revealed different levels of dwarfing under different genetic background and environments (plant height variation data not shown). The petiole length of *dsp* was less than 2 cm, showing a difference from previously described short-petiole mutants, for instance D76-1609 (8.59 ± 0.93 cm) [10], NJ90L-1SP (7–13 cm) [12], and LN89-3502TP (5–12 cm) [13]. Moreover, *dsp* was relatively more valuable for progenies selection compared with NJ90-1SP and LN89-3502TP, which have inferior agronomic characters including abnormal leaf and pulvinus trait. Considering the ideal soybean architecture model proposed as compact plants with a small stature as well one or two branches [1], *dsp* mutant provides a valuable genetic resource for the development of a soybean ideal plant-type.

Figure 1. Canopy architecture and petiole morphology of the *dsp* mutant (**a–b**). Aerial view of the *dsp* mutant (**a**) and BW-2 wild-type plant (**b**). (**c**) The morphology and petiole length of wild type and mutant. Scale bars represent 2cm in all pictures.

3.2. Genetic Analysis of Petiole Length in the dsp Mutant

Two crosses were made between the common short-petiole *dsp* mutant and two soybean cultivars with normal petiole, HDS-1 and BW-2, respectively. All the F_1 plants obtained from the two populations (HDS-1 × *dsp* and BW-2 × *dsp*) had normal petioles, indicating that the short petiole of *dsp* mutant was recessive to the normal petiole. This result was consistent with that of previous genetic studies based on short-petiole lines, which also revealed the recessive nature of short-petiole length in soybean [10–12]. However, the results of Cary and Nickell (1999) [13] were in contrast. They reported that a short petiole was controlled by a single gene with incomplete dominance.

In the F_2 population from the cross BW-2 × *dsp*, eight out of 238 F_2 individuals had the mutant-type phenotype the same as the *dsp* parent (MT plants). However, the ratio of wild-type plants (WT plants) relative to mutant-type plants (MT plants) was not significantly different from a 15:1 ratio ($\chi^2 = 2.91$, $p = 0.09$) (Table 1). In the case of HDS-1 × *dsp* cross, segregation for petiole-length trait was observed in 87 of the 184 $F_{2:3}$ rows derived from 184 F_2 WT plants, which was not significantly different from a 8:7 ratio for segregating and nonsegregating in long-petiole rows ($\chi^2 = 2.47$, $p = 0.12$) (Table 1). According to subsequent mapping results of genes responsible for *dsp*, F_2 WT plants with genotype as *Dsp1dsp1dsp2dsp2* or *dsp1dsp1Dsp2dsp2* from HDS-1 × *dsp* cross were screened out based on flanking markers of two mapping regions (Figures 2 and 3), and random 81 and 91 $F_{2:3}$ WT plants were grouped to validate the inheritance of *dsp* mutant. As a result, both ratios between those segregating and homozygous (nonsegregating) F_3 individuals fit a 2:1 ratio based on $F_{3:4}$ families ($\chi^2 = 0.01$, $p = 0.91$; $\chi^2 = 0.07$, $p = 0.80$) (Table 1).

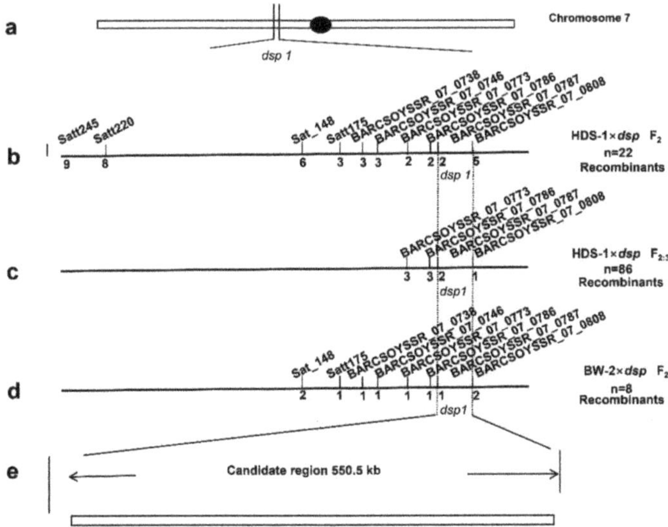

Figure 2. Mapping of the *dsp1* locus. (**a–d**) Two F_2 populations of HDS-1 × *dsp* (**b**) and BW-2 × *dsp* (**d**) were used here. One $F_{2:3}$ population derived from *Dsp1dsp1dsp2dsp2* F_2 plants of HDS-1 × *dsp* cross was used to verify the mapping result of *dsp1* (**c**). The *dsp1* locus was mapped to a 550.5-kb region nearby centromere on chromosome 07 with BARCSOYSSR_07_0787 and BARCSOYSSR_07_0808 as flanking markers (**e**). Black spot represents centromere. Number below every SSR marker means recombinant individual number.

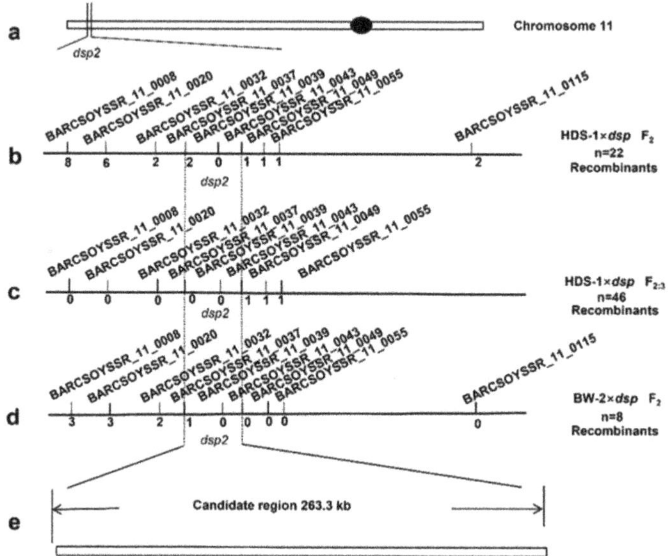

Figure 3. Mapping of the *dsp2* locus. (**a–d**) Two F_2 populations of HDS-1 × *dsp* (**b**) and BW-2 × *dsp* (**d**) were used here. One $F_{2:3}$ population derived from *dsp1dsp1Dsp2dsp2* F_2 plants of HDS-1 × *dsp* was used to verify the mapping result of *dsp2* (**c**). The *dsp2* locus was mapped between BARCSOYSSR_11_0037 and BARCSOYSSR_11_0043, a 263.3-kb region in the front end of chromosome 11 (**e**). Black spot represents centromere. Number below every SSR marker means recombinant individual number.

These results demonstrated that the short-petiole trait of *dsp* mutant is controlled by two recessive genes even though the segregation ratio in F_2 population of HDS-1 × *dsp* was significantly different from a 15:1 ratio ($\chi^2 = 11.16$, $p = 0.00$) in the 2014 field experiment. That is probably attributed to poor emergence by poor seed quality of mutant plants (Table 1). This is somewhat similar to the case of genetic analysis for SS98206SP, a short-petiole line controlled by a single recessive gene and fit a 15:1 ratio of F_2 progenies between long and short petioles based on bad seed quality in 2006 [11].

3.3. Mapping Genes dsp1 and dsp2 with SSR Markers

Out of a total 1015 SSR markers screened for polymorphism, only 67 markers distributed to 12 chromosomes were found to be polymorphic between WT and MT DNA pools derived from F_2 population of HDS-1 × *dsp* cross. Finally, polymorphic markers on chromosome 07 (linkage group M) and chromosome 11 (linkage group B1) were detected to be linked with the short-petiole mutant phenotype of *dsp*. These genomic regions governing the mutant phenotype of *dsp* was named as *dsp1* and *dsp2*, respectively.

A total of 22 and 8 F_2 MT plants from HDS-1 × *dsp* and BW-2 × *dsp*, respectively, were genotyped using linked markers screened from chromosome 07 and chromosome 11. New SSR markers from Song (2010) [23] were synthesized to narrow the mapping regions of *dsp1* and *dsp2* loci. Eventually, the *dsp1* locus was mapped to a 550.5-Kb region on chromosome 07 with flanking markers as BARCSOYSSR_07_0787 and BARCSOYSSR_07_0808 (Figure 2). The mapping region is near to centromere according to SoyBase database [21]. A total of 36 gene models were present within this region (Glyma.Wm82.a1.v1.1) (Table 2). The *dsp2* locus was mapped within a 263.3-Kb region between BARCSOYSSR_11_0037 and BARCSOYSSR_11_0049 markers on the front of chromosome 11, harboring 33 gene models (Figure 3, Table 3). Furthermore, these $F_{2:3}$ MT plants from *Dsp1dsp1dsp2dsp2* or *dsp1dsp1Dsp2dsp2* F_2 individuals were used to confirm the mapping regions. Among them, 86 $F_{2:3}$ MT plants from *Dsp1dsp1 dsp2dsp2* F_2 plants were utilized to validate the mapping result of *dsp1*.

Then *dsp1* was mapped to the same genomic region as identified earlier using F_2 populations (Figure 2). Similarly, 46 $F_{2:3}$ MT plants from *dsp1dsp1 Dsp2dsp2* F_2 plants confirmed the locus *dsp2*. The right boundary of *dsp2* is definite with BARCSOYSSR_11_0049, which is also consistent with the mapping result of *dsp2* in F_2 populations (Figure 3). The petiole length of *dsp* mutant is very similar to another SS98206SP line, while the *lps3* locus underlying short petiole of SS98206SP was reported to be mapped on chromosome 13 [14]. Hence, *dsp* is a novel short-petiole line different from SS98206SP.

Table 2. List of gene models in the 550.5-Kb physical interval of the *derived short-petiole1* (*dsp1*) locus.

Glyma.Wm82. a1.v1.1 Gene Models	Glyma.Wm82. a2.v1 Gene Models	Description
Glyma07g16731	*Glyma.07g139000*	Mitochondrial carrier protein
Glyma07g16740	*Glyma.07g139100*	Major Facilitator Superfamily
Glyma07g16750	*Glyma.07g139200*	Uncharacterized protein
Glyma07g16760	*Glyma.07g139300*	Uncharacterized protein
Glyma07g16770	*Glyma.07g139400*	Probable lipid transfer
Glyma07g16790	*Glyma.07g139500*	PRONE (Plant-specific Rop nucleotide exchanger)
Glyma07g16800	*Glyma.07g139600*	Glutathione S-transferase, C-terminal domain
Glyma07g16810	*Glyma.07g139700*	Glutathione S-transferase, C-terminal domain
Glyma07g16830	*Glyma.07g139800*	Glutathione S-transferase, C-terminal domain
Glyma07g16840	*Glyma.07g139900*	Glutathione S-transferase, C-terminal domain
Glyma07g16850	*Glyma.07g140000*	Glutathione S-transferase, C-terminal domain
Glyma07g16860	*Glyma.07g140200*	Glutathione S-transferase, C-terminal domain
Glyma07g16876	*Glyma.07g140300*	Glutathione S-transferase, C-terminal domain
Glyma07g16893	-	Cytochrome P450 CYP4/CYP19/CYP26 subfamilies
Glyma07g16910	*Glyma.07g140400*	Glutathione S-transferase, C-terminal domain
Glyma07g16925	*Glyma.07g140500*	Glutathione S-transferase, C-terminal domain
Glyma07g16940	*Glyma.07g140700*	Glutathione S-transferase, C-terminal domain
Glyma07g16950	*Glyma.07g140800*	Universal stress protein family
Glyma07g16970	*Glyma.07g141000*	Uncharacterized protein
Glyma07g16980	*Glyma.07g141100*	Myb-like DNA-binding domain
Glyma07g16990	*Glyma.07g141200*	Ctr copper transporter family
Glyma07g17000	*Glyma.07g141300*	Probable lipid transfer
Glyma07g17010	*Glyma.07g141400*	Non-specific serine/threonine protein kinase.
Glyma07g17030	*Glyma.07g141500*	Probable lipid transfer
Glyma07g17060	*Glyma.07g141600*	Ctr copper transporter family
Glyma07g17080	*Glyma.07g141700*	Hsp20/alpha crystallin family
Glyma07g17090	*Glyma.07g141800*	Uncharacterized protein
Glyma07g17101	*Glyma.07g141900*	Domain of unknown function (DUF3527)
Glyma07g17110	*Glyma.07g142000*	PAP2 superfamily C-terminal
Glyma07g17116	*Glyma.07g142100*	Ubiquitinyl hydrolase 1.
Glyma07g17130	*Glyma.07g142300*	RING-variant domain
Glyma07g17140	*Glyma.07g142400*	Multicopper oxidase
Glyma07g17150	*Glyma.07g142500*	Multicopper oxidase
Glyma07g17170	*Glyma.07g142600*	Multicopper oxidase
Glyma07g17180	*Glyma.07g142700*	Fructose-1-6-bisphosphatase
Glyma07g17190	*Glyma.07g142800*	Uncharacterized protein

Table 3. List of gene models in the 263.3-Kb physical interval of the *dsp2* locus.

Glyma.Wm82. a1.v1.1 Gene Models	Glyma.Wm82. a2.v1 Gene Models	Description
Glyma11g01300	*Glyma.11G011000*	RNA-binding proteins
Glyma11g01310	*Glyma.11G011100*	Uncharacterized protein
Glyma11g01320	*Glyma.11G011200*	NADH: ubiquinone oxidoreductase, B17.2 subunit
Glyma11g01330	*Glyma.11G011300*	E3 ubiquitin ligase
Glyma11g01340	*Glyma.11G011400*	Translin-associated protein X
Glyma11g01350	*Glyma.11G011500*	Chalcone and stilbene synthases
Glyma11g01360	*Glyma.11G011600*	PPR repeat
Glyma11g01370	*Glyma.11G011700*	Nuclear transport receptor CRM1/MSN5 (importin beta superfamily)
Glyma11g01380	*Glyma.11G011800*	GTP-binding ADP-ribosylation factor Arf1
Glyma11g01390	*Glyma.11G011900*	Plant protein of unknown function (DUF946)
Glyma11g01405	*Glyma.11G012000*	Guanosine-3′,5′-bis(diphosphate)3′-pyrophosphohydrolase
Glyma11g01420	*Glyma.11G012100*	Ribonuclease P
Glyma11g01430	*Glyma.11G012200*	DEAD/DEAH box helicase/Helicase conserved C-terminal domain
Glyma11g01441	*Glyma.11G012300*	Pwwp domain-containing protein
Glyma11g01450	*Glyma.11G012400*	Cell division cycle 20 (CDC20) (Fizzy)-related
Glyma11g01460	*Glyma.11G012500*	Putative u4/u6 small nuclear ribonucleoprotein
Glyma11g01470	*Glyma.11G012700*	Mitochondrial outer membrane protein
Glyma11g01480	*Glyma.11G012800*	Galactosyltransferases
Glyma11g01491	*Glyma.11G012900*	Aspartyl proteases
Glyma11g01501	*Glyma.11G013000*	Aspartyl proteases
Glyma11g01510	*Glyma.11G013100*	Aspartyl proteases
Glyma11g01520	*Glyma.11G013200*	Uncharacterized protein
Glyma11g01530	*Glyma.11G013300*	PLAC8 family
Glyma11g01536	*Glyma.11G013400*	DYW family of nucleic acid deaminases (DYW_deaminase)
Glyma11g01543	*Glyma.11G013500*	PPR repeat (PPR)
Glyma11g01550	*Glyma.11G013600*	PPR repeat (PPR)
Glyma11g01570	*Glyma.11G013700*	leucine-rich PPR motif-containing protein, mitochondrial (LRPPRC)
Glyma11g01580	*Glyma.11G013800*	Complex 1 protein (LYR family)
Glyma11g01595	*Glyma.11G013900*	KH domain containing RNA binding protein
Glyma11g01610	*Glyma.11G014000*	Protein Phosphatase methylesterase-1 related
Glyma11g01620	*Glyma.11G014100*	Cytochrome c
Glyma11g01640	*Glyma.11G014200*	Ethylene-responsive transcription factor ERF021
Glyma11g01650	*Glyma.11G014300*	Nuclear transport factor 2 (NTF2) domain

Based on the synteny information obtained from SoyBase [21], the mapping region of *dsp1* and *dsp2* belongs to two nonhomologous fragments on chromosome 07 and chromosome 11, respectively (Figure 4). Furthermore, homologous protein information retrieved from Phytozome [22] revealed that none of the encoding proteins within both regions displayed homology (Tables 2 and 3). Hence, these two above loci identified to be responsible for short-petiole phenotype of the *dsp* mutant did not function as duplicated genes.

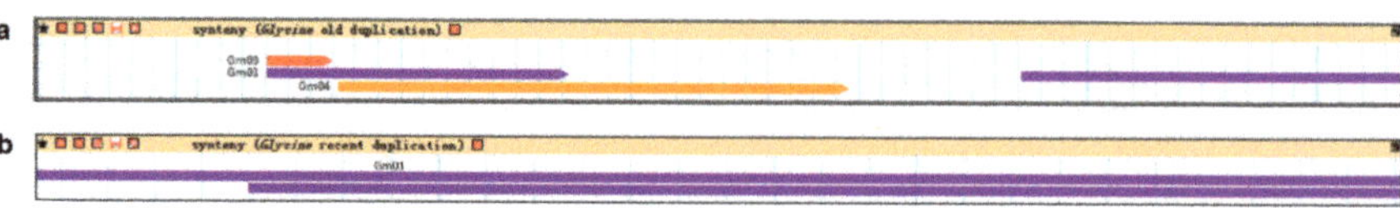

Figure 4. The duplicated regions of *dsp1* and *dsp2* locus from SoyBase. (**a**) The duplicated regions of the *dsp1* locus. (**b**) The duplicated regions of the *dsp2* locus.

As a paleopolyploid, the genome of soybean contains 70.3% duplicate regions due to two whole genome duplication (WGD) events [24–27]. Duplicated genes may undergo pseudogenization, sub-functionalization, or neo-functionalization [28], and the divergence of duplicated genes is thought to provide the basis for adaptive evolution [29]. For example, among the four homologous genes of the *Arabidopsis terminal flower* gene (*TFL1*) in soybean, only one has been found to control growth habit; the other copies may have additional functions because they have been reported to show different transcriptional patterns [30]. A disease-like rugose leaf phenotype in soybean was attributed to two recessive duplicated loci of *rl1* and *rl2* [31]. In our study, all genes located within the mapping regions of *dsp1* and *dsp2* loci have one or more duplicated copies in other chromosomes. The duplicated effect of *dsp1* and *dsp2* candidate genes on the short petiole of *dsp* mutant indicated the possible functional differentiation or genetic interaction during petiole development. Therefore, the short petiole of *dsp* mutant is a complex trait, and the identification of candidate genes underlying *dsp1* and *dsp2* loci will greatly help to clarify the mechanism of petiole development.

4. Conclusions

In summary, we identified a novel short-petiole mutant line *"dsp"* that was demonstrated to be controlled by two recessive gene designated as *dsp1* and *dsp2*. The mapping of *dsp1* and *dsp2* revealed a redundant function between two nonhomologous loci on the formation of short petiole. Hence, the present study provides potential genetic resources, linked markers as well as genes governing the short petiole in *dsp* mutant, and these valuable genes will be in turn used for rapid introgression into elite soybean backgrounds for developing cultivars with short petiole and high yield via marker-assisted breeding. Therefore, the availability of these resources could greatly facilitate the dream of developing soybean ideotype.

Author Contributions: Conceptualization: M.L. and T.Z.; Data curation: M.L. and Y.W.; Formal analysis: J.A.B. and Y.L.; Funding acquisition: T.Z.; Methodology: Y.L. and J.K.; Software: Y.W.; Writing—original draft: M.L.; Writing—review and editing: T.Z. and J.G.

Funding: This research was funded by the National Natural Science Foundation of China (Grant Nos. 31571691, 31871646), the Fundamental Research Funds for the Central Universities (KYT201801), the MOE Program for Changjiang Scholars and Innovative Research Team in University (PCSIRT_17R55), the MOE 111 Project (B08025), and the Jiangsu Collaborative Innovation Center for Modern Crop Production (JCIC-MCP) Program.

Conflicts of Interest: The authors declare no conflict of interest.

References

1. Chavarria, G.; Caverzan, A.; Müller, M.; Rakocevic, M. Soybean architecture plants: From solar radiation interception to crop protection. In *Soybean-The Basis of Yield, Biomass and Productivity*; InTechOpen: London, UK; p. 15.
2. Mwanamwenge, J.; Siddique, K.H.M.; Sedgley, R.H. Canopy development and light absorption of grain legume species in a short season mediterranean-type environment. *J. Agron. Crop Sci.* **1997**, *179*, 1–7. [CrossRef]
3. Isoda, A.; Mori, M.; Matsumoto, S.; Li, Z.; Wang, P. High yielding performance of soybean in Northern Xinjiang, China. *Plant Prod. Sci.* **2015**, *9*, 401–407. [CrossRef]
4. Liu, X.; Jin, J.; Wang, G.; Herbert, S.J. Soybean yield physiology and development of high-yielding practices in Northeast China. *Field Crop Res.* **2008**, *105*, 157–171. [CrossRef]
5. Muraoka, H.; Takenaka, A.; Tang, Y.; Koizumi, H.; Washitani, I. Flexible leaf orientations of Arisaema heterophyllum maximize light capture in a forest understorey and avoid excess irradiance at a deforested site. *Ann. Bot.* **1998**, *82*, 297–307. [CrossRef]
6. Junior, C.P.; Kawakami, J.; Bridi, M.; Conte, M.V.D.; Michalovicz, L. Phenological and quantitative plant development changes in soybean cultivars caused by sowing date and their relation to yield. *Afr. J. Agric. Res.* **2015**, *10*, 515–523. [CrossRef]
7. Kokubun, M. Design and evaluation of soybean ideotypes. *Bull. Tohoku Natl. Agric. Exp. Stn.* **1988**, *77*, 77–142.

8. Niinemets, Ü.; Fleck, S. Petiole mechanics, leaf inclination, morphology and investment in support in relation to light availability in the canopy of Liriodendron tulipifera. *Oecologia* **2002**, *132*, 21–33. [CrossRef]
9. Thomas, J.F.; Raper, C.D. Internode and petiole elongation of soybean in response to photoperiod and end-of-day light quality. *Bot. Gaz.* **1985**, *146*, 495–500. [CrossRef]
10. Jun, T.H.; Kang, S.T.; Moon, J.K.; Seo, M.J.; Yun, H.T.; Lee, S.K.; Lee, Y.H.; Kim, S.J. Genetic analysis of new short petiole gene in soybean. *J. Crop Sci. Biotechnol.* **2009**, *12*, 87–89. [CrossRef]
11. Kilen, T. Inheritance of a short petiole trait soybean. *Crop Sci.* **1983**, *23*, 1208–1210. [CrossRef]
12. You, M.; Zhao, T.; Gai, J.; Yen, Y. Genetic analysis of short petiole and abnormal pulvinus in soybean. *Euphytica* **1998**, *102*, 329–333. [CrossRef]
13. Jun, T.H.; Kang, S.T. Genetic map of *lps3*: A new short petiole gene in soybeans. *Genome* **2012**, *55*, 140–146. [CrossRef] [PubMed]
14. Cary, T.R.; Nickell, C.D. Genetic analysis of a short petiole-type soybean, LN89-3502TP. *J. Hered.* **1999**, *90*, 300–301. [CrossRef]
15. Pearson, K.X. On the criterion that a given system of deviations from the probable in the case of a correlated system of variables is such that it can be reasonably supposed to have arisen from random sampling. *Philos. Mag.* **1900**, *50*, 157–175. [CrossRef]
16. Yates, F. Contingency tables involving small numbers and the χ^2 test. *J. R. Stat. Soc.* **1934**, *1*, 217–235.
17. Michelmore, R.W.; Paran, I.; Kesseli, R.V. Identification of markers linked to disease-resistance genes by bulked segregant analysis: A rapid method to detect markers in specific genomic regions by using segregating populations. *Proc. Natl. Acad. Sci. USA* **1991**, *88*, 9828–9832. [CrossRef]
18. Lander, E.S.; Green, P.; Abrahamson, J.; Barlow, A.; Daly, M.J.; Lincoln, S.E.; Newburg, L. MAPMAKER: An interactive computer package for constructing primary genetic linkage maps of experimental and natural populations. *Genomics* **1987**, *1*, 174–181. [CrossRef]
19. Kosambi, D.D. The estimation of ma distance from recombination values. *Ann. Eugen.* **1944**, *12*, 172–175. [CrossRef]
20. Song, Q.J.; Marek, L.F.; Shoemaker, R.C.; Lark, K.G.; Concibido, V.C.; Delannay, X.; Specht, J.E.; Cregan, P.B. A new integrated genetic linkage map of the soybean. *Theor. Appl. Genet.* **2004**, *109*, 122–128. [CrossRef]
21. Grant, D.; Nelson, R.T.; Cannon, S.B.; Shoemaker, R.C. SoyBase, the USDA-ARS soybean genetics and genomics database. *Nucleic Acids Res.* **2010**, *38*, D843–D846. [CrossRef]
22. Goodstein, D.M.; Shu, S.; Howson, R.; Neupane, R.; Hayes, R.D.; Fazo, J.; Mitros, T.; Dirks, W.; Hellsten, U.; Putnam, N.; et al. Phytozome: A comparative platform for green plant genomics. *Nucleic Acids Res.* **2012**, *40*, D1178–D1186. [CrossRef] [PubMed]
23. Song, Q.; Jia, G.; Zhu, Y.; Grant, D.; Nelson, R.T.; Hwang, E.Y.; Hyten, D.L.; Cregan, P.B. Abundance of SSR motifs and development of candidate polymorphic SSR markers (BARCSOYSSR_1.0) in soybean. *Crop Sci.* **2010**, *50*, 1950–1960. [CrossRef]
24. Chan, C.; Qi, X.; Li, M.W.; Wong, F.L.; Lam, H.M. Recent developments of genomic research in soybean. *J. Genet. Genom.* **2012**, *39*, 317–324. [CrossRef] [PubMed]
25. Schlueter, J.A.; Lin, J.Y.; Schlueter, S.D.; Vasylenko-Sanders, I.F.; Deshpande, S.; Yi, J.; O'bleness, M.; Roe, B.A.; Nelson, R.T.; Scheffler, B.E.; et al. Gene duplication and paleopolyploidy in soybean and the implications for whole genome sequencing. *BMC Genom.* **2007**, *8*, 330. [CrossRef]
26. Schmutz, J.; Cannon, S.B.; Schlueter, J.; Ma, J.; Mitros, T.; Nelson, W.; Hyten, D.L.; Song, Q.; Thelen, J.J.; Cheng, J.; et al. Genome sequence of the palaeopolyploid soybean. *Nature* **2010**, *463*, 178. [CrossRef]
27. Shoemaker, R.C.; Schluete, J.; Doyle, J.J. Paleopolyploidy and gene duplication in soybean and other legumes. *Curr. Opin. Plant Biol.* **2006**, *9*, 104–109. [CrossRef]
28. Innan, H.; Kondrashov, F. The evolution of gene duplications: Classifying and distinguishing between models. *Nat. Rev. Genet.* **2010**, *11*, 97–108. [CrossRef]
29. Flagel, L.E.; Wendel, J.F. Gene duplication and evolutionary novelty in plants. *New Phytol.* **2009**, *183*, 557–564. [CrossRef]

30. Tian, Z.; Wang, X.; Lee, R.; Li, Y.; Specht, J.E.; Nelson, R.L.; McClean, P.E.; Qiu, L.; Ma, J. Artificial selection for determinate growth habit in soybean. *Proc. Natl. Acad. Sci. USA* **2010**, *107*, 8563–8568. [CrossRef]
31. Wang, Y.; Chen, W.; Zhang, Y.; Liu, M.; Kong, J.; Yu, Z.; Jaffer, A.M.; Gai, J.; Zhao, T. Identification of two duplicated loci controlling a disease-like rugose leaf phenotype in soybean. *Crop Sci.* **2016**, *56*, 1611. [CrossRef]

 *agronomy*

Article

Assessment of Water Absorption Capacity and Cooking Time of Wild Under-Exploited *Vigna* Species towards their Domestication

Difo Voukang Harouna [1,2], Pavithravani B. Venkataramana [2,3], Athanasia O. Matemu [1] and Patrick Alois Ndakidemi [2,3,*]

[1] Department of Food Biotechnology and Nutritional Sciences, Nelson Mandela African Institution of Science and Technology (NM-AIST), Arusha P.O. Box 447, Tanzania
[2] Centre for Research, Agricultural Advancement, Teaching Excellence and Sustainability in Food and Nutrition Security (CREATES-FNS), Nelson Mandela African Institution of Science and Technology, Arusha P.O. Box 447, Tanzania
[3] Department of Sustainable Agriculture, Biodiversity and Ecosystems Management, Nelson Mandela African Institution of Science and Technology, Arusha P.O. Box 447, Tanzania
* Correspondence: Patrick.ndakidemi@nm-aist.ac.tz; Tel.: +255-757-744-772

Received: 16 July 2019; Accepted: 26 August 2019; Published: 4 September 2019

Abstract: Some phenotypic traits from wild legumes are relatively less examined and exploited towards their domestication and improvement. Cooking time for instance, is one of the most central factors that direct a consumer's choice for a food legume. However, such characters, together with seed water absorption capacity are less examined by scientists, especially in wild legumes. Therefore, this study explores the cooking time and the water absorption capacity upon soaking on 84 accessions of wild *Vigna* legumes and establishes a relationship between their cooking time and water absorbed during soaking for the very first time. The accessions were grown in two agro-ecological zones and used in this study. The Mattson cooker apparatus was used to determine the cooking time of each accession and 24 h soaking was performed to evaluate water absorbed by each accession. The two-way analysis of variance revealed that there is no interaction between the water absorption capacity and cooking time of the wild *Vigna* accessions with their locations or growing environments. The study revealed that there is no environment × genotype interaction with respect to cooking time and water absorption capacity as phenotypic traits while genotype interactions were noted for both traits within location studied. Furthermore, 11 wild genotypes of *Vigna* accessions showed no interaction between the cooking time and the water absorption capacity when tested. However, a strong negative correlation was observed in some of the wild *Vigna* species which present phenotypic similarities and clusters with domesticated varieties. The study could also help to speculate on some candidates for domestication among the wild *Vigna* species. Such key preliminary information could be of vital consideration in breeding, improvement, and domestication of wild *Vigna* legumes to make them useful for human benefit as far as cooking time is concerned.

Keywords: non-domesticated legumes; *Vigna racemosa*; *Vigna ambacensis*; *Vigna reticulata*; *Vigna vexillata*; wild food legumes; legumes; *Vigna* species; cooking time; water absorption; domestication

---

1. Introduction

Legumes (family: Fabaceae), the third largest family among flowering plants, grouping about 650 genera and 20,000 species, represent the second most valuable plant source of nutrients for both humans and animals [1]. Their importance in human life through positive impact in global food security is uncontestable due to the contribution of some of the domesticated commercialized legumes such as soybeans, cowpeas, and common beans [2]. Yet, their production rate remains unsatisfying compared with their consumption rate due to several challenges ranging from agronomic constraints to policy issues through farmers' and consumers' acceptability [3,4]. These challenges have directed the interest of some scientists towards investigating novel alternatives by screening the hitherto wild non-domesticated species within the little-known genera of legumes in order to find important traits that fit consumers' acceptance and desire without necessarily genetically engineering them [4].

Generally, taste and smell are the first senses that come to peoples' mind whenever they think about the consumption of food and drink [5]. However, consumers' responses to food depend on several factors not only limited to sensory characteristics of the product and their physiological status. They also depend on other factors, such as previous information acquired about the product, their past experience, and their attitudes and beliefs [6].

Soaking is usually a processing technique performed prior to cooking of grains and legumes. Hence, the evaluation of water absorption characteristics of different seeds during soaking is an important parameter that is well considered by researchers who have proven that grains show different water absorption rates and water absorption capacities in different soaking conditions [7]. Understanding water absorption in legumes during soaking is a very important aspect because it affects succeeding processes such as the cooking time and the quality of the final product [8]. Water absorption of seeds during soaking mainly depends on soaking time, water temperature, and some seeds' physical characteristics like hardness and seed coat thickness, and may be related to cooking time for a specific type of grain or legume. This is one of the gaps that this study is attempting to address.

Cooking time, a sign of cooking quality, is one of the most central factors that direct a consumer's choice for a food legume as longer cooking time is one of the foremost limitations that make legumes uneconomical and unacceptable to consumers [9]. Cooking usually implies heat application that causes physicochemical changes like gelatinization of starch, denaturation of proteins, solubilization of some of the polysaccharides, softening and breakdown of the middle lamella, a cementing material found in the cotyledon [9]. Cooking also inactivates or reduces the levels of anti-nutrients such as trypsin inhibitors and flatulence-causing oligosaccharides, resulting in improved nutritional and sensory qualities [10].

Cooking time is also one of the phenotypic characters assessed by many breeding programs using the Mattson Bean Cooker as the recommended equipment for measuring the variable [11]. The cooking time of legumes depend on their genera, species, and varieties [7,9].

Common beans (*Phaseolus vulgaris*) and cowpea (*Vigna unguiculata*) are the mainly cultivated and consumed varieties of legumes worldwide that belong to two different genera, the *Phaseolus* and *Vigna* respectively [2,12,13]. It is reported that fewer domesticated edible species as compared with the numerous non-domesticated wild species exist in most legumes' genera. Domesticated and semi-domesticated species are denoted as neglected and underutilized species due to little attention being paid to them or the complete ignorance of their existence by agricultural researchers, plant breeders, and policymakers [14].

The genus *Vigna*, of the present study, is a large collection of vital legumes consisting of more than 200 species [15]. It comprises several species of agronomic, economic, and environmental importance. The most common domesticated ones include the mung bean (*V. radiata* (L.) Wilczek), urd bean (*V. mungo* (L.) Hepper), cowpea (*V. unguiculata* (L.) Walp.), azuki bean (*V. angularis* (Willd.) Ohwi & Ohashi), bambara groundnut (*V. subterranea* (L.) Verdc.), moth bean (*V. aconitifolia* (Jacq.) Maréchal), and rice bean (*V. umbellata* (Thunb.) Ohwi & Ohashi). Many of these species are valued as forage, green manure, and cover crops, besides their value as high protein grains. Moreover, the genus also comprises more than 100 wild species that do not possess common names apart from their scientific appellation yet [16]. They are simply known as underexploited wild *Vigna* species, or non-domesticated *Vigna* species [2,15]. This could be some of the reasons as to why very little or almost no scientific attention has been given to them, especially concerning the human domestic utilization such as consumption, cookability, functional, and processing characteristics such as water absorption capacity and soaking.

Therefore, this study evaluates the cooking time and water absorption capacity upon soaking on 84 accessions of wild *Vigna* legumes and establishes a relationship between their cooking time and water absorbed during soaking.

2. Materials and Methods

2.1. Sample Collection and Preparation

The 84 accessions of wild *Vigna* species (seeds) used in this study were obtained from gene banks as presented in Table 1. All the accessions were planted in an experimental plot following the augmented block design arrangement [17] and allowed to grow until complete maturity before harvesting in order to have enough seeds for analysis. After harvesting, the seeds were sun-dried to maintain a uniform moisture content of grains to 10%–14% using a moisture grain tester (DICKEY-JOHN, model: MINIGAC 1, Minneapolis, MN, USA) [18]. An illustration of the seeds of some of the samples is also shown in Figure 1. In addition, three domesticated *Vigna* legumes—that is, cowpea (*V. unguiculata*), rice bean (*V. umbellata*), and a semi-domesticated landrace (*V. vexillata*)—were used as checks. The checks were obtained from the Genetic Resource Center (GRC-IITA), Nigeria (cowpea), the National Bureau of Plant Genetic Resources (NBPGR), India (rice bean), and the Australian Grain Genebank (AGG), Australia (semi-domesticated landrace *V. vexillata*).

Figure 1. Photographs illustrating seed morphology of wild *Vigna* species. Four seeds per accession were pictured under the same conditions to give an image of the morphology and the relative size. Distances of lines in the background are 1 cm in the vertical and horizontal directions. Source: Authors based on seeds requested from the Australian Grain Genebank (AGG) (**a–e,q–t**) and the Genetic Resources Center, International Institute of Tropical Agriculture, (IITA), Ibadan, Nigeria (**f–p**).

Table 1. Wild *Vigna* species collected from the gene banks/self.

Vigna Species	Genebank/Number of Accession			Total
	GRC, IITA Ibadan, Nigeria	**AGG Horsham, Victoria**	**Self-Collected**	
Vigna racemosa	-	4	-	4
Vigna reticulata	30	1	-	31
Vigna vexillata	29	6	-	35
Vigna ambacensis	11	0	-	11
Unknown V. racemosa Accession (Nigeria)	-	-	1	1
Unknown V. reticulata Accession (Nigeria)	-	-	1	1
Unknown Vigna (Tanzania)	-	-	1	1
Total	70	11	3	84

GRC, IITA: Genetic Resource Center, Germplasm Health Unit, International Institute of Tropical Agriculture (IITA), Headquarters, PMB 5320, Oyo Road, Idi-Oshe, Ibadan, Nigeria. AGG: Australian Grain Genebank, Department of Economic Development, Jobs, Transport and Resources, Private Bag 260, Horsham, Victoria 3401.

2.2. Sample Cultivation (Multiplication) Process: Experimental Design and Study Site.

The collected seeds were planted in two agro-ecological zones located at two agricultural research stations in Tanzania during the main cropping season (March–September, 2018). The first site was at the Tanzania Coffee Research Institute (TaCRI), located at Hai District, Moshi, Kilimanjaro region (latitude 3°13′59.59″ S, longitude 37°14′54″ E) which is at a high altitude (1681 m) a.s.l. The second site was at the Tanzania Agricultural Research Institute (TARI), Selian, Arusha, which is at a mid-altitude agroecological zone. TARI-Selian lies at latitude 3°21′50.08″ N and longitude 36°38′06.29″ E at an elevation of 1390 m a.s.l.

A total of 160 accessions of wild *Vigna* legumes were planted in an augmented block design following the randomization generated by the statistical tool on the website [19] with three checks. The seeds were planted in eight blocks of 26 lines each with every line containing 10 seeds of each accession. Each check was replicated two times in a block as generated by the statistical tool. The field was monitored and maintained in good conditions from germination to complete maturity before harvesting and seeds were prepared for further analysis. Eighty-four accessions were then selected based on the availability of seeds after maturity for this study.

2.3. Seed Soaking Process

The soaking method adopted from McWatters and modified by Shafaei was used with a slight modification [7,20].

Ten seeds of each accession were randomly selected and weighed, then placed in glass beakers containing 200 mL distilled water and allowed to stand at room temperature (25 °C) for 24 h. The weight of water absorbed by various seeds was measured after 24 h, as it is the soaking time generally practiced by most consumers at home. After reaching required time, the soaked samples were removed from the beakers and placed on a blotter paper to eliminate the excess water, and then weighed. A precision electronic balance (Model GF400, accuracy ± 0.001 g A&D Company Ltd, Taunton, MA, USA) was used to measure weight of sample before and after immersion. All tests were performed in triplicate. The weight of water absorbed was determined using the formula below [7,20]:

$$Wa = (Wf - Wi)/Wi,$$

where, *Wa* is the water absorption, *Wf* is weight of seeds after immersion (g), and *Wi* is weight of seeds before immersion (g).

2.4. Cooking Process on a Mattson Bean Cooker

A Mattson Bean Cooker (MBC) apparatus was used to record the mean cooking time of each accession of wild *Vigna* legume. The apparatus consists of 25 plungers and a cooking rack with 25 reservoir-like perforated saddles, each of which holds a grain and a plunger calibrated to a specific weight. Each plunger weighs 90 g and terminates in a stainless-steel probe of 1.0 mm in diameter [21]. The cooking proceeded by immersing MBC in a beaker with boiling water (98 °C) over a hotplate. The 50% cooked point, indicated by plungers dropping and penetrating 13 (approximately 50% of the 25 individual seeds) of the individual beans, corresponds to the sensory preferred degree of cooking, according to methodology adapted from Proctor and Watts [11,22]. A digital chronometer was used to record the cooking time during the process.

2.5. Yield per Plant Data Collection and Evaluation

The yield per plant parameter was evaluated by the method adopted from Adewale [23] and converted to the unit used by Bisht [17]. A total of 10 seeds of each accession were planted in eight blocks of 26 lines as earlier described in Section 2.2. Matured pods from the 10 plants of each accession were harvested, threshed, sun-dried, and weighed. The weight of total seeds for each plant was then recorded and the mean seed weight for all the plants harvested on a line (plot) was evaluated as yield per plant. Similarly, the mean seed weight for all the accessions of the same species was evaluated as the yield per plant for the species.

2.6. Data Analysis

The values for water absorption capacity and cooking time were recorded in triplicate and presented as mean ± standard error using XLSTAT. The data were subjected to two-way analysis of variance (ANOVA), correlation coefficients, and Tukey's test. $p < 0.05$ was considered statistically significant. Agglomerative Hierarchical Clustering (AHC) analysis was performed to examine similarities between accessions. Descriptive statistics for the yield traits as well as cooking time and water absorption capacity were also computed using XLSTAT. All the data were entered in an excel sheet and analyzed using XLSTAT-Base version 21.1.57988.0.

3. Results

3.1. Cooking Time and Water Absorption Capacity of Domesticated Legumes

The water absorptions and the cooking time for a landrace of *Vigna vexillata* (check 1), cowpea (check 2), and rice bean (check 3) used here were harvested from two different agro-ecological zones, as shown in Table 2a.

The values for both water absorption and cooking time showed no significant difference between agroecological zones and between the three species and therefore no environment × species interaction (Table 2a). A detailed presentation of the interactions between species (*V. vexillata* landrace, *V. unguiculata*, and *V. umbellate*) as replicated within locations is shown in Table 2b,c. It shows that there is no replicate interaction effect between species within locations for the water absorption capacity trait in all the tested combinations. However, replicate interaction effects were significant ($p < 0.05$) when tested within locations between species for the cooking time trait except when tested across locations (Table 2c).

The values for cooking time showed significant differences between the three domesticated varieties ($p < 0.05$). Pearson correlation analysis shows that there is no correlation between the water absorption capacity and cooking time considering only the three seed varieties ($r = -0.030$ for site A, 0.029 for site B) (Figure 2). Cowpea has a higher cooking time than rice bean which also cook longer than the landrace of *V. vexillata*.

Table 2. Results of the cooking time and water absorption capacity for the domesticated legume seeds. (**a**) Means, analysis of variance and type III sum of square analysis for the cooking time and water absorption capacity traits of domesticated legume seeds. (**b**) Details of interactions within locations effects for water absorption capacity trait. (**c**) Details of interactions within locations effects for cooking time trait.

(**a**)

	Water Absorption Capacity		Cooking Time (min)	
Checks	Site A	Site B	Site A	Site B
Landrace of *Vigna vexillata*	1.33 ± 0.11 [a]	1.32 ± 0.13 [a]	10.24 ± 0.15 [a]	10.26 ± 0.15 [a]
Cowpea (*Vigna unguiculata*)	1.27 ± 0.08 [a]	1.27 ± 0.08 [a]	16.29 ± 0.15 [c]	16.31 ± 0.15 [c]
Rice Bean (*Vigna umbellata*)	1.16 ± 0.06 [a]	1.16 ± 0.06 [a]	13.20 ± 0.12 [b]	13.23 ± 0.12 [b]

Analysis of Variance (ANOVA)

	Water Absorption Capacity					Cooking Time (min)				
Source	DF	Sum of Squares	Mean Squares	F	p	DF	Sum of Squares	Mean Squares	F	p
Model	5	1.263	0.253	1.134	0.343	5	1582.515	316.503	356.710	<0.0001
Error	258	57.475	0.223			258	228.919	0.887		
Corrected Total	263	58.738				263	1811.434			

Type III Sum of Squares Analysis

Source	DF	Sum of Squares	Mean Squares	F	p	DF	Sum of Squares	Mean Squares	F	p
Location (Site) Effect	1	0.001	0.001	0.004	0.950	1	0.044	0.044	0.050	0.823
Species Effect	2	1.262	0.631	2.833	0.061	2	1582.470	791.235	891.749	<0.0001
Location X Species	2	0.000	0.000	0.000	1.000	2	0.000	0.000	0.000	1.000

Results are represented as the mean value of triplicates ± standard error. Different letters in the same column represent statistically different mean values ($p = 0.05$). Site A: TARI-Selian; Site B: TaCRI. *DF*: Degree of freedom; *F*: *F*-ratio; *p*: *p*-value.

(b)

Location × Species/Tukey (HSD)/Analysis of the Differences between the Categories with a Confidence Interval of 95% (Water Absorption Capacity)

Contrast	Difference	Standardized Difference	Critical value	Pr > Diff	Significant
Location-Site A × Species-Check 1 vs. Location-Site B × Species-Check 3	0.173	1.675	2.871	0.550	No
Location-Site A × Species-Check 1 vs. Location-Site A × Species-Check 3	0.172	1.661	2.871	0.559	No
Location-Site A × Species-Check 1 vs. Location-Site B × Species-Check 2	0.063	0.620	2.871	0.990	No
Location-Site A × Species-Check 1 vs. Location-Site A × Species-Check 2	0.059	0.584	2.871	0.992	No
Location-Site A × Species-Check 1 vs. Location-Site B × Species-Check 1	0.006	0.054	2.871	1.000	No
Location-Site B × Species-Check 1 vs. Location-Site B × Species-Check 3	0.167	1.619	2.871	0.587	No
Location-Site B × Species-Check 1 vs. Location-Site A × Species-Check 3	0.166	1.605	2.871	0.596	No
Location-Site B × Species-Check 1 vs. Location-Site B × Species-Check 2	0.057	0.563	2.871	0.993	No
Location-Site B × Species-Check 1 vs. Location-Site A × Species-Check 2	0.054	0.527	2.871	0.995	No
Location-Site A × Species-Check 2 vs. Location-Site B × Species-Check 3	0.114	1.160	2.871	0.855	No
Location-Site A × Species-Check 2 vs. Location-Site A × Species-Check 3	0.112	1.145	2.871	0.862	No
Location-Site A × Species-Check 2 vs. Location-Site B × Species-Check 2	0.004	0.038	2.871	1.000	No
Location-Site B × Species-Check 2 vs. Location-Site B × Species-Check 3	0.110	1.122	2.871	0.872	No
Location-Site B × Species-Check 2 vs. Location-Site A × Species-Check 3	0.108	1.107	2.871	0.878	No
Location-Site A × Species-Check 3 vs. Location-Site B × Species-Check 3	0.001	0.014	2.871	1.000	No
Tukey's d critical value			4.061		

Check 1: Landrace of *Vigna* vexillata; Check 2: Cowpea (*Vigna unguiculata*); Check 3: Rice Bean (*Vigna umbellata*).

(c)

Location × Species/Tukey (HSD)/Analysis of the Differences between the Categories with a Confidence Interval of 95% (Cooking Time)

Contrast	Difference	Standardized Difference	Critical value	Pr > Diff	Significant
Location-Site B × Species-Check 2 vs. Location-Site A × Species-Check 1	6.074	29.913	2.871	<0.0001	Yes
Location-Site B × Species-Check 2 vs. Location-Site B × Species-Check 1	6.048	29.785	2.871	<0.0001	Yes
Location-Site B × Species-Check 2 vs. Location-Site A × Species-Check 3	3.109	15.908	2.871	<0.0001	Yes
Location-Site B × Species-Check 2 vs. Location-Site B × Species-Check 3	3.083	15.775	2.871	<0.0001	Yes
Location-Site B × Species-Check 2 vs. Location-Site A × Species-Check 2	0.026	0.135	2.871	1.000	No
Location-Site A × Species-Check 2 vs. Location-Site A × Species-Check 1	6.048	29.785	2.871	<0.0001	Yes
Location-Site A × Species-Check 2 vs. Location-Site B × Species-Check 1	6.022	29.657	2.871	<0.0001	Yes
Location-Site A × Species-Check 2 vs. Location-Site A × Species-Check 3	3.083	15.775	2.871	<0.0001	Yes
Location-Site A × Species-Check 2 vs. Location-Site B × Species-Check 3	3.057	15.642	2.871	<0.0001	Yes
Location-Site B × Species-Check 3 vs. Location-Site A × Species-Check 1	2.991	14.514	2.871	<0.0001	Yes
Location-Site B × Species-Check 3 vs. Location-Site B × Species-Check 1	2.965	14.388	2.871	<0.0001	Yes
Location-Site B × Species-Check 3 vs. Location-Site A × Species-Check 3	0.026	0.131	2.871	1.000	No
Location-Site A × Species-Check 3 vs. Location-Site A × Species-Check 1	2.965	14.388	2.871	<0.0001	Yes
Location-Site A × Species-Check 3 vs. Location-Site B × Species-Check 1	2.939	14.262	2.871	<0.0001	Yes
Location-Site B × Species-Check 1 vs. Location-Site A × Species-Check 1	0.026	0.122	2.871	1.000	No
Tukey's d critical value			4.061		

Check 1: Landrace of *Vigna vexillata*; Check 2: Cowpea (*Vigna unguiculata*); Check 3: Rice Bean (*Vigna umbellata*).

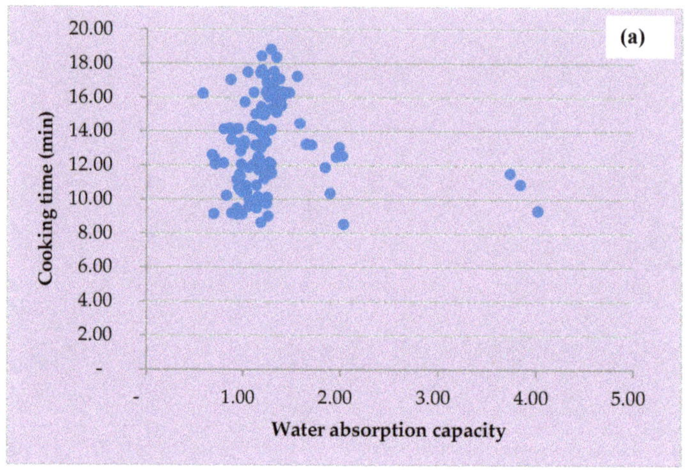

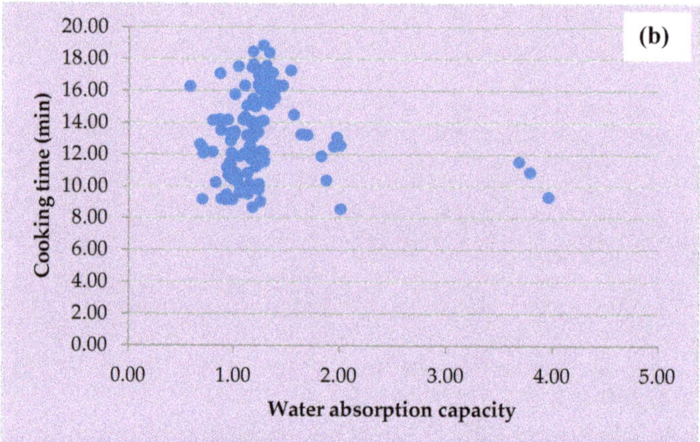

Figure 2. Correlation between water absorption and cooking time for the three checks. (**a**) Plotted with data from Site A; (**b**) plotted with data from Site B.

3.2. Cooking Time and Water Absorption Capacity of Vigna ambacensis Accessions

The water absorption capacities and the cooking times for 11 accessions of wild *Vigna ambacensis* are presented in Table 3.

The values for water absorption capacity and cooking time showed no significant difference ($p > 0.05$) when compared with the values of their corresponding accession harvested in the other agro-ecological zone (Table 3).

Considering the water absorption capacity, all the wild accessions exhibited significantly low values as compared with all three checks. The water absorption capacity of wild accessions varied from 0.08 ± 0.01 to 0.47 ± 0.01 (Table 3) in both site A and B. Accession TVNu342 showed no significant difference in water absorption capacity with three checks and with accession TVNu219.

The cooking time of the wild accessions varied from 23.02 ± 0.50 to 24.26 ± 0.07 min in both sites (Table 3). All the wild accessions possessed significantly higher cooking time values compared with the three checks (Table 3). None of the accessions cooked faster than the checks.

Table 3. Cooking time and water absorption of *Vigna ambacensis* accessions.

Species/Accession Number	Water Absorption Capacity		Cooking Time (min)	
	Site A	Site B	Site A	Site B
Landrace of *Vigna vexillata*	1.33 ± 0.11 [a]	1.32 ± 0.13 [a]	10.24 ± 0.15 [a]	10.26 ± 0.15 [a]
Cowpea (*Vigna unguiculata*)	1.27 ± 0.08 [a]	1.27 ± 0.08 [a]	16.29 ± 0.15 [b]	16.31 ± 0.15 [b]
Rice bean (*Vigna umbellata*)	1.16 ± 0.06 [a]	1.16 ± 0.06 [a]	13.20 ± 0.12 [c]	13.23 ± 0.12 [c]
TVNu1699	0.14 ± 0.01 [c]	0.13 ± 0.01 [c]	24.26 ± 0.07 [d]	23.87 ± 0.10 [d]
TVNu342	0.47 ± 0.01 [a,b]	0.45 ± 0.01 [a,b]	23.34 ± 0.16 [d]	23.35 ± 0.18 [d]
TVNu877	0.22 ± 0.01 [c]	0.21 ± 0.01 [c]	24.10 ± 0.19 [d]	23.71 ± 0.22 [d]
TVNu223	0.21 ± 0.01 [c]	0.21 ± 0.02 [c]	23.35 ± 0.55 [d]	23.36 ± 0.50 [d]
TVNu720	0.22 ± 0.01 [c]	0.20 ± 0.01 [c]	23.02 ± 0.50 [d]	23.03 ± 0.45 [d]
TVNu219	0.28 ± 0.02 [b,c]	0.26 ± 0.01 [b,c]	24.06 ± 0.49 [d]	24.08 ± 0.50 [d]
TVNu1840	0.11 ± 0.01 [c]	0.10 ± 0.01 [c]	23.36 ± 0.21 [d]	23.37 ± 0.30 [d]
TVNu1804	0.09 ± 0.01 [c]	0.08 ± 0.01 [c]	23.55 ± 0.52 [d]	23.56 ± 0.50 [d]
TVNu1792	0.23 ± 0.01 [c]	0.09 ± 0.01 [c]	23.28 ± 0.22 [d]	23.30 ± 0.30 [d]
TVNu1644	0.09 ± 0.01 [c]	0.21 ± 0.01 [c]	23.12 ± 0.10 [d]	23.13 ± 0.15 [d]
TVNu1185	0.12 ± 0.01 [c]	0.11 ± 0.01 [c]	23.34 ± 0.33 [d]	23.35 ± 0.30 [d]

Analysis of Variance (ANOVA)

	Water Absorption Capacity					Cooking Time (min)				
Source	DF	Sum of Squares	Mean Squares	F	p	DF	Sum of Squares	Mean Squares	F	p
Model	27	60.707	2.248	11.756	<0.0001	27	6864.480	254.240	313.317	<0.0001
Error	302	57.761	0.191			302	245.057	0.811		
Corrected Total	329	118.469				329	7109.537			

Type III Sum of Squares Analysis

Source	DF	Sum of squares	Mean squares	F	p	DF	Sum of Squares	Mean Squares	F	p
Location (Site)	1	0.002	0.002	0.011	0.916	1	0.007	0.007	0.008	0.929
Genotype (Accessions)	13	60.704	4.670	24.414	<0.0001	13	6864.433	528.033	650.730	<0.0001
Location × Genotype	13	0.001	0.000	0.001	1.000	13	0.002	0.000	0.000	1.000

Results are represented as the mean value of triplicates ± standard error. Mean values without any letter in common within each column are significantly different ($p = 0.05$). Site A: TARI-Selian; Site B: TaCRI. *DF*: Degree of freedom; *F*: *F*-ratio; *p*: *p*-value.

Additionally, there was no correlation between the water absorption capacity and cooking time considering only the 11 accessions studied ($r = -0.025$ for site A and $r = -0.024$ for site B) (Figure 3).

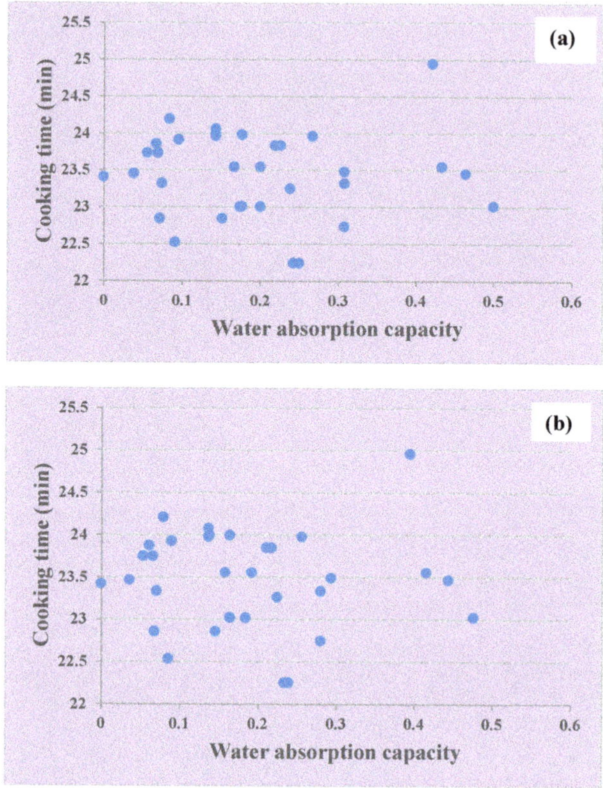

Figure 3. Correlation between water absorption and cooking time for the *Vigna ambacensis* accessions. (**a**) Plotted with data from Site A; (**b**) plotted with data from Site B.

3.3. Cooking Time and Water Absorption Capacity of Vigna vexillata Accessions

The result for water absorption capacity and cooking time for 35 accessions of wild *Vigna vexillata* is shown in Table 4. The values for water absorption capacity and cooking time show no significant difference ($p > 0.05$) when compared with the values of their corresponding accessions harvested in the other agro-ecological zone.

The Water Absorption Capacity in all the wild accessions with exception of TVNu781 and TVNu837 showed significant low values compared with the three checks (Table 4). The water absorption capacity of the wild *V. vexillata* accessions varied from 0.04 ± 0.00 to 1.10 ± 0.03 in both site A and B (Table 4).

Considering the cooking time, there is a high diversity in differences among the accessions. The cooking time varied from 16.22 ± 0.23 to 31.04 ± 0.33 min in site A and from 16.24 ± 0.20 to 31.06 ± 0.31 min in site B (Table 4). Accessions TVNu781, AGG308107WVIG2, AGG308097WVIG1, and TVNU1624 exhibited relatively similar cooking time with check 2 (Cowpea) (Table 4). Conversely, cooking time for all other remaining accessions was significantly higher than all the checks. Pearson correlation analysis shows that there is a weak negative correlation between the water absorption capacity and cooking time considering the wild *V. vexillata* tested ($r = -0.31$ for site A and $r = -0.32$). Furthermore, the regression analysis shows that the water absorption capacity and cooking time are related by the equation: $Y = -5.12x + 27.15$ with $R^2 = 0.094$ (Figure 4).

Table 4. Cooking time and water absorption capacity of *Vigna vexillata* accessions.

Species/Accession Number	Water Absorption Capacity		Cooking Time (min)	
	Site A	Site B	Site A	Site B
Landrace of *Vigna vexillata*	1.33 ± 0.11 [a]	1.32 ± 0.13 [a]	10.24 ± 0.15 [n]	10.26 ± 0.15 [n]
Cowpea (*Vigna unguiculata*)	1.27 ± 0.08 [a,b,c]	1.27 ± 0.08 [a,b,c]	16.29 ± 0.15 [l]	16.31 ± 0.15 [l]
Rice Bean (*Vigna umbellata*)	1.16 ± 0.06 [a,b,c]	1.16 ± 0.06 [a,b,c]	13.20 ± 0.12 [m]	13.23 ± 0.12 [m]
TVNu781	1.10 ± 0.02 [abcd]	1.10 ± 0.01 [abcd]	31.04 ± 0.33 [a,b]	31.06 ± 0.31 [a,b]
TVNu837	1.07 ± 0.01 [abcd]	1.05 ± 0.01 [abcd]	29.34 ± 0.32 [a,b,c,d,e,f,g]	29.35 ± 0.01 [a,b,c,d,e,f,g]
TVNu1582	0.73 ± 0.01 [abcd]	0.67 ± 0.02 [abcd]	16.25 ± 0.24 [l]	16.26 ± 0.30 [l]
TVNu1358	0.57 ± 0.01 [abcd]	0.53 ± 0.01 [abcd]	17.37 ± 0.26 [l]	17.38 ± 0.28 [l]
AGG308107WVIG2	0.43 ± 0.01 [abcd]	0.41 ± 0.01 [abcd]	26.32 ± 0.49 [f,g,h,i,j]	26.33 ± 0.51 [f,g,h,i,j]
TVNu1593	0.42 ± 0.01 [abcd]	0.38 ± 0.01 [bcd]	16.22 ± 0.23 [l]	16.24 ± 0.20 [l]
TVNu1591	0.41 ± 0.01 [abcd]	0.38 ± 0.01 [bcd]	26.38 ± 0.40 [f,g,h,i,j]	26.39 ± 0.43 [f,g,h,i,j]
TVNu120	0.40 ± 0.01 [abcd]	0.38 ± 0.01 [bcd]	31.10 ± 0.31 [a]	30.71 ± 0.34 [a]
TVNu333	0.40 ± 0.02 [abcd]	0.37 ± 0.02 [bcd]	26.28 ± 0.40 [f,g,h,i,j]	26.30 ± 0.35 [f,g,h,i,j]
TVNu1546	0.39 ± 0.02 [bcd]	0.37 ± 0.01 [bcd]	29.07 ± 0.13 [a,b,c,d,e,f]	29.08 ± 0.15 [a,b,c,d,e,f]
AGG308101WVIG1	0.37 ± 0.01 [bcd]	0.34 ± 0.01 [cd]	29.36 ± 0.50 [a,b,c,d,e]	29.37 ± 0.47 [a,b,c,d,e]
TVNu1701	0.35 ± 0.01 [bcd]	0.33 ± 0.01 [cd]	24.59 ± 0.50 [j]	24.60 ± 0.57 [j]
AGG308096 WVIG2	0.34 ± 0.01 [cd]	0.32 ± 0.01 [cd]	26.47 ± 0.59 [f,g,h,i,j]	26.49 ± 0.60 [f,g,h,i,j]
TVNu1629	0.33 ± 0.01 [cd]	0.32 ± 0.02 [cd]	28.19 ± 1.15 [b,c,d,e,f,g,h,i]	28.20 ± 1.20 [b,c,d,e,f,g,h,i]
TVNu293	0.33 ± 0.01 [cd]	0.31 ± 0.01 [cd]	26.32 ± 0.28 [f,g,h,i,j]	26.33 ± 0.30 [f,g,h,i,j]
TVNu832	0.32 ± 0.01 [cd]	0.30 ± 0.01 [cd]	25.46 ± 0.36 [h,i,j]	25.47 ± 0.36 [h,i,j]
TVNu1796	0.32 ± 0.01 [cd]	0.30 ± 0.01 [cd]	27.38 ± 0.48 [c,d,e,f,g,h,i,j]	27.39 ± 0.50 [c,d,e,f,g,h,i,j]
TVNu1529	0.32 ± 0.01 [cd]	0.30 ± 0.01 [cd]	27.29 ± 0.64 [c,d,e,f,g,h,i,j]	27.30 ± 0.64 [c,d,e,f,g,h,i,j]
TVNu1628	0.30 ± 0.01 [cd]	0.28 ± 0.01 [cd]	26.30 ± 0.36 [f,g,h,i,j]	26.31 ± 0.33 [f,g,h,i,j]
TVNu1344	0.29 ± 0.01 [cd]	0.28 ± 0.01 [cd]	30.03 ± 0.44 [a,b,c]	30.04 ± 0.44 [a,b,c]
TVNu1632	0.29 ± 0.01 [cd]	0.28 ± 0.01 [cd]	29.41 ± 0.52 [a,b,c,d]	28.25 ± 0.50 [a,b,c,d]
TVNu1370	0.28 ± 0.01 [cd]	0.26 ± 0.02 [cd]	28.23 ± 0.39 [l]	29.43 ± 0.40 [l]
TVNu1360	0.28 ± 0.01 [cd]	0.25 ± 0.01 [cd]	27.05 ± 0.71 [d,e,f,g,h,i,j]	27.60 ± 0.72 [d,e,f,g,h,i,j]
TVNu1624	0.25 ± 0.01 [cd]	0.23 ± 0.01 [d]	26.28 ± 0.46 [f,g,h,i,j]	26.29 ± 0.46 [f,g,h,i,j]
TVNu1621	0.25 ± 0.01 [cd]	0.23 ± 0.01 [d]	17.24 ± 0.47 [l]	17.26 ± 0.48 [l]
AGG62154WVIG_1	0.20 ± 0.01 [d]	0.19 ± 0.01 [d]	21.33 ± 0.17 [k]	21.34 ± 0.17 [k]
TVNu1092	0.19 ± 0.01 [d]	0.18 ± 0.01 [d]	29.02 ± 0.23 [a,b,c,d,e,f,g]	29.04 ± 0.55 [a,b,c,d,e,f,g]
TVNu479	0.18 ± 0.01 [d]	0.17 ± 0.01 [d]	26.50 ± 0.56 [e,f,g,h,i,j]	26.52 ± 0.20 [e,f,g,h,i,j]
AGG308097WVIG 1	0.17 ± 0.01 [d]	0.16 ± 0.01 [d]	28.56 ± 0.50 [a,b,c,d,e,f,g]	28.58 ± 0.50 [a,b,c,d,e,f,g]
TVNu178	0.17 ± 0.01 [d]	0.16 ± 0.01 [d]	17.03 ± 0.54 [l]	16.64 ± 0.01 [l]
TVNu955	0.11 ± 0.01 [d]	0.11 ± 0.01 [d]	25.28 ± 0.47 [i,j]	25.29 ± 0.47 [i,j]
TVNu1378	0.11 ± 0.01 [d]	0.10 ± 0.00 [d]	16.29 ± 0.45 [l]	16.31 ± 0.47 [l]
TVNu1586	0.06 ± 0.00 [d]	0.05 ± 0.01 [d]	28.39 ± 0.29 [a,b,c,d,e,f,g]	28.41 ± 0.30 [a,b,c,d,e,f,g]
TVNu381	0.04 ± 0.00 [d]	0.042 ± 0.00 [d]	25.41 ± 0.63 [h,i,j]	25.42 ± 0.64 [h,i,j]
AGG308099WVIG2	0.042 ± 0.01 [d]	0.04 ± 0.01 [d]	26.16 ± 0.48 [h,i,j]	26.17 ± 0.50 [h,i,j]

Table 4. *Cont.*

Species/Accession Number	Water Absorption Capacity					Cooking Time (min)				
	Site A	Site B		Site A				Site B		
	Analysis of Variance (ANOVA)									
	Water Absorption Capacity					Cooking Time (min)				
Source	DF	Sum of Squares	Mean Squares	F	p	DF	Sum of Squares	Mean Squares	F	p
Model	75	111.003	1.480	9.649	<0.0001	75	22,437.582	299.168	368.513	<0.0001
Error	398	61.050	0.153			398	323.106	0.812		
Corrected Total	473	172.052				473	22,760.688			
Type III Sum of Squares Analysis										
Source	DF	Sum of Squares	Mean Squares	F	p	DF	Sum of Squares	Mean Squares	F	p
Location (Site)	1	0.018	0.018	0.117	0.732	1	0.012	0.012	0.015	0.903
Genotype (Accessions)	37	110.978	2.999	19.554	<0.0001	37	22,437.529	606.420	746.983	<0.0001
Location × Genotype	37	0.013	0.000	0.002	1.000	37	0.005	0.000	0.000	1.000

Results are represented as the mean value of triplicates ± standard error. Mean values without any letter in common within each column are significantly different ($p = 0.05$). Site A: TARI-Selian; Site B: TaCRI. *DF*: Degree of freedom; *F*: *F*-ratio; *p*: *p*-value.

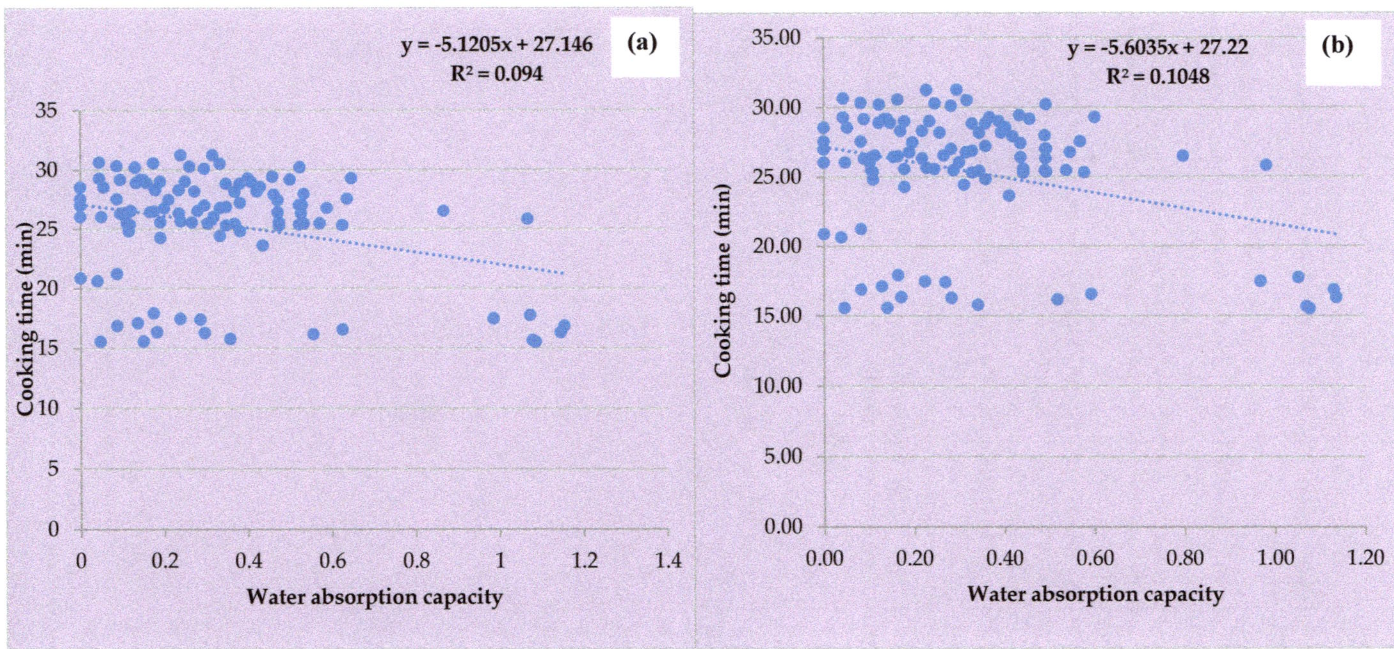

Figure 4. Correlation between water absorption and cooking time for the *Vigna vexillata* accessions. (**a**) Plotted with data from Site A; (**b**) plotted with data from Site B.

3.4. Cooking Time and Water Absorption capacity of Vigna reticulata Accessions

Table 5 shows the various values for water absorption capacity and cooking time for 32 accessions of wild *Vigna reticulata*. The values for water absorption capacity and cooking time showed no significant difference ($p > 0.05$) when compared with the values of their corresponding accessions harvested in the other agro-ecological zone.

All the wild accessions showed significantly low water absorption capacity values compared with the checks except for TVNu1520, and TVNu325 (Table 5). The water absorption capacity of the wild *V. reticulata* accessions varied from 0.06 ± 0.01 to 1.27 ± 0.08 in site A and from 0.06 ± 0.01 to 1.32 ± 0.13 in site B (Table 5). No location and genotype $\times$ location interactions ($p > 0.05$) were observed for both water absorption capacity and cooking time traits in these accessions. However, only genotype interaction was observed for both traits ($p < 0.05$).

Regarding cooking time, there is a high diversity in differences of means among the accessions. Twenty-five accessions showed significant higher cooking time values. Check 2 showed no significant difference in cooking time with TVNu325 and the unknown *V. reticulata* accession only. The cooking times for all accessions varied from 17.41 ± 0.44 to 30.25 ± 0.41 min in site A and from 17.42 ± 0.45 to 30.26 ± 0.42 min in site B (Table 5).

Pearson correlation analysis shows that there is a weak negative correlation between the water absorption and cooking time considering the wild *V. reticulata* tested ($r = -0.43$ for site A and $r = -0.45$) (Figure 5. Furthermore, the regression analysis shows that the water absorption and cooking time are related by the equation: $Y = -2.57x + 27.77$ with $R^2 = 0.18$ (Figure 5).

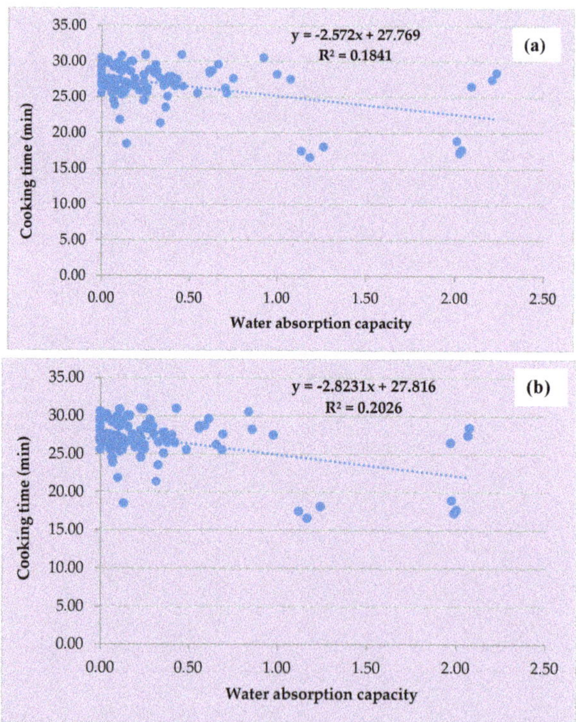

Figure 5. Correlation between water absorption and cooking time for the *Vigna reticulata* accessions. (a) Plotted with data from Site A; (b) plotted with data from Site B.

Table 5. Cooking time and water absorption capacity of *Vigna reticulata* accessions.

Species/Accession Number	Water Absorption Capacity		Cooking Time (min)	
	Site A	Site B	Site A	Site B
Landrace of *Vigna vexillata*	1.33 ± 0.11 [a,b,c]	1.32 ± 0.13 [a,b,c]	10.24 ± 0.15 [h]	10.26 ± 0.15 [h]
Cowpea (*Vigna unguiculata*)	1.27 ± 0.08 [a,b,c,d]	1.27 ± 0.08 [a,b,c,d]	16.29 ± 0.15 [f]	16.31 ± 0.15 [f]
Rice Bean (*Vigna umbellata*)	1.16 ± 0.06 [a,b,c,d]	1.16 ± 0.06 [a,b,c,d]	13.20 ± 0.12 [g]	13.23 ± 0.12 [g]
TVNu324	0.49 ± 0.02 [c,d]	0.47 ± 0.02 [c,d]	26.51 ± 0.47 [a,b,c,d]	26.53 ± 0.48 [a,b,c,d]
TVNu325	2.03 ± 0.02 [a,b]	1.99 ± 0.01 [a,b]	17.92 ± 0.51 [f]	17.93 ± 0.52 [f]
Unknown _*Vigna reticulata*	1.20 ± 0.02 [a,b,c,d]	1.18 ± 0.02 [a,b,c,d]	17.41 ± 0.44 [f]	17.42 ± 0.45 [f]
TVNu343	0.19 ± 0.01 [c,d]	0.18 ± 0.01[c,d]	30.25 ± 0.41 [a]	30.26 ± 0.42 [a]
TVNu767	0.12 ± 0.01 [c,d]	0.11 ± 0.01 [c,d]	29.18 ± 0.99 [a,b,c,d]	29.20 ± 1.00 [a,b,c,d]
TVNu1520	2.18 ± 0.03 [a]	2.04 ± 0.03 [a]	27.46 ± 0.91 [a,b,c,d]	27.48 ± 0.92 [a,b,c,d]
TVNu349	0.31 ± 0.02 [c,d]	0.29 ± 0.01 [c,d]	29.14 ± 0.74 [a,b,c,d]	28.76 ± 0.75 [a,b,c,d]
TVNu379	0.77 ± 0.01 [c,d]	0.71 ± 0.02 [c,d]	29.38 ± 0.46 [a,b,c,d]	28.99 ± 0.44 [a,b,c,d]
TVNu524	0.17 ± 0.01 [c,d]	0.17 ± 0.01 [c,d]	25.57 ± 0.57 [c,d,e]	25.58 ± 0.58 [c,d,e]
TVNu1698	0.12 ± 0.01 [c,d]	0.11 ± 0.01 [c,d]	26.34 ± 0.56 [b,c,d,e]	26.36 ± 0.57 [b,c,d,e]
TVNu1191	0.22 ± 0.01 [c,d]	0.21 ± 0.01 [c,d]	25.38 ± 1.00 [d,e]	25.39 ± 0.99 [d,e]
TVNu1394	0.82 ± 0.02 [b, c,d]	0.75 ± 0.01 [b, c,d]	28.21 ± 0.99 [a,b,c,d]	27.82 ± 0.97 [a,b,c,d]
TVNu-224	0.19 ± 0.01 [c,d]	0.18 ± 0.01 [c,d]	25.55 ± 0.51 [c,d,e]	25.57 ± 0.52 [c,d,e]
TVNu739	0.15 ± 0.01 [c,d]	0.14 ± 0.01 [c,d]	28.50 ± 0.46 [a,b,c,d]	28.52 ± 0.47 [a,b,c,d]
TVNu56	0.24 ± 0.02 [c,d]	0.22 ± 0.02 [c,d]	27.01 ± 2.73 [a,b,c,d]	26.62 ± 2.70 [a,b,c,d]
TVNu1405	0.29 ± 0.02 [c,d]	0.26 ± 0.02 [c,d]	30.03 ± 0.64 [a,b]	29.64 ± 0.62 [a,b]
TVNu607	0.08 ± 0.01 [d]	0.08 ± 0.01 [d]	27.33 ± 0.49 [a,b,c,d]	26.38 ± 0.47 [a,b,c,d]
TVNu916	0.12 ± 0.01 [c,d]	0.11 ± 0.01 [c,d]	26.37 ± 0.52 [b,c,d,e]	27.84 ± 0.55 [b,c,d,e]
AGG17856WVIG 1	0.16 ± 0.01 [c,d]	0.15 ± 0.01 [c,d]	28.23 ± 1.00 [a,b,c,d]	27.35 ± 0.97 [a,b,c,d]
TVNu1790	0.32 ± 0.02 [c,d]	0.29 ± 0.02 [c,d]	28.43 ± 0.47 [a,b,c,d]	28.44 ± 0.47 [a,b,c,d]
TVNu491	0.15 ± 0.01 [c,d]	0.14 ± 0.01 [c,d]	28.44 ± 0.93 [a,b,c,d]	28.45 ± 0.92 [a,b,c,d]
TVNu1808	0.16 ± 0.01 [c,d]	0.15 ± 0.01 [c,d]	29.36 ± 0.42 [a,b,c]	29.38 ± 0.43 [a,b,c]
TVNu738	0.12 ± 0.01 [c,d]	0.12 ± 0.01 [c,d]	26.42 ± 0.39 [a,b,c,d]	26.43 ± 0.40 [a,b,c,d]
TVNu1779	0.19 ± 0.02 [c,d]	0.17 ± 0.01 [c,d]	26.12 ± 2.04 [c,d,e]	25.74 ± 2.01 [c,d,e]
TVNu605	0.42 ± 0.02 [c,d]	0.36 ± 0.02 [c,d]	29.16 ± 0.51 [a,b,c,d]	29.18 ± 0.51 [a,b,c,d]
TVNu57	0.06 ± 0.01 [d]	0.06 ± 0.01 [d]	28.00 ± 0.55 [a,b,c,d]	27.61 ± 0.48 [a,b,c,d]
TVNu138	0.23 ± 0.01 [c,d]	0.21 ± 0.01 [c,d]	27.19 ± 0.62 [a,b,c,d]	26.80 ± 0.60 [a,b,c,d]
TVNu161	0.18 ± 0.01 [c,d]	0.16 ± 0.01 [c,d]	30.02 ± 0.77 [a,b]	29.64 ± 0.76 [a,b]
TVNu758	0.16 ± 0.01 [c,d]	0.15 ± 0.01 [c,d]	27.10 ± 0.30 [a,b,c,d]	27.11 ± 0.30 [a,b,c,d]
TVNu1825	0.25 ± 0.02 [c,d]	0.23 ± 0.02 [c,d]	25.50 ± 0.91 [d,e]	25.51 ± 0.91 [d,e]
TVNu1522	0.19 ± 0.01 [c,d]	0.17 ± 0.01 [c,d]	22.56 ± 0.57 [e]	22.57 ± 0.57 [e]
TVNu1388	0.18 ± 0.01 [c,d]	0.16 ± 0.01 [c,d]	26.53 ± 0.69 [a,b,c,d]	26.54 ± 0.70 [a,b,c,d]

Table 5. *Cont.*

Species/Accession Number	Water Absorption Capacity					Cooking Time (min)				
	Site A		Site B			Site A		Site B		
Analysis of Variance (ANOVA)										
	Water Absorption Capacity					Cooking Time (min)				
Source	*DF*	**Sum of Squares**	**Mean Squares**	*F*	*p*	*DF*	**Sum of Squares**	**Mean Squares**	*F*	*p*
Model	69	131.740	1.909	11.989	<0.0001	69	22845.864	331.099	225.891	<0.0001
Error	386	61.473	0.159			386	565.779	1.466		
Corrected Total	455	193.213				455	23411.643			
Type III Sum of Squares Analysis										
Source	*DF*	**Sum of Squares**	**Mean Squares**	*F*	*p*	*DF*	**Sum of Squares**	**Mean Squares**	*F*	*p*
Location (Site)	2	0.025	0.013	0.080	0.924	2	0.052	0.026	0.018	0.982
Genotype (Accessions)	34	88.722	2.609	16.385	<0.0001	34	12987.598	381.988	260.610	<0.0001
Location × Genotype	33	0.033	0.001	0.006	1.000	33	0.000	0.000	0.000	1.000

Results are represented as the mean value of triplicates ± standard error. Mean values without any letter in common within each column are significantly different (p = 0.05). Site A: TARI-Selian; Site B: TaCRI. *DF*: Degree of freedom; *F*: *F*-ratio; *p*: *p*-value.

3.5. Cooking Time and Water Absorption of Vigna racemosa Accessions

The results for water absorption capacity and cooking time for accessions of wild *Vigna racemosa* are shown in Table 6. The values for water absorption capacity and cooking time tested showed no significant difference ($p > 0.05$) when compared with the values of their corresponding accession harvested in the other agro-ecological zone through two-way analysis of variance (ANOVA).

The water absorption capacity of some of the wild accessions showed significant difference to each other and to the three checks. The unknown *Vigna racemosa* and unknown *Vigna* legume accessions displayed significantly low values similar to the three checks (Table 6). The water absorption capacity of the wild *V. racemosa* accessions varied from 0.08 ± 0.01 to 1.35 ± 0.03 in site A and from 0.08 ± 0.00 to 1.32 ± 0.13 in site B (Table 6).

On the other hand, non-significant difference in cooking time between AGG51603WVIG1, AGG52867WVIG1 accessions and check 1 was observed. Besides, they were all significantly different from check 2, check 3, and the other accessions (Table 6). Generally, AGG53597WVIG1 exhibited superior low cooking time compared with the three checks. The cooking time for all accessions varied from 8.26 ± 0.42 to 30.33 ± 0.48 min in site A and from 7.87 ± 0.40 to 30.34 ± 0.50 min in site B (Table 6).

Pearson correlation analysis shows that there is a strong negative correlation between the water absorption and cooking time considering the wild *V. racemosa* accessions tested ($r = -0.91$ for site A and $r = -0.92$ for site B). Furthermore, the regression analysis shows that the water absorption capacity and cooking time are related by the equation: $Y = -17.17x + 32.10$ with $R^2 = 0.84$ (Figure 6)

3.6. Water Absorption Capacity, Cooking Time, and Clustering Analysis of the Four Vigna species for Domestication and Crop Improvement

The Figure 7 below shows the pattern of evolution of water absorption as a function of cooking time to depict the existing relationship between the two parameters for the eighty four accessions from the four wild *Vigna* species (*V. ambacensis*, *V. reticulata*, *V. vexillata*, and *V. racemosa*) and three domesticated species. It shows that the relationship is a strong negative correlation (-0.69 for site A and -0.70 for site B) between the water absorption and the cooking time which follows the equation: $Y_A = -7.99X + 26.52$ ($R^2 = 0.48$) or $Y_B = -8.21X + 26.57$ ($R^2 = 0.50$) (Figure 7).

Agglomerative Hierarchical Clustering (AHC) analysis performed on all the four *Vigna* species taking water absorption capacity, cooking time, and their individual weights before any processing as variable traits revealed seven classes (Figure 8). Details of various accessions belonging to each class are provided in Table 7. Class 1 consists of nineteen accessions of *V. reticulata*, sixteen accessions of *V. vexillata*, and all the eleven accessions of *V. ambacensis*. Class 2 consists of only eight accessions of *V. reticulata* and ten accessions of *V. vexillata* while class 3 consists of two accessions of *V. reticulata*, one accession of *V. vexillata*, three accessions of *V. racemosa* and check 2 and 3. The class 4 consists of one accession of *V. vexillata* and check 3 only, while one accession makes up class 5. Class 6 is made up of four accessions of *V. vexillata* and class 7 of two *V. reticulata* and two *V. vexillata*.

Table 6. Cooking time and water absorption capacity of *Vigna racemosa* accessions.

Species/Accession Number	Water Absorption		Cooking Time (min)	
	Site A	Site B	Site A	Site B
Landrace of *Vigna vexillata*	1.33 ± 0.11 [a]	1.32 ± 0.13 [a]	10.24 ± 0.15 [d]	10.26 ± 0.15 [d]
Cowpea (*Vigna unguiculata*)	1.27 ± 0.08 [a]	1.27 ± 0.08 [a]	16.29 ± 0.15 [b]	16.31 ± 0.15 [b]
Rice Bean (*Vigna umbellata*)	1.16 ± 0.06 [a]	1.16 ± 0.06 [a]	13.20 ± 0.12 [c]	13.23 ± 0.12 [c]
AGG53597WVIG1	1.35 ± 0.03 [a]	1.33 ± 0.02 [a]	8.26 ± 0.42 [d]	7.87 ± 0.40 [d]
AGG51603WVIG1	1.29 ± 0.01 [a]	1.27 ± 0.02 [a]	10.15 ± 0.22 [d,e]	10.17 ± 0.25 [d,e]
AGG52867WVIG1	1.04 ± 0.04 [a]	1.02 ± 0.00 [a]	11.27 ± 0.41 [d]	11.28 ± 0.42 [d]
Unknown *Vigna* legume	0.43 ± 0.01 [a,b]	0.39 ± 0.02 [a,b]	29.35 ± 0.31 [a]	28.97 ± 0.30 [a]
Unknown *Vigna racemosa*	0.08 ± 0.01 [b]	0.08 ± 0.00 [b]	30.33 ± 0.48 [a]	30.34 ± 0.50 [a]

Analysis of Variance (ANOVA)

	Water Absorption Capacity					Cooking Time (min)				
Source	DF	Sum of Squares	Mean Squares	F	p	DF	Sum of Squares	Mean Squares	F	p
Model	15	13.441	0.896	4.279	<0.0001	15	4957.993	330.533	386.632	<0.0001
Error	278	58.223	0.209			278	237.663	0.855		
Corrected Total	293	71.664				293	5195.656			

Type III Sum of Squares Analysis

Source	DF	Sum of Squares	Mean Squares	F	p	DF	Sum of Squares	Mean Squares	F	p
Location (Site)	1	0.004	0.004	0.017	0.896	1	0.006	0.006	0.007	0.934
Genotypes (Accessions)	7	13.436	1.919	9.165	<0.0001	7	4957.947	708.278	828.489	<0.0001
Location × Genotype	7	0.003	0.000	0.002	1.000	7	0.001	0.000	0.000	1.000

Results are represented as the mean value of triplicates ± standard error. Mean values without any letter in common within each column are significantly different ($p = 0.05$). Site A: TARI-Selian; Site B: TaCRI. *DF*: Degree of freedom; *F*: F-ratio; *p*: p-value.

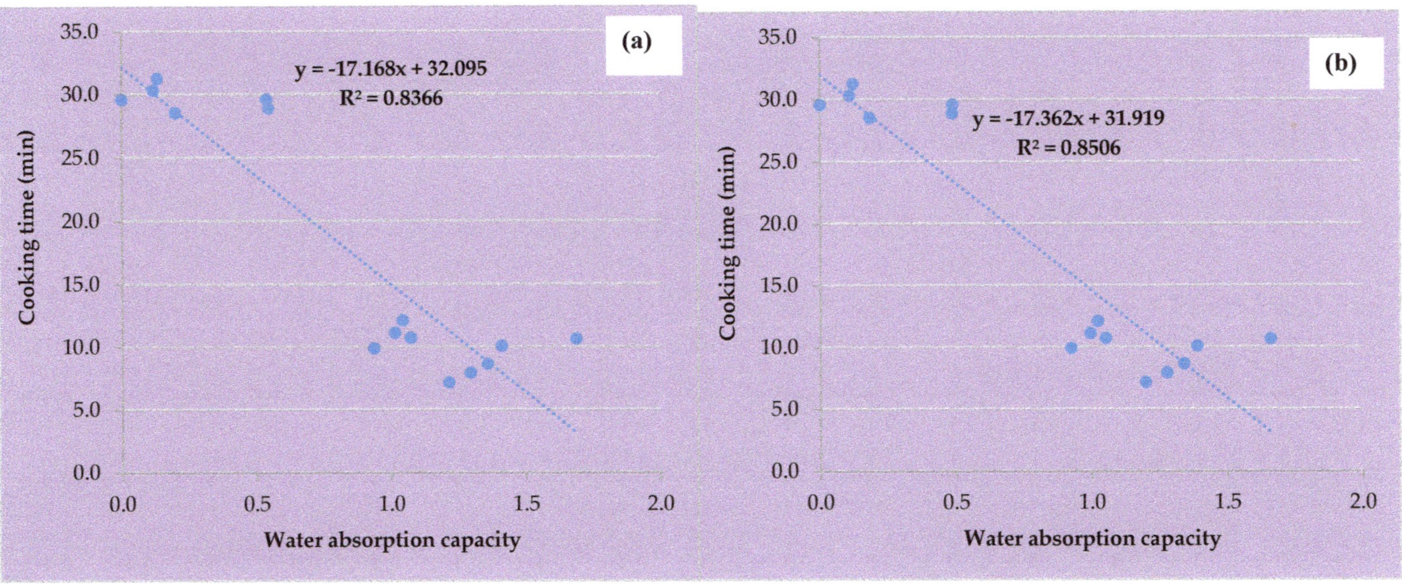

Figure 6. Correlation between water absorption and cooking time for the *Vigna racemosa* accessions. (**a**) Plotted with data from Site A; (**b**) plotted with data from Site B.

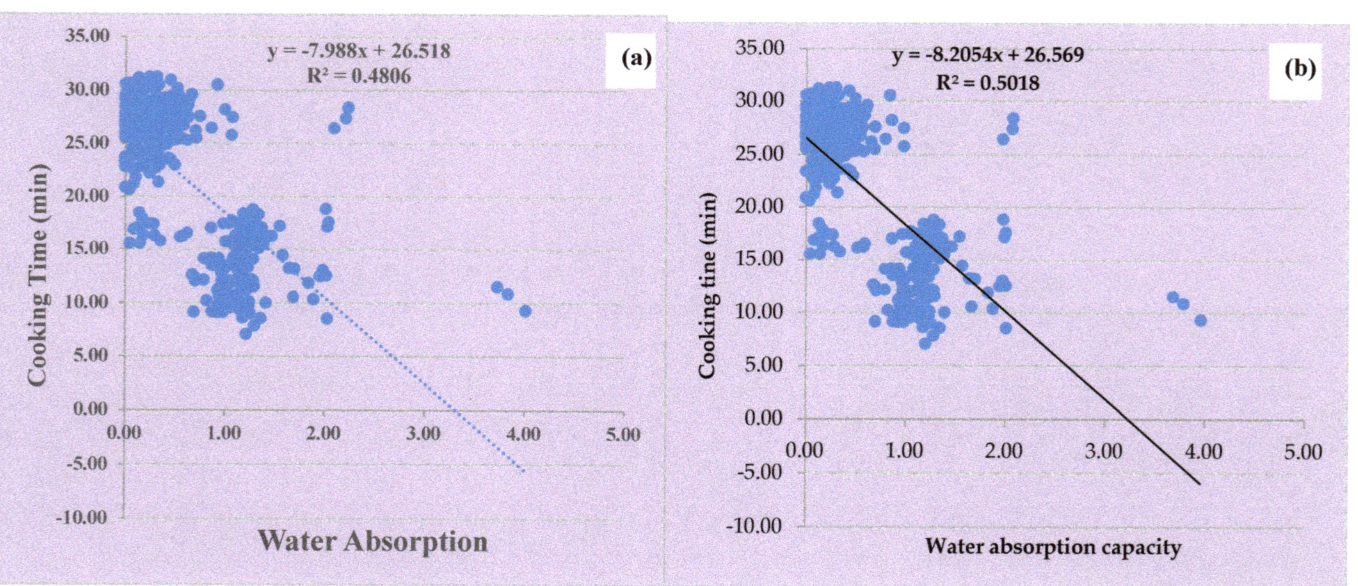

Figure 7. Correlation between water absorption and cooking time for the *Vigna* species studied. (**a**) Plotted with data from Site A; (**b**) plotted with data from Site B.

Table 7. Details of classes from the dendrogram *.

Class	1	2	3	4	5	6	7
Object	47	20	8	2	1	4	4
	TVNu324_VRe	TVNu1632_VV	TVNu325_VRe	Check 3	TVNu1520_VRe	AGG308107WVIG2_VV	TVNu379_VRe
	TVNu342_VA	TVNu1701_VV	Check 2	TVNu781_VV		AGG62154WVIG_1_VV	TVNu1582_VV
	AGG308101WVIG1 _VV	TVNu1629_VV	Unknown _*Vigna reticulata*			TVNu1624_VV	TVNu1358_VV
	TVNu1344_VV	TVNu767_VRe	AGG51603WVIG1_VRa			AGG308097WVIG1_VV	TVNu1394_VRe
	AGG308096 WVIG2_VV	TVNu343_VRe	AGG53597WVIG1_VRa				
	TVNu120 _VV	TVNu333_VV	Check 1				
	TVNu1529_VV	TVNu1370_VV	TVNu837_VV				
	TVNu720_VA	TVNu349_VRe	AGG52867WVIG1_VRa				
	TVNu223_VA	TVNu1378_VV					
	TVNu1546_VV	TVNu1405_VRe					
	TVNu1698_VRe	TVNu1593_VV					
	TVNu877_VA	Unknown *Vigna*					
	TVNu524_VRe	TVNu381_VV					
	TVNu1699_VA	TVNu479_VV					
	TVNu1191_VRe	TVNu605 _VRe					
	TVNu1621_VV	TVNu1360_VV					
	TVNu607_VRe	TVNu1790_VRe					
	TVNu56_VRe	TVNu1808_VRe					
	TVNu- 224_VRe	Unknown_*Vigna_racemosa*					
	TVNu739_VRe	TVNu161_VRe					
	TVNu916_VRe						
	TVNu955_VV						
	TVNu1092 _VV						
	TVNu1591_VV						
	TVNu178_VV						
	TVNu293_VV						
	TVNu1840_VA						

Table 7. *Cont.*

Class	1	2	3	4	5	6	7
Object	**47**	**20**	**8**	**2**	**1**	**4**	**4**
	AGG17856WVIG_1_VRe						
	TVNu738_VRe						
	TVNu1796_VV						
	TVNu1792_VA						
	TVNu832_VV						
	TVNu219_VA						
	TVNu491_VRe						
	TVNu1628_VV						
	TVNu1779_VRe						
	TVNu138_VRe						
	AGG308099WVIG2_VV						
	TVNu1804_VA						
	TVNu1586_VV						
	TVNu57_VRe						
	TVNu1825_VRe						
	TVNu1644_VA						
	TVNu758_VRe						
	TVNu1388_VRe						
	TVNu1522_VRe						
	TVNu1185_VA						

* Abbreviations put beside the accession names serves to identify species: VA stands for *V. ambacensis*, VV for *V.vexillata*, VRe for *V. reticulata*, and VRa for *V. racemos*.

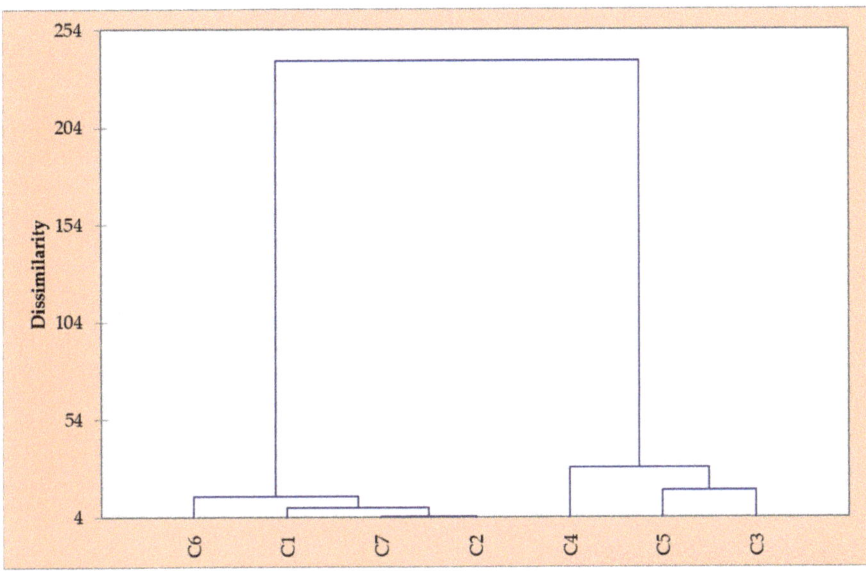

Figure 8. Dendrograms showing relationship among 84 accessions of the four wild *Vigna* species and three domesticated varieties regarding their weights before soaking, water absorption and cooking time.

3.7. Descriptive Statistics and Yield Traits of the Wild Vigna Species

Table 8 shows results of the means values for water absorption capacity, cooking time, and yield traits of the four wild species studied. *Vigna ambacensis* present mean values of 0.20, 23.45 min, and 1.74 g for water absorption capacity, cooking time, and yield per plant, respectively, in site A, while in site B the mean values are 0.18, 23.43 min, and 0.78 g for water absorption capacity, cooking time, and yield per plant, respectively. In *Vigna vexillata*, the values of 0.34, 25.42 min, and 16.84 g were found for water absorption capacity, cooking time, and yield per plant, respectively, in site A and 0.32, 25.40 min, and 12.54 g for water absorption capacity, cooking time, and yield per plant, respectively, in site B. For *Vigna reticulata*, the mean values are 0.39, 26.77 min, and 10.60 g for water absorption capacity, cooking time, and yield per plant, respectively, in site A, while in site B the mean values are 0.37, 26.78 min, and 6.78 g for water absorption capacity, cooking time, and yield per plant, respectively. Finally, *Vigna racemosa* present mean values of 0.84, 17.70 min, and 28.25g for water absorption capacity, cooking time, and yield per plant, respectively, in site A, while in site B the mean values are 0.81, 17.72 min, and 18.28 g for water absorption capacity, cooking time, and yield per plant, respectively. The yield values varied from 13.45 g (*V. vexillata* landrace) to 86.04 g (rice bean) in site A, while it varied from 7.62 g (*V. vexillata* landrace) to 61.92 g (rice bean) in site B for the domesticated legumes. For the wild legumes, it varied from 1.74 g (*Vigna ambacensis*) to 28.25 g (*Vigna racemosa*) in site A and from 0.78 g (*Vigna ambacensis*) to 18.28 g (*Vigna racemosa*) in site B.

Table 8. Descriptive statistic and yield traits of the wild *Vigna* species.

Species	Descriptive Parameters	Water Absorption Capacity		Cooking Time (min)		Yield per Plant (g)	
		Site A	Site B	Site A	Site B	Site A	Site B
Landrace of *Vigna vexillata*	Mean	1.33	1.32	10.24	10.26	13.45	7.62
	CV (%)	9.50	9.37	1.46	1.45	4.59	6.34
	Range	0.69–4.01	0.70–3.96	8.56–11.89	8.59–11.91	9.00–26.55	4.94–17.31
Cowpea (*Vigna unguiculata*)	Mean	1.27	1.27	16.29	16.31	52.690	26.657
	CV (%)	1.85	1.82	0.93	0.93	5.48	5.42
	Range	0.58–1.58	0.58–1.57	14.06–18.84	14.09–18.87	28.80–106.08	14.71–53.35
Rice Bean (*Vigna umbellata*)	Mean	1.16	1.16	13.20	13.23	86.04	61.92
	CV (%)	4.05	4.02	0.92	0.91	2.378	2.361
	Range	0.67–2.02	0.68–2.00	11.73–14.98	11.76–15.01	60.27–109.76	43.51–78.86
Vigna ambacensis	Mean	0.20	0.18	23.45	23.43	1.74	0.78
	CV (%)	11.21	11.20	0.44	0.42	22.36	14.25
	Range	0.00–0.50	0.00–0.58	22.25–24.95	22.26–24.96	0.72–5.36	0.43–1.65
Vigna vexillata	Mean	0.34	0.32	25.42	25.40	16.84	12.54
	CV (%)	7.80	7.95	1.73	1.70	9.48	6.77
	Range	0.00–1.15	0.00–1.13	15.54–31.28	15.55–31.30	9.48–63.00	7.61–35.26
Vigna reticulata	Mean	0.39	0.37	26.77	26.78	10.60	6.78
	CV (%)	13.83	14.04	1.20	1.27	10.55	10.62
	Range	0.00–2.24	0.00–2.08	16.60–30.98	16.58–40.00	4.32–30.36	2.58–17.69
Vigna racemosa	Mean	0.84	0.81	17.70	17.72	28.25	18.28
	CV (%)	16.70	17.06	14.83	14.81	37.02	38.37
	Range	0.00–1.69	0.00–1.66	7.11–31.19	7.12–31.22	2.08–49.00	1.21–34.70

CV: Coefficient of variation; Range (Minimum–Maximum).

4. Discussion

The values for water absorption capacity and cooking time showed no significant difference when compared with the values of their corresponding accessions harvested in the other agroecological zone for all the accessions tested. This could be due to the existence of a very slight difference in the characteristics of the two agroecological zones that could not significantly affect the genetic performance of the *Vigna* genus regarding the weight, water absorption, and cooking time. This is further justified by the fact that the interaction effect (location × genotype) showed that the differences observed for cooking time and water absorption capacities do not depend on location in all the accessions tested in this study (Tables 2a, 3, 4, 5 and 6). In the same line, a recent report revealed that the agroecological conditions could affect some nutrients like amino acids, protein, and minerals in quinoa but have no effect on their saponin and fiber content [24]. Furthermore, this study also demonstrates that the replication of the same species within the same location does not depend on the other species for the water absorption capacity trait (Table 2b), while for cooking time trait, there is an interaction with other species within the same location (Table 2c). This could be an important characteristic to be exploited in breeding programs.

The non-significant or significant changes observed in the mean seed weights of some accessions when compared before and after soaking depicted here by their water absorption capacity values could be explained by the fact that some accessions possess a seed coat more water permeable than others (Tables 2–6). The seed coat water permeability of seeds as a phenotype possesses a crucial role in legumes cooking properties and germination [25]. However, the development of legume seed coat has not yet been characterized at a molecular level to strongly support its genetic implication [26]. A study involving legume showed that the water absorption of dry beans differs between varieties [27].

Looking at the *V. ambacensis* species, all the wild accessions exhibited significantly lower water absorption capacity values as compared with all three checks (Table 3). The accession TVNu342, with a water absorption capacity not significantly different from the checks exhibited a higher cooking time. This could imply that not only the water absorption capacity is directly or indirectly linked to cooking times of legumes and requires further physiological investigation. The genus *Vigna* possess a very large number of species in which very few have been studied extensively. The *V. ambacensis* is among the non-studied species [2]. The very first comprehensive web genomic resource of the genus *Vigna* has just recently been published and that covered only three commercially domesticated species [28]. Taxonomic rearrangements are also still under investigation [29] and efforts to domesticate some of the selected wild *Vigna* species is in progress [2,4]. Pearson correlation analysis shows that there is no correlation between the water absorption and cooking time considering only the three domesticated species ($r = -0.025$). This could be due to some individual physiological differences or similarities among the tested accessions which requires further examination at molecular level as reports on *V. ambacensis* studies are very scanty and need to be addressed for proper exploitation of its full potential towards domestication [2].

For the *V. vexillata* species, Table 4 proved that there are some phenotypic similarities between the wild accessions with the cowpea and rice bean with regard to their water absorption capacity values as many accessions show no significant different values with those two checks. Henceforth, it requires further investigations at molecular level involving phylogenetic analysis to establish a strong relationship between the accessions. In this regard, it is noted that the genetic diversity and structure of *V. vexillata* as well as many wild *Vigna* legumes are still under investigation [2,29–31]. The idea is also supported by an earlier report that stipulated that domestication of the commercial *V. vexillata* (zombie pea) is not certain and it took place more than once in different regions [32]. Concerning the cooking time (Table 4), there is a high diversity in differences among the accessions. This could also explain why there is a weak negative correlation between the water absorption and cooking time considering the wild *V. vexillata* tested ($r = -0.31$) (Figure 4).

Wild *V. reticulata* species revealed that there is no significant difference between accessions regarding water absorption capacity (Table 5) with cowpea and rice bean except for TVNu1520, TVNu325 and the *V. vexillata* landrace. This demonstrates a considerable variability among the accessions as far as water absorption capacity is concerned as a phenotypic trait. Considering the cooking time (Table 5), a high diversity in differences of means among the accessions is noticed. Twenty-five accessions show no significant difference to each other but significantly different from the *V. vexillata* landrace and rice bean, while cowpea showed no significant difference to TVNu325 and the unknown *V. reticulata* accession. Curiously, scanty information about *V. reticulata* is also noticed. The genotype interactions in both water absorption capacity and cooking time phenotypic traits simply demonstrate the phenotypic diversity of these accessions which is very important in breeding.

Though very few accessions were included in this study, *V. racemosa* species present more phenotypic similarities with the *V. vexillata* landrace and cowpea with respect to the water absorption capacity and cooking time traits studied. It was revealed from the results that there is no significant difference between the means of the following wild accessions (AGG51603WVIG1, AGG53597WVIG1, AGG52867WVIG1) regarding their weights before soaking and the *V. vexillata* landrace and cowpea. The weights taken after the soaking process revealed a similar phenomenon while the water absorption shows closeness to rice bean. In the case of cooking time, two accessions seem to be related to the *V. vexillata* landrace. All these assumptions need further investigations as *V. racemosa* also suffer from scanty information.

This study also showed that there is a strong negative correlation between the water absorption and the cooking time with a correlation coefficient of $r = -0.69$ which follows the equation: $Y = -7.99x + 26.52$ ($R^2 = 0.48$) for site A and $Y = -8.21x + 26.57$ ($R^2 = 0.50$) for the site B (Figure 7). This result is in line with previous reports. For example, an early report proved that the cooking time was longer in bean varieties without prior soaking [33]. A similar result was found within classes of oriental noodle, in which cooking time was significantly shortened with increase in water absorption [34]. This could be an important parameter to guide the breeding of legumes with regards to cooking time when knowing their water absorption capacity.

Agglomerative hierarchical clustering (AHC) analysis performed on all the four species revealed the existence of seven classes when weight of accessions before the soaking process, water absorption capacity, and cooking time are taken as parameters (Figure 8). Details of various accessions belonging to each class are provided in Table 7. The analysis shows that some accessions between the four species can be grouped together in the same cluster as they present similar traits or relationship. It is in line with what the first comparison based on Tukey analysis showed in this study. For example, class 1 consists of *V. vexillata, ambacensis* and *reticulata* accessions while class 2 is mainly *V. vexillata* with few *V. reticulata*. It is also noted that all *V. ambacensis* are grouped in class 1. This can simply imply that there are phenotypic trait similarities of the accessions within species with each other and with checks. However, further molecular investigations are needed to fully investigate assumptions of any genetic relationship within and between species. The classifications of the *Vigna* species remain a continuous and evolving process as their origin are still subject of speculations. For example, it is reported that the Asian *Vigna* were still belonging to the genus *Phaseolus* until 1970 [30]. It is generally speculated that the *Vigna* might have originated from Africa and evolved from the African genus *Wajira* as it is basal compared with *Vigna* and *Phaseolus* [30]. Although, little attention has been paid to the conservation of the African wild *Vigna* species as more than 20 species are apparently not conserved in any ex-situ collection despite their several ethnobotanical uses [30]. Therefore, it could be speculated from this study that accessions in groups 3, 4, and 5 are likely candidates for domestication since these groups contain the check lines, though further investigations are required.

Based on a general assessment view, the values of yield per plant for the wild *Vigna* species studied here are lower than those of the domesticated species, especially cowpea and rice bean (Table 8). A similar finding was reported by an earlier report [7]. However, it might be important to note that the yield per plant for these wild legume accessions may be influenced by their seed characteristics because some of them could have a high number of seeds per plant with a surprising low weight as compared with the domesticated ones that produced fewer numbers of seed. The low seed weights in wild accessions could be attributed to their small seed sizes compared to domesticated ones with bigger seed sizes. The domesticated species here could have certainly acquired bigger seed sizes during the domestication process. Seed size is one of the important domestication traits [35] that should be considered by breeders in the course of improvement and domestication of these wild legumes as they all presented smaller seed sizes by mere looking (Figure 1). From this study, yield, water absorption capacity, and cooking time are apparently not related, though they are very important traits that need to be considered in breeding and selection of wild candidates for domestication. This may be due to the fact that yield mainly depends on seed physical characteristics such as seed size, seed weight, and seed number, while cooking time and water absorption capacity depends on seed physiological characteristics such as seed coat biosynthesis [25]. This could also be supported by the high variation in yield between locations as compared with low variations in cooking time and water absorption capacity (Table 8). In the same vein, it is also noted that the domesticated legumes possess high values of yield per plant in addition to their low cooking time and high water absorption capacity values as compared with the wild ones. Such characteristics might be among the factors that hinders their utilization as earlier reported [4]. Yield is a very important trait in crop domestication. However, these wild legumes with multipurpose utilizations as suggested by farmers in our earlier investigation [4] fit well as candidates for domestication considering the domestication criteria established by researchers recently [35]. Crop domestication of novel species is becoming one of the potential alternatives to mitigate the global food security challenge.

5. Conclusions

Despite their under-exploitation for human benefits, the wild *Vigna* legumes possess important cooking characteristics comparable with the domesticated ones. The present study revealed that the cooking time and water absorption capacity of wild legumes do not depend on their cultivation environment. Furthermore, it proved that there is a strong negative correlation between the water absorption capacity and cooking time in wild *Vigna* species. The study also revealed that some wild *Vigna* species present no significant difference in their cooking times with domesticated species which could be a positive acceptability trait to consumers. However, they might require considerable improvement in terms of seed physical characteristics to impact on their yield. Such key preliminary information could be of vital consideration in breeding, improvement, and domestication of wild *Vigna* legumes to make them useful for human benefit as far as cooking time is concerned. Investigations of nutritional and biochemical composition of these under-exploited legumes will also be of great importance to both scientists (breeders) and consumers for achieving food variety addition.

Author Contributions: P.A.N. and A.O.M conceived and designed the experiments; D.V.H. performed the experiments, collected data, analyzed the data, and made the first draft of the manuscript; P.B.V. and A.O.M. supervised the research and internally reviewed the manuscript; and P.A.N. made the final internal review and revised the final draft manuscript.

Funding: This research was partially funded by the Centre for Research, Agricultural Advancement, Teaching Excellence and Sustainability in Food and Nutrition Security (CREATES-FNS) through the Nelson Mandela African Institution of Science and Technology (NM-AIST). The research also received funding support from the International Foundation for Science (IFS) through the grant number I-3-B-6203-1.

Acknowledgments: This research was partially funded by the Centre for Research, Agricultural Advancement, Teaching Excellence and Sustainability in Food and Nutrition Security (CREATES-FNS) through the Nelson Mandela African Institution of Science and Technology (NM-AIST). The authors acknowledge the additional funding support from the International Foundation for Science (IFS) through the grant number I-3-B-6203-1. The authors also acknowledge the Genetic Resources Center, International Institute of Tropical Agriculture (IITA), Ibadan-Nigeria as well as the Australian Grains Genebank (AGG) for providing detail information and seed materials for illustration. The authors are thankful to Mary Mdachi and Lameck Makoye of the Tanzania Agricultural Research Institute (TARI), Selian-Arusha, Tanzania for their technical support in carrying out the Mattson Cooking experiment. The statistical advice from Usman Rachid of the Health and Rehabilitation Research Center, Auckland University of Technology, New Zealand is also appreciated.

Conflicts of Interest: The authors declare no conflict of interest.

References

1. Bhat, R.; Karim, A.A. Exploring the nutritional potential of wild and underutilized legumes. *Compr. Rev. Food Sci. Food Saf.* **2009**, *8*, 305–331. [CrossRef]

2. Harouna, D.V.; Venkataramana, P.B.; Ndakidemi, P.A.; Matemu, A.O. Under-exploited wild *Vigna* species potentials in human and animal nutrition: A review. *Glob. Food Secur.* **2018**, *18*, 1–11. [CrossRef]

3. Ojiewo, C.; Monyo, E.; Desmae, H.; Boukar, O.; Mukankusi-Mugisha, C.; Thudi, M.; Pandey, M.K.; Saxena, R.K.; Gaur, P.M.; Chaturvedi, S.K.; et al. Genomics, genetics and breeding of tropical legumes for better livelihoods of smallholder farmers. *Plant Breed.* **2018**, *138*, 487–499. [CrossRef]

4. Harouna, D.V.; Venkataramana, P.B.; Matemu, A.O.; Ndakidemi, P.A.; Harouna, D.V.; Venkataramana, P.B.; Matemu, A.O.; Ndakidemi, P.A. Wild *Vigna* legumes: farmers' perceptions, preferences, and prospective uses for human exploitation. *Agronomy* **2019**, *9*, 284. [CrossRef]

5. Wang, Q.J.; Mielby, L.A.; Junge, J.Y.; Bertelsen, A.S.; Kidmose, U.; Spence, C.; Byrne, D.V. The role of intrinsic and extrinsic sensory factors in sweetness perception of food and beverages: A review. *Foods* **2019**, *8*, 211. [CrossRef] [PubMed]

6. Costell, E.; Tárrega, A.; Bayarri, S. Food acceptance: The role of consumer perception and attitudes. *Chemosens. Percept.* **2010**, *3*, 42–50. [CrossRef]

7. Shafaei, S.M.; Masoumi, A.A.; Roshan, H. Analysis of water absorption of bean and chickpea during soaking using Peleg model. *J. Saudi Soc. Agric. Sci.* **2016**, *15*, 135–144. [CrossRef]

8. Turhan, M.; Sayer, S.; Gunasekaran, S. Application of Peleg model to study water absorption in bean and chickpea during soaking. *J. Food Eng.* **2002**, *53*, 153–159. [CrossRef]

9. Hamid, S.; Muzaffar, S.; Wani, I.A.; Masoodi, F.A.; Bhat, M.M. Physical and cooking characteristics of two cowpea cultivars grown in temperate Indian climate. *J. Saudi Soc. Agric. Sci.* **2016**, *15*, 127–134. [CrossRef]

10. Difo, V.H.; Onyike, E.; Ameh, D.A.; Njoku, G.C.; Ndidi, U.S. Changes in nutrient and antinutrient composition of *Vigna* racemosa flour in open and controlled fermentation. *J. Food Sci. Technol.* **2015**, *52*, 6043–6048. [CrossRef]

11. Siqueira, B.D.S.; Vianello, R.P.; Fernandes, K.F.; Bassinello, P.Z. Hardness of carioca beans (*Phaseolus vulgaris* L.) as affected by cooking methods. *LWT Food Sci. Technol.* **2013**, *54*, 13–17. [CrossRef]

12. Garcia, E.H.; Pena-Valdivia, C.B.; Aguirre, J.R.R.; Muruaga, J.S.M. Morphological and agronomic traits of a wild population and an improved cultivar of Common Bean (*Phaseolus vulgaris* L.). *Ann. Bot.* **1997**, *79*, 207–213. [CrossRef]

13. Gepts, P. Origins of plant agriculture and major crop plants. In *Our fragile world, forerunner volumes to the Encyclopedia of Life-Supporting Systems*; Tolba, M.K., Ed.; eolss publishers: Oxford, UK, 2001; pp. 629–637.

14. Padulosi, S.; Thompson, J.; Rudebjer, P. *Fighting Poverty, Hunger and Malnutrition with Neglected and Underutilized Species (NUS): Needs, Challenges and the Way Forward*; Bioversity International: Rome, Italy, 2013; ISBN 9789290439417.

15. Pratap, A.; Malviya, N.; Tomar, R.; Gupta, D.S.; Jitendra, K. Vigna. In *Alien Gene Transfer in Crop Plants, Volume 2: Achievements and Impacts*; Pratap, A., Kumar, J., Eds.; Springer Science+Business Media, LLC: Berlin/Heidelberg, Germany, 2014; pp. 163–189. ISBN 9781461495727.

16. Tomooka, N.; Naito, K.; Kaga, A.; Sakai, H.; Isemura, T.; Ogiso-Tanaka, E.; Iseki, K.; Takahashi, Y. Evolution, domestication and neo-domestication of the genus Vigna. *Plant Genet. Resour. Characterisation Util.* **2014**, *12*, S168–S171. [CrossRef]

17. Bisht, I.S.; Bhat, K.V.; Lakhanpaul, S.; Latha, M.; Jayan, P.K.; Biswas, B.K.; Singh, A.K. Diversity and genetic resources of wild *Vigna* species in India. *Genet. Resour. Crop Evol.* **2005**, *52*, 53–68. [CrossRef]

18. Bassett, A.N.; Cichy, K.A.; Ambechew, D.; Bassett, A.N.; Cichy, K.A.; Ambechew, D. Cooking time and sensory analysis of a dry bean diversity panel. *Annu. Rep. Bean Improv. Coop.* **2017**, *60*, 154–156.

19. IASRI Augmented Block Designs. Available online: http://www.iasri.res.in/design/AugmentedDesigns/home.htm (accessed on 18 January 2019).

20. MCWatters, K.H.; CHinnan, M.S.; PHillips, R.D.; Beuchat, L.R.; Reid, L.B.; Mensa-Wilmot, Y.M. Functional, nutritional, mycological, and Akara- making properties of stored cowpea meal. *J. Food Sci.* **2002**, *67*, 2229–2234. [CrossRef]

21. Wang, N.; Daun, J.K. Determination of cooking times of pulses using an automated Mattson cooker apparatus. *J. Sci. Food Agric.* **2005**, *85*, 1631–1635. [CrossRef]

22. Proctor, J.R.; Watts, B.M. Development of a modified Mattson bean cooker procedure based on sensory panel cookability evaluation. *Can. Inst. Food Sci. Technol. J.* **2013**, *20*, 9–14. [CrossRef]

23. Adewale, B.D.; Okonji, C.; Oyekanmi, A.A.; Akintobi, D.A.C.; Aremu, C.O. Genotypic variability and stability of some grain yield components of Cowpea. *Afr. J. Agric. Res.* **2010**, *5*, 874–880.

24. Reguera, M.; Conesa, C.M.; Gil-Gómez, A.; Haros, C.M.; Pérez-Casas, M.Á.; Briones-Labarca, V.; Bolaños, L.; Bonilla, I.; Álvarez, R.; Pinto, K.; et al. The impact of different agroecological conditions on the nutritional composition of quinoa seeds. *PeerJ* **2018**, *6*, e4442. [CrossRef]

25. Smykal, P.; Vernoud, V.; Blair, M.W.; Soukup, A.; Thompson, R.D. The role of the testa during development and in establishment of dormancy of the legume seed. *Front. Plant Sci.* **2014**, *5*, 1–19.

26. Thompson, R.; Burstin, J.; Gallardo, K. Post-genomics studies of developmental processes in legume seeds: Figure 1. *Plant Physiol.* **2009**, *151*, 1023–1029. [CrossRef] [PubMed]

27. Zamindar, N.; Baghekhandan, M.S.; Nasirpour, A.; Sheikhzeinoddin, M. Effect of line, soaking and cooking time on water absorption, texture and splitting of red kidney beans. *J. Food Sci. Technol.* **2013**, *50*, 108–114. [CrossRef] [PubMed]

28. Jasrotia, R.S.; Yadav, P.K.; Iquebal, M.A.; Arora, V.; Angadi, U.B.; Tomar, R.S.; Rai, A.; Kumar, D. VigSatDB: Genome-wide microsatellite DNA marker database of three species of *Vigna* for germplasm characterization and improvement. *Database* **2019**, *2019*, baz055. [PubMed]

29. Gore, P.G.; Tripathi, K.; Pratap, A.; Bhat, K.V.; Umdale, S.D.; Gupta, V.; Pandey, A. Delineating taxonomic identity of two closely related *Vigna* species of section Aconitifoliae: V. trilobata (L.) Verdc. and V. stipulacea (Lam.) Kuntz in India. *Genet. Resour. Crop Evol.* **2019**, *66*, 1155–1165. [CrossRef]

30. Tomooka, N.; Kaga, A.; Isemura, T.; Vaughan, D. Vigna. In *Wild Crop Relatives: Genomic and Breeding Resources: Legume Crops and Forages*; Kole, C., Ed.; Springer: Berlin/Heidelberg, Germany; London, UK, 2011; pp. 1–321. ISBN 9783642143878.

31. Karuniawan, A.; Iswandi, A.; Kale, P.R.; Heinzemann, J.; Grüneberg, W.J. *Vigna vexillata* (L.) A. Rich. cultivated as a root crop in Bali and Timor. *Genet. Resour. Crop Evol.* **2006**, *53*, 213–217. [CrossRef]

32. Dachapak, S.; Somta, P.; Poonchaivilaisak, S.; Yimram, T.; Srinives, P. Genetic diversity and structure of the zombi pea (*Vigna vexillata* (L.) A. Rich) gene pool based on SSR marker analysis. *Genetica* **2017**, *145*, 189–200. [CrossRef] [PubMed]

33. Corrêa, M.M.; Jaeger De Carvalho, L.M.; Nutti, M.R.; Luiz, J.; De Carvalho, V.; Neto, A.R.H.; Gomes Ribeiro, E.M. Water Absorption, Hard Shell and Cooking Time of Common Beans (Phaseolus vulgaris L.). *Afr. J. Food Sci. Technol.* **2010**, *1*, 13–20.

34. Hatcher, D.W.; Kruger, J.E.; Anderson, M.J. Influence of water absorption on the processing and quality of oriental noodles. *Cereal Chem.* **1999**, *76*, 566–572. [CrossRef]

35. Schlautman, B.; Barriball, S.; Ciotir, C.; Herron, S.; Miller, A.J. Perennial grain legume domestication Phase I: Criteria for candidate species selection. *Sustainability* **2018**, *10*, 730. [CrossRef]

 *agronomy*

Article

Trypsin Inhibitor Assessment with Biochemical and Molecular Markers in a Soybean Germplasm Collection and Hybrid Populations for Seed Quality Improvement

Kulpash Bulatova [1,*], Shynar Mazkirat [1], Svetlana Didorenko [1], Dilyara Babissekova [1], Mukhtar Kudaibergenov [1], Perizat Alchinbayeva [1], Sholpan Khalbayeva [1] and Yuri Shavrukov [2] 

[1] Kazakh Scientific Research Institute of Agriculture and Plant Growing, Almaty Region, Kazakhstan; shynarbek.mazkirat@gmail.com (S.M.); svetl_did@mail.ru (S.D.); janeka__88@mail.ru (D.B.); muhtar.sarsenbek@mail.ru (M.K.); piko1206@mail.ru (P.A.); sholpan_2706@mail.ru (S.K.)

[2] College of Science and Engineering, Biological Sciences, Flinders University, Bedford Park 5042, Australia; yuri.shavrukov@flinders.edu.au

* Correspondence: bulatova_k@rambler.ru; Tel.: +77273883925

Received: 12 January 2019; Accepted: 9 February 2019; Published: 11 February 2019

Abstract: A soybean germplasm collection was studied for the identification of accessions with low trypsin inhibitor content in seeds. Twenty-nine accessions, parental plants, and two hybrid populations were selected and analyzed using genetic markers for alleles of the *Ti3* locus, encoding Kunitz trypsin inhibitor (KTI). Most of the accessions had high or very high KTI (49.22–73.53 Trypsin units inhibited (TUI/mg seeds), while the two local Kazakh cultivars, Lastochka and Ivushka, were found to have a moderately high content of KTI, at 54.16–54.87 TUI/mg. In contrast, two soybean cultivars from Italy, Hilario and Ascasubi, showed the lowest levels of trypsin units inhibited, at 25.47–27.87 TUI/mg. Electrophoresis of seed proteins in a non-denaturing system showed a simple discrimination pattern and very clear presence/absence of bands corresponding to KTI. The SSR marker *Satt228* was the most effective diagnostic marker among the three examined, and it confirmed the presence of the homozygous null-allele *ti3/ti3* in cultivars Ascasubi and Hilario, which were used for hybridization with the local cv. Lastochka. Heterozygote F_1 hybrid plants and homozygous *ti3/ti3* lines in F_2 segregating populations were successfully identified using *Satt228*. Finally, through marker-assisted selection with *Satt228*, prospective homozygous *ti3/ti3* lines were produced for further application in the breeding program aimed at improving soybean seed quality in Kazakhstan.

Keywords: Kunitz trypsin inhibitor; molecular markers; non-denaturing electrophoresis; seed quality; seed storage proteins; soybean; SSR (Simple sequence repeat) markers; *Ti* gene

---

1. Introduction

Soybean [*Glycine max* (L.) Merrill] is one of the most important crops for protein and oil production in Kazakhstan, and the processed seed products are mainly used for the purpose of food and fodder. Soybean breeding programs and seed production have been conducted in Kazakhstan for over 40 years, where 16 cultivars have been developed and introduced into local agribusiness. The area under soybean crop is increasing annually and now accounts for more than 120,000 hectares. However, this is still not enough to provide sufficient raw materials for oil processing and livestock and poultry farms in Kazakhstan.

The soybean germplasm collection and breeding lines (750 accessions) have been assessed using the main breeding indicators for seed productivity, quality and length of the growing season. Seeds were evaluated for quality indicators including protein and oil content. However, the content of anti-nutritional compounds in seeds was not previously taken into consideration by domestic breeders. As a result, parents with high levels of anti-nutritional compounds were unwittingly used for hybridization, which has significantly affected the resulting quality of the processed soybean products. The presence of proteinase inhibitors in seeds is known to be one of the main obstacles in the development and expansion of commercial soybean products. Soybean seeds contain two classes of major proteinase inhibitors, Kunitz trypsin inhibitor (KTI) and Bowman-Birk proteinase inhibitor [1,2]. Trypsin is an essential digestive enzyme found in the vertebrate small intestine that catalyzes the degradation of proteins, enabling their absorption into the bloodstream. KTI contributes up to 80% of the total trypsin inhibitor activity in soybean seeds [3].

Despite the moderate content of KTI in total soybean proteins, feeding of such seeds to livestock inhibits their growth and weight gain [4] and causes pancreatic hypertrophy [5]. For soybean seeds with high KTI, a preliminary heat treatment is required to inactivate the trypsin inhibitor enzyme before it can be fed to poultry and livestock. This heat treatment affectively destroys the undesirable anti-nutritional component, but also leads to an increase in the cost of the final product as well as decreased levels of available amino acids [6].

Initially, a single gene, *Ti*, encoding a high content of KTI in seeds of soybean, *G. max*, was described [7], where three dominant alleles, Ti^a, Ti^b, and Ti^c, were reported for trypsin inhibitor A_2, and the allele Ti^a was very common and widespread [8]. The multiple alleles were further extended during the study of 1368 germplasm accessions of Korean wild soybean (*G. soja* Sieb. & Zucc.), where the additional two rare dominant alleles, Tib^{i7-1} and Tib^{i5}, were reported [9]. The *Ti/ti* locus was mapped to linkage group 9, with a genetic distance of 16.2 cM from the acid phosphatase locus, *Ap* [7], and 15.3 cM from the leucine aminopeptidase locus, *Lap1* [10]. Later, a comprehensive molecular analysis revealed 10 genes encoding KTI in soybean, but only four of them, *Ti1*, *Ti2*, *Ti3*, and *Ti4*, were transcribed into mRNA [11]. It was confirmed that only one gene, *Ti3* (GenBank ID: S45092.1), represents the major KTI, since its expression level is much more prevalent in soybean seeds compared to the other genes. The *Ti3* gene was reported to be transcribed at the seed maturation stage, producing most of the KTI protein found in soybean seeds [12]. This KTI polypeptide represents a soybean storage protein with a molecular weight of 21–21.5 kDa, with specific activity for trypsin inhibition [13,14]. The sequence of *Ti3* was identical to the common allele Ti^a discovered earlier [11]. Two other genes, *Ti1* and *Ti2*, displaying a lack of KTI activity, are mostly expressed at different stages of leaf, stem, and root development, and contribute a much smaller level of mRNA during embryogenesis and seed development [11]. The remaining *Ti* genes do not encode proteins with KTI activity and are assumed to be 'Pseudogenes'.

A single recessive null-allele, *ti*, resulted in the absence of KTI [8], caused by three nucleotide differences within the *Ti3* coding region [12]. The null-allele was initially originated from soybean germplasm accessions M91-212006, and the segregating analysis of progenies from mapping populations F_2 and F_3 was reported [15]. It was suggested that the absence of a 21.5 kDa protein in the total spectrum of seed storage proteins was related to the null-allele *ti*. Soybean germplasm accessions PI157440 and PI196168 were reported as additional sources of the null-allele *ti*. The breeding lines, based on introgression of the null-allele *ti* into commercial soybean cultivars using a conventional backcrossing program, revealed low KTI and improved seed quality for feeding chickens and young pigs [8]. In a further study, the soybean accession PI542044 with pedigree origin from PI157440 and carrying the null-allele *ti* was backcrossed with recurrent parents. The introgression of the *ti* null-allele at BC_2F_2 and BC_3F_2 generations was controlled by Marker-assisted selection (MAS) with *ti* allele-specific primers and linked SSR markers. Nine and six breeding lines with genetic backgrounds of JS97-52 and DS9712 / DS9814 recurrent parents, respectively, were obtained in Indian breeding programs [16,17]. The introgressed null-allele *ti* was confirmed in the breeding lines and the KTI content was reduced by 69–84% [16] and an additional seed yield improvement was reported [17].

The aims of this research were: (1) to study a germplasm collection of soybean for the identification of genotypes with low KTI content in seeds; (2) to evaluate diagnostic genetic markers for the null-allele *ti3* in parental forms and produce hybrid populations from crosses between the lowest KTI genotype identified and elite Kazakh soybean cultivars; (3) to employ marker-assisted selection for low KTI genotypes in hybrid populations for seed quality improvement in the Kazakh breeding program.

2. Materials and Methods

The soybean germplasm collection comprising 29 cultivars was selected and received from the Kazakh Scientific Research Institute of Agriculture and Plant Growing, Almaty region, Kazakhstan (Listed in Table 2). Two hybrid combinations, Lastochka × Ascasubi and Lastochka × Hilario, F_1, F_2, and F_3, generations, were produced from manual crosses using a method described recently in the patent [18].

KTI activity was determined according to the method described by Kakade et al. [19], using casein as a substrate and expressed in trypsin units inhibited per milligram of soybean meal.

Isolation of glycinin storage proteins was carried out in phosphate buffer (pH 6.9), with subsequent electrophoretic separation in accordance with the protocol published earlier [20] in the presence of a molecular mass marker set ranging from 10 to 130 kDa (Thermo Fisher Scientific, Vilnius, Lithuania). Fixation and staining of the protein bands was carried out in 12.5% trichloroacetic acid. Quantitative measurement of the spectra components, their molecular mass, and relative mobility was carried out by means of the Quantum ST4 Gel documenting system (Vilber, Collégien, France), and the relative percentage of each band was calculated as a ratio of all components using the 'Quantum Capt' computer software supplemented to the equipment. Electrophoresis in a non-denaturing system was carried out according to the protocol published earlier [21]. Proteins were extracted from 10 mg of flour using 62.5 mM Tris HCl (pH 8.1) buffer for 40 min. The protein probe was prepared by mixing 100 µl of supernatant with 100 µl of 62.5 mM Tris HCl (pH 6.9) buffer mixture containing 20% glycerin and bromophenol blue marker dye. Fixation and staining of the protein bands was performed as mentioned above.

Genomic DNA was extracted using the CTAB method [22] from the first true leaves of individual seedlings grown in greenhouse conditions. Leaf tissue samples, about 200 mg each, were transferred to 2-ml test tubes with 800 µl of CTAB extraction buffer containing 1.0% polyvinylpyrrolidone (PVP40) and 0.2% β-mercaptoethanol, and homogenized using a stainless steel pestle. Extracted and precipitated DNA was re-dissolved in 400 µl of 1 M NaCl solution and treated with 2 µl (10 mg/mL) of RNase A (Thermo Scientific, Waltham, MA, USA) at 37 °C for 30 min. DNA was precipitated with cold 100% ethanol and washed with 70% ethanol. Isolated DNA was then dissolved in 100 µl of sterile water. The concentration and quality of DNA samples was determined at 260 and 280 nm using a spectrophotometer Jenway 6715 (Jenway, Staffordshire, UK). DNA samples were diluted with sterile water to a concentration of 100 ng/µl for use in further experiments.

PCR was performed in a total volume of 15 µL containing a cocktail with the following final concentrations: 1×PCR buffer, 2.5 mM $MgCl_2$, 200 µM each of dNTPs, 0.5 µM of each forward and reverse primers, BSA (2 mg/mL), and 0.5 U of *Taq* DNA polymerase (GeneLab, Astana, Kazakhstan). The PCR was conducted in a thermocycler (Bio-Rad, iCycler, Portland, ME, USA), where the amplification conditions were as follows: initial denaturation at 94 °C for 5 min, followed by 35 cycles of 94 °C for 20 s, 55 °C for 20 s, 72 °C for 30 s, and with a final extension of 10 min at 72 °C.

The amplification products were separated in polyacrylamide gel (8% acrylamide, 1×TBE buffer), and gels were stained with ethidium bromide for digital imaging by the Quantum ST4 Gel documenting system (Vilber, Collégien, France), as indicated above. The dimensional characteristics of PCR products were determined using the computer software 'Quantum Capt' (Vilber, Collégien, France) to determine the length and intensity of DNA fragments.

Three SSR markers tightly linked to the *Ti3* gene and distinguishing between alleles were selected based on published data [23], and primer sequences are presented in Table 1.

IBM SPSS Statistical software Desktop 25.0.0.0 (IBM, Armonk, NY, USA). was used to calculate and analyze means and standard error. Welch's ANOVA test was applied for comparison of accessions with low and high KTI due to different standard deviations and heteroscedasticity. Observed and expected segregations were analyzed using Chi-square test. One-way ANOVA and post-hoc Tukey–Kramer test with the minimum significant difference were applied for calculation of significant difference among genotypes of parent and hybrids with different KTI.

Table 1. SSR markers and corresponding primer sequences used for the identification of soybean *Ti3/ti3* genotypes.

Markers		Primer Sequences
Satt228	F	TCATAACGTAAGAGATGGTAAAACT
	R	CATTATAAGAAAACGTGCTAAAGAG
Satt409	F	CCTTAGACCATGAATGTCTCGAAGAA
	R	CTTAAGGACACGTGGAAGATGACTAC
Ti/ti, gene specific marker	F	CTTTTGTGCCTTCACCACCT
	R	GAATTCATCATCAGAAACTCTA

3. Results

Biochemical analyses of soybean seeds in the germplasm collection were carried out for the purpose of characterizing the activity of anti-nutritional components by measuring the inhibition of trypsin by KTI. Our results show that two cultivars originating from Italy—Ascasubi and Hilario—with the lowest trypsin units inhibited, TUI (27.87 and 25.47 units/mg of dry ground seeds, respectively, Table 2), were significantly different to other studied soybean accessions ($p < 0.001$, using Welch's ANOVA test). The best local soybean cultivars from Kazakhstan were Lastochka and Ivushka, with moderately high KTI and showing 54.16 and 54.87 units/mg of TUI, respectively, but these were still significantly higher than Hilario and Ascasubi. (Table 2).

Table 2. Characterization of soybean collection samples according to the activity of the Kunitz trypsin inhibitor. TUI, trypsin units inhibited $\pm$ SE ($n = 3$).

Entry	Sample Name	Country of origin	TUI/mg	Group
1	Hilario	Italy	25.47 $\pm$ 2.44	Low
2	Ascasubi	Italy	27.87 $\pm$ 0.85	Low
3	Vilana	Russia	49.22 $\pm$ 1.56	High
4	Selekta 301	Russia	51.05 $\pm$ 0.57	High
5	Blamcos	Italy	51.62 $\pm$ 0.57	High
6	Lastochka	Kazakhstan	54.16 $\pm$ 0.56	High
7	Ivushka	Kazakhstan	54.87 $\pm$ 1.28	High
8	Sava	Serbia	54.87 $\pm$ 1.84	High
9	Triumph	Serbia	56.14 $\pm$ 0.28	High
10	Galina	Ukraine	56.85 $\pm$ 0.15	High
11	Slaviia	Russia	57.41 $\pm$ 0.14	High
12	Korsak	Ukraine	57.41 $\pm$ 0.42	High
13	Luna	Italy	57.98 $\pm$ 0.15	High

Table 2. *Cont.*

Entry	Sample Name	Country of origin	TUI/mg	Group
14	Dekabig	USA	57.98 ± 0.70	High
15	Pamiat IuGK	Kazakhstan	59.24 ± 0.56	High
16	Zhansaya	Kazakhstan	59.96 ± 0.14	High
17	Fora	Russia	61.65 ± 1.72	High
18	Harbin	China	61.93 ± 2.12	High
19	Voevodzhanka	Serbia	62.36 ± 0.85	High
20	Ayzere	Kazakhstan	63.77 ± 0.84	High
21	Perizat	Kazakhstan	64.91 ± 0.56	High
22	Sabira	Kazakhstan	66.18 ± 0.98	High
23	Safrana	France	66.60 ± 2.83	High
24	Delta	Russia	67.45 ± 0.29	High
25	Akky	Kazakhstan	67.59 ± 1.84	High
26	Santana	France	68.30 ± 0.56	High
27	Birlik KV	Kazakhstan	69.85 ± 1.27	High
28	Zen	Switzerland	71.69 ± 1.13	High
29	Atlantic	Italy	73.53 ± 0.71	High

Electrophoresis of seed storage proteins from the studied soybean cultivars revealed the presence of a 21 kDa molecular component, corresponding to the Kunitz trypsin inhibitor, in all analyzed genotypes. Importantly, only two cultivars, Hilario and Ascasubi, with the lowest amounts of KTI, showed the absence of the 21 kDa molecular component. Figure 1a shows the comparison of the KTI spectra in Hilario with three other cultivars as an example of the absence or presence of the KTI band. The relative percentage of KTI component in total seed storage proteins established by densitometry in the cultivar Triumph was only 1.03%, which was clearly absent in Hilario (Figure 1b).

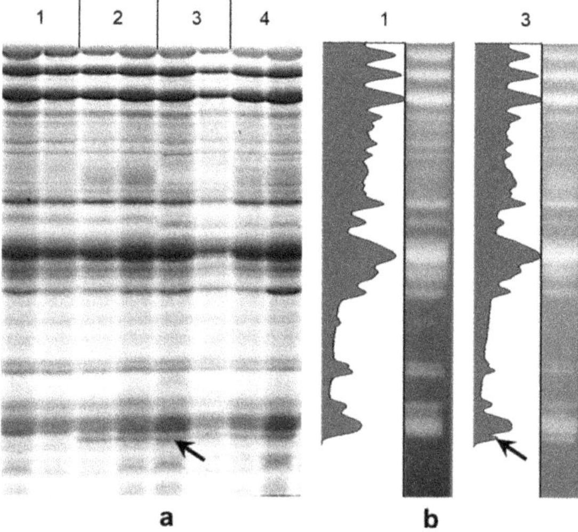

Figure 1. Electrophoregram of polyacrylamide gel (**a**) and densitograms (**b**) of soybean seed storage proteins. Germplasms are labelled above the image as follows: 1, Hilario; 2, Atlantic; 3, Triumph; and 4, Luna. The band for the 21 kDa molecular component KTI is indicated by arrows.

Molecular analysis of *Ti3* locus encoding KTI was carried out with three SSR markers. The markers *Satt228*, *Satt409*, and the gene-specific marker *Ti/ti* are tightly linked to the *Ti3* locus, and they can be perfectly used as diagnostic markers for MAS of genotypes with the null-allele *ti3*. Amplification products of these markers during PCR analysis with DNA from two soybean cultivars with low KTI (Hilario and Ascasubi) and one local cultivar Lastochka with high KTI are shown in Figure 2. The presented data confirmed that both Italian cultivars (Hilario and Ascasubi) have null-allele *ti3* while Kazakh cultivar has dominant allele *Ti3* in the locus.

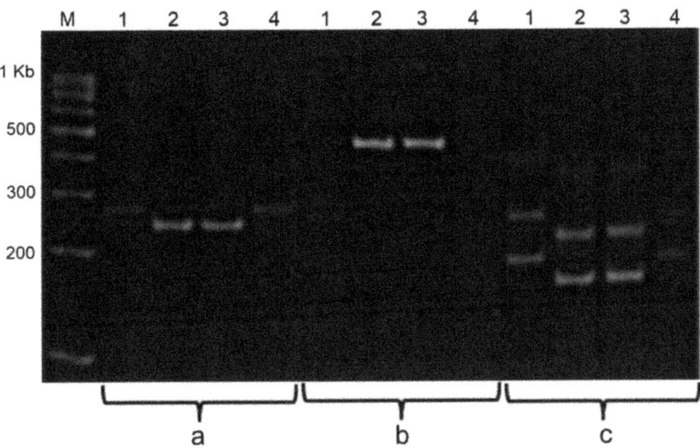

Figure 2. Amplification products of three SSR markers linked to the soybean *Ti3* gene. M, Marker 100 bp; 1 and 4, Lastochka; 2, Hilario; and 3, Ascasubi. The individual marker used is indicated at the bottom as follows: (**a**) *Satt228*; (**b**) *Ti/ti*-gene specific marker; (**c**) *Satt409*.

All markers clearly distinguish between soybean genotypes on the basis of the *Ti3* alleles detected, but the *Ti/ti*-gene-specific marker does not allow for the identification of homo- and heterozygote genotypes with dominant alleles *Ti3*, which makes it difficult to apply this marker in segregating populations. In contrast, the SSR marker *Satt228* was a much more suitable diagnostic marker for the further screening of plants in segregating populations, for the identification of homozygote progenies with null-allele *ti3* and production of the best non-segregating breeding lines with low KTI in seeds.

The cultivars Hilario and Ascasubi were identified as the genotypes with the lowest amounts of KTI, and they were selected in the results of the biochemical and PCR analyses with the confirmed null-allele *ti3*. Hybrid populations were produced with the null-allele *ti3* from Hilario and Ascasubi introgressed into the genetic background of local Kazakh cultivars.

In the hybrid combinations, Lastochka × Ascasubi and Lastochka × Hilario, PCR analysis of DNA from the F_1 plants and SSR marker *Satt228* confirmed the presence of the null-allele *ti3* in Ascasubi and Hilario (Data not shown). All F_1 plants were heterozygous for the *Ti3* locus with high KTI and were used for further production of the F_2 populations. Biochemical analysis for the spectrum of seed storage proteins using non-denaturing electrophoresis was applied for analyses of Kazakh varieties and F_2 segregating populations originating from the crosses, Lastochka × Ascasubi and Lastochka x Hilario. All local varieties and the parental plants from cultivars Ascasubi and Hilario, as well as segregants with null-alleles, *ti3/ti3*, showed very clear presence/absence of bands in a simple discrimination pattern. This method can be used for the selection of genotypes with the null allele of the KTI locus in seeds. However, homo- and heterozygote genotypes with dominant alleles, *Ti3/Ti3* and *Ti3/ti3*, could not be separated using polyacrylamide gel electrophoresis (Figure 3).

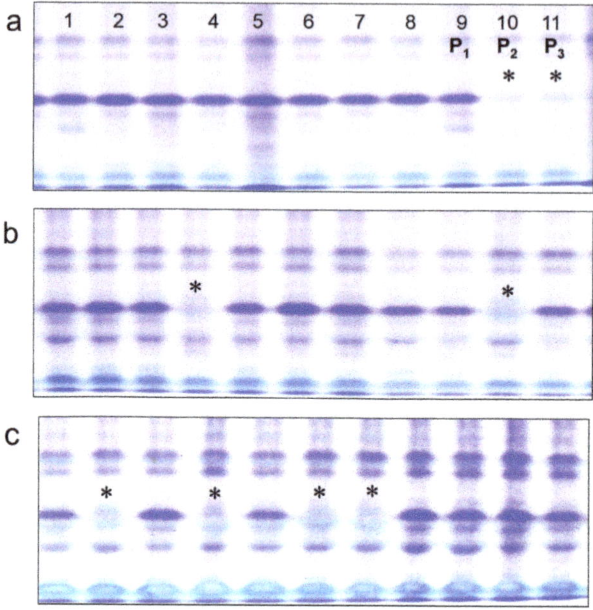

Figure 3. Polyacrylamide gel patterns of soybean seed storage proteins: (**a**) local varieties (1–8) and (9) parental cultivar Lastochka (P₁) with genotypes *Ti3/Ti3*; two other parental forms (10 and 11), Ascasubi (P₂) and Hilario (P₃), respectively, with genotype *ti3/ti3*. Examples of the F₂ segregating populations: (**b**) Lastochka × Ascasubi; and (**c**) Lastochka × Hilario. Homozygote genotypes, *ti3/ti3*, are indicated by asterisks (*).

In contrast, application of the SSR marker *Satt228* could identify all three types of genotypes at the *Ti3* locus. An example of genotyping of soybean plants using this SSR marker is presented in the second cross, Lastochka × Hilario. PCR analysis of the F₂ of hybrid plants in the presence of both parents confirmed that plants of cv. Lastochka have genotypes with the dominant alleles *Ti3/Ti3*, while plants of cv. Hilario are homozygotes with the recessive null-allele *ti3/ti3*. Among progenies of the F₂ hybrid population, all genotypes with the dominant and recessive alleles of the *Ti3* locus were identified (Figure 4).

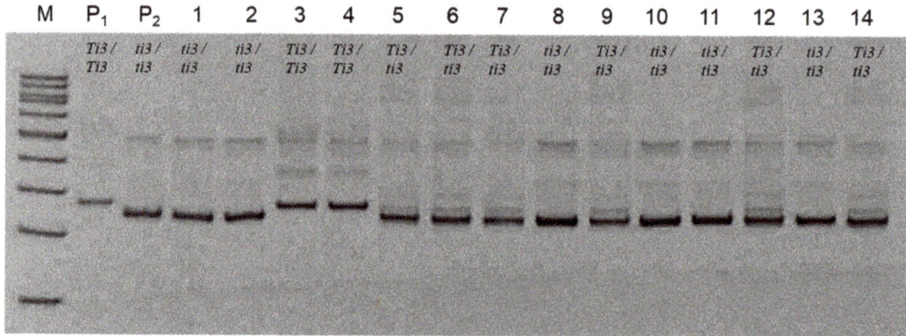

Figure 4. PCR products of parental forms and the F₂ hybrid population of Lastochka × Hilario amplified with *Satt228*. M, Marker 100 bp; P₁, Lastochka; P₂, Hilario; 1–14, Hybrid F₂ plants. Genotypes for the *Ti3* locus are shown for each plant.

The comparison of segregation analyses using seed storage proteins and the molecular SSR marker *Satt228* revealed full consensus and confirmed the Mendelian monogenic-type inheritance in both studied F_2 populations of Lastochka × Ascasubi and Lastochka × Hilario (Table 3).

Table 3. Segregation analyses of genotypes of the *Ti3* locus using seed storage proteins and the molecular SSR marker *Satt228* in F_2 populations of Lastochka × Ascasubi and Lastochka × Hilario. The asterisks (* and **) indicate no significant differences ($p < 0.05$ and $p < 0.01$, respectively) between observed and expected segregations for simple Mendelian ratio (3:1 for seed storage proteins and 1:2:1 for SSR marker *Satt228*), compared to the Chi-square distribution table.

	Genotypes			Total Number of Plants	$\chi^2 < \chi^2_{(Table1)}$
	Ti3/Ti3	*Ti3/ti3*	*ti3/ti3*		
Lastochka × Ascasubi					
Seed storage proteins, observed segregation	46		24	70	2.78 * <
Expected segregation	52.5		17.5	70	$3.84_{(df = 1)}$
SSR marker *Satt228*, observed segregation	15	31	24	70	3.23 ** <
Expected segregation	17.5	35	17.5	70	$4.61_{(df = 2)}$
Lastochka × Hilario					
Seed storage proteins, observed segregation	38		15	53	0.31 ** <
Expected segregation	39.75		13.25	53	$2.71_{(df = 1)}$
SSR marker *Satt228*, observed segregation	12	26	15	53	0.36 ** <
Expected segregation	13.25	26.5	13.25	53	$4.61_{(df = 2)}$

MAS was applied for segregants with homozygote null-allele genotypes, *ti3/ti3*, based on the results of screening using the SSR marker *Satt228* on F_2 hybrid progenies of Lastochka × Ascasubi and Lastochka × Hilario. The selected homozygote F_2 plants, both with recessive and dominant alleles at the *Ti3* locus, were grown and seeds from F_3 families were finally analyzed for KTI content based on TUI results (presented in Table 4).

Table 4. Kunitz trypsin inhibitor analysis in homozygote F_3 families originating from two hybrid populations after MAS with SSR marker *Satt228* for trypsin units inhibited (TUI) $\pm$ SE ($n = 3$). Minimum significant differences ($p > 0.01$) based on one-way ANOVA with post-hoc Tukey–Kramer test are shown by different letters. No statistical differences are shown by identical letters.

Parent/Progeny	Name	Genotype *Ti3/ti3*	TUI/mg of Dry Seeds	Statistical Differences
Lastochka × Ascasubi				
♀P_1	Lastochka	*Ti3/Ti3*	53.2 ± 0.2	a
♂P_2	Ascasubi	*ti3/ti3*	25.2 ± 0.1	c
F_3 family 1	(Lastochka × Ascasubi) − 1	*Ti3/Ti3*	43.0 ± 0.6	b
F_3 family 2	(Lastochka × Ascasubi) − 2	*ti3/ti3*	23.7 ± 1.3	c

Table 4. *Cont.*

Parent/Progeny	Name	Genotype *Ti3/ti3*	TUI/mg of Dry Seeds	Statistical Differences
		Lastochka × Hilario		
$♀P_1$	Lastochka	*Ti3/Ti3*	53.2 ± 0.2	a
$♂P_3$	Hilario	*ti3/ti3*	23.2 ± 0.1	c
F_3 family 3	(Lastochka × Hilario) − 3	*Ti3/Ti3*	45.5 ± 0.6	b
F_3 family 4	(Lastochka × Hilario) − 4	*ti3/ti3*	17.4 ± 1.3	d

The presented results show consistent inheritance and very significant differences in KTI in homozygote F_3 families originating from recombinants in both hybrid combinations. In the first hybrid (Lastochka × Ascasubi), the F_3 family showed KTI content statistically similar to the parental form Ascasubi, while in the second hybrid combination (Lastochka × Hilario), statistically reduced KTI was found in the F_3 family (Table 4).

4. Discussion

Soybean is a very widely used legume crop, but improvement of seed quality for lower KTI content is a very important and challenging task representing a critical step for soybean breeding. Therefore, the identification of soybean germplasm resources with low KTI and their introgression in hybridization programs promotes the development of promising soybean breeding lines with improved protein composition.

Among studied local Kazakh soybean cultivars, only two, Ivushka and Lastochka, were found to have more moderate levels of KTI. However, two Italian cultivars, Hilario and Ascasubi, were identified as having the lowest KTI, with significantly less than other studied soybean germplasms at 25.47 and 27.87 units of trypsin inhibited, respectively. Nevertheless, these cultivars with the lowest content of anti-nutritional factors are relatively old, having been developed in Italy in the early 1990s. Therefore, they are most valuable as donors of genetic resources to the modern soybean breeding program.

Biochemical screening of the germplasm collection for KTI and electrophoretic analysis of the seed protein composition allow for the evaluation of a range of soybean cultivars produced both locally and overseas, thus making it possible to select the most suitable parental forms for crossings and hybrid production. Homozygote progenies with the null-allele *ti3/ti3* can be successfully identified and novel breeding lines can be produced from hybrid populations. However, this process must be improved and significantly sped up via the use of MAS in the initial steps of selections, using a suitable diagnostic marker strongly associated with the null-allele *ti3* to produce new soybean cultivars with improved seed quality.

The presented results show the effective application of both biochemical and molecular markers for the null-allele *ti3* in the studied soybean germplasm collection, and in two segregating populations. Biochemical analysis of seed storage proteins using polyacrylamide gel electrophoresis is very accurate and can be used at the seed stage whilst also saving the viable part of the *ti3/ti3* line for multiplication. Therefore, it is important to enrich the initial pool of recombinant lines with the null allele of the *Ti3* locus because a large part of the lines can be removed at the early stage of the breeding process. The use of the diagnostic SSR marker *Satt228* is based on regular PCR and is very simple and quick. SSR markers are well known as being polymorphic, with codominant inheritance, and therefore can be widely used for genotype identification and selection of desired traits [23]. However, the distance between the SSR marker and the gene of interest can vary depending on the type of population [6,24]. For example, the genetic distance between *Satt228* and *Ti3* varied between 0 and 3.7 cM in two different populations [23]. Therefore, the number of recombinant genotypes in F_2 hybrids of Lastochka

× Ascasubi and Lastochka × Hilario (Table 3) is very small and can be estimated as 0–2.6 and 0–2.0 plants, respectively. Another codominant SSR marker *Satt409* could also be successfully used instead (Figure 2), but it was mapped to a region more genetically distant from the *Ti3* locus at 4.5–21.9 cM [23], meaning that there would be a greater chance of unwanted recombinants. Therefore, either native electrophoresis of storage proteins or SSR markers can support the reliable selection of new promising breeding lines, and the choice will depend on the cost or convenience as preferred by researchers. The applied MAS with *Satt228* was very effective in our experiments with both segregating populations, where homozygote genotypes *ti3/ti3* were identified, isolated, and propagated. The simple Mendelian-type inheritance of the *Ti3* locus has been confirmed [12,15] and helps to estimate the ratio for homozygotes with the null-allele *ti3*. Finally, segregants with *ti3/ti3* genotypes were successfully verified for low KTI content and propagated for further yield analysis and development of prospective breeding lines. Our results were similar to those published earlier on the introgression of the null-allele *ti3* and MAS to select recombinant genotypes with low KTI content in an Indian soybean breeding program [16,17]. Hybridization and transference of the beneficial *ti3* null-alleles determining low KTI in seeds using MAS is very important to enhance the market value and overall soybean seed quality in Kazakhstan.

5. Conclusions

Biochemical analysis of seeds of 29 varieties from soybean germplasm collection revealed only two cultivars, Hilario and Ascasubi, with the lowest activity of the Kunitz trypsin inhibitor (KTI). In contrast, all soybean cultivars currently grown in Kazakhstan have a KTI enzyme activity ranging from 54.16 to 69.85 of trypsin units inhibited per mg of seeds, for Lastochka and Birlik KV, respectively. Using a method employing protein analysis and molecular markers, the null-allele *ti3* was confirmed in the cultivars Hilario and Ascasubi. These genotypes were then used in crosses with domestic soybean cultivars. The SSR marker *Satt228* was identified as the most effective diagnostic marker, confirming the heterozygosity of the F_1 generation and helping to select homozygous lines with the null allele *ti3* in F_2 and in F_3 segregated populations to enable the reduction of KTI and the improvement of seed quality.

Author Contributions: Conceptualization and supervision: K.B.; data collecting and analysis: S.M.; soybean resources and hybrid populations design: S.D.; molecular genotyping: D.B.; agronomic performance: M.K.; protein electrophoresis: P.A.; trypsin inhibitor activity assay: S.K.; review and editing: Y.S.

Funding: This research was funded by the Ministry of Agriculture (Kazakhstan), Scientific Technical Program, BR06249214 (M.K.).

Acknowledgments: We thank Carly Schramm for her critical comments and English editing.

Conflicts of Interest: The authors declare no conflict of interest.

References

1. Laskowski, M., Jr.; Kato, I. Protein inhibitors of proteinases. *Annu. Rev. Biochem.* **1980**, *49*, 593–629. [CrossRef] [PubMed]
2. Tan-Wilson, A.L.; Chen, J.C.; Duggan, M.C.; Chapman, C.; Obach, R.S.; Wilson, K.A. Soybean Bowman-Birk trypsin isoinhibitors: Classification and report of a glycine-rich trypsin inhibitor class. *J. Agric. Food Chem.* **1987**, *35*, 974–981. [CrossRef]
3. Kumar, V.; Rani, A.; Rawal, R. Deployment of gene specific marker in development of Kunitz trypsin inhibitor free soybean genotypes. *Indian J. Exp. Biol.* **2013**, *51*, 1125–1129. [PubMed]
4. Palacios, M.; Easter, R.A.; Soltwedel, K.T.; Parsons, C.M.; Douglas, M.W.; Hymowitz, T. Effect of soybean variety and processing on growth performance of young chicks and pigs. *J. Anim. Sci.* **2004**, *82*, 1108–1114. [CrossRef] [PubMed]
5. Liencer, I.E. Possible adverse effects of soybean anticarcinogens. *J. Nutr.* **1995**, *125*, 744–750.

6. Chung, J. Identification and confirmation of SSR marker tightly linked to the *Ti* locus in soybean [*Glycine max* (L.) Merr.]. In *Soybean—Genetics and Novel Techniques for Yield Enhancement*; Krezhova, D., Ed.; InTech: Rijeka, Croatia, 2011; pp. 201–212.

7. Hildebrand, D.F.; Orf, J.H.; Hymowitz, T. Inheritance of an acid phosphatase and its linkage with the Kunitz trypsin inhibitor in seed protein of soybeans. *Crop Sci.* **1980**, *20*, 83–85. [CrossRef]

8. Hymowitz, T. Genetics and breeding of soybeans lacking the Kunitz trypsin inhibitor. In *Nutritional and Toxicological Significance of Enzyme Inhibitors in Foods. Advances in Experimental Medicine and Biology*; Friedman, M., Ed.; Springer: Boston, MA, USA, 1986; Volume 199, pp. 291–298.

9. Krishnamurthy, P.; Singh, R.J.; Tsukamoto, C.; Park, J.H.; Lee, J.D.; Chung, G. Kunitz trypsin inhibitor polymorphism in the Korean wild soybean (*Glycine soja* Sieb. & Zucc.). *Plant Breed.* **2013**, *132*, 311–316.

10. Klang, Y.T.; Chiang, Y.C. Genetic linkage of a leucine aminopeptidase locus with the Kunitz trypsin inhibitor locus in soybeans. *J. Hered.* **1986**, *77*, 128–129. [CrossRef]

11. Jofuku, K.D.; Goldberg, R.B. Kunitz trypsin inhibitor genes are differentially expressed during the soybean life cycle and in transformed tobacco plants. *Plant Cell* **1989**, *1*, 1079–1093. [CrossRef] [PubMed]

12. Jofuku, K.D.; Schipper, R.D.; Goldberg, R.B. A frameshift mutation prevents Kunitz trypsin inhibitor mRNA accumulation in soybean embryos. *Plant Cell* **1989**, *1*, 427–435. [CrossRef] [PubMed]

13. Ryan, C.A. Proteinase inhibitors. In *Biochemistry of Plants*; Marcus, A., Ed.; Academic Press: San Diego, CA, USA, 1981; pp. 351–370.

14. Kim, S.H.; Hara, S.; Hase, S.; Ikenaka, T.; Toda, H.; Kitamura, K.; Kaizuma, N. Comparative study on amino acid sequences of Kunitz-type soybean trypsin inhibitors, *Ti, Tib,* and *Tic. J. Biochem.* **1985**, *98*, 435–448.

15. Vollmann, J.; Schausberger, H.; Bistrich, H.; Lelley, T. The presence or absence of the soybean Kunitz trypsin inhibitor as a quantitative trait locus for seed protein content. *Plant Breed.* **2002**, *121*, 272–274. [CrossRef]

16. Kumar, V.; Rani, A.; Rawal, R.; Mourya, V. Marker assisted accelerated introgression of null allele of Kunitz trypsin inhibitor in soybean. *Breed. Sci.* **2015**, *65*, 447–452. [CrossRef] [PubMed]

17. Maranna, S.; Verma, K.; Talukdar, A.; Lal, S.K.; Kumar, A.; Mukherjee, K. Introgression of null allele of Kunitz trypsin inhibitor through marker-assisted backcross breeding in soybean (*Glycine max* L. Merr.). *BMC Genetics* **2016**, *17*, 106. [CrossRef] [PubMed]

18. Didorenko, S.V.; Karyagin, Y.G.; Bulatova, K.M. Method of soybean hybridization. Submitted by Kazakh Scientific Research Institute of Agriculture and Plant Growing, Almaty region, Kazakhstan. Republic of Kazakhstan Patent No. 31427, 21 July 2016.

19. Kakade, M.L.; Simons, N.; Liener, I.E. An evaluation of natural vs synthetic substances for measuring the antitryptic activity of soybean samples. *Cereal Chem.* **1974**, *46*, 518–526.

20. Laemmli, U.K. Cleavage of structural proteins during the assembly of the head of bacteriophage T4. *Nature* **1970**, *227*, 680–685. [CrossRef] [PubMed]

21. Davis, B.J. Description of discontinuous buffer system for non-denaturing gels in electrophoresis. *Ann. New York Acad. Sci.* **1964**, *209*, 373–381.

22. Murray, M.G.; Thompson, W.F. Rapid isolation of high molecular weight plant DNA. *Nucleic Acids Res.* **1980**, *8*, 4321–4325. [CrossRef] [PubMed]

23. Kim, M.S.; Park, M.J.; Jeong, W.H.; Nam, K.C.; Chung, J.I. SSR marker tightly linked to the *Ti* locus in soybean [*Glycine max* (L.) Merr.]. *Euphytica* **2006**, *152*, 361–366. [CrossRef]

24. Cregan, P.B.; Jarvik, T.; Bush, A.L.; Shoemaker, R.C.; Lark, K.G.; Kahler, A.L.; Kaya, N.; VanToai, T.T.; Lohnes, D.G.; Chung, J.; et al. An integrated genetic linkage map of the soybean genome. *Crop Sci.* **1999**, *39*, 1464–1490. [CrossRef]

Article

Seedling Growth and Transcriptional Responses to Salt Shock and Stress in *Medicago sativa* L., *Medicago arborea* L., and Their Hybrid (Alborea)

Eleni Tani [1], Efi Sarri [1], Maria Goufa [1], Georgia Asimakopoulou [1], Maria Psychogiou [2], Edwin Bingham [3], George N. Skaracis [1] and Eleni M. Abraham [4,*] 

[1] Department of Crop Science, Laboratory of Plant Breeding and Biometry, Agricultural University of Athens, Iera Odos 75, 11855 Athens, Greece; etani@aua.gr (E.T.); efisarri.18@gmail.com (E.S.); marog@aua.gr (M.G.); georgia_asim@hotmail.com (G.A.); gskaracis@aua.gr (G.N.S.)

[2] Department of Natural Resources Management and Agricultural Engineering, Laboratory of Agricultural Hydraulics, Agricultural University of Athens, Iera Odos 75, 11855 Athens, Greece; mariapsy1@gmail.com

[3] Agronomy Department, University of Wisconsin-Madison, Madison, WI 53706, USA; ebingham@wisc.edu

[4] Faculty of Agriculture, Forestry and Natural Environment, School of Forestry and Natural Environment, Aristotle University of Thessaloniki, 54124 Thessaloniki, Greece

*   Correspondence: eabraham@for.auth.gr; Tel.: +30-231-099-2301

Received: 16 September 2018; Accepted: 15 October 2018; Published: 18 October 2018

Abstract: Salinity is a major limiting factor in crop productivity worldwide. *Medicago sativa* L. is an important fodder crop, broadly cultivated in different environments, and it is moderately tolerant of salinity. *Medicago arborea* L. is considered a stress-tolerant species and could be an important genetic resource for the improvement of *M. sativa's* salt tolerance. The aim of the study was to evaluate the seedling response of *M. sativa*, *M. arborea*, and their hybrid (Alborea) to salt shock and salt stress treatments. Salt treatments were applied as follows: salt stress treatment at low dose (50 mM NaCl), gradual acclimatization at 50–100 and 50–100–150 mM NaCl, and two salt shock treatments at 100 and 150 mM NaCl. Growth rates were evaluated in addition to transcriptional profiles of representative genes that control salt uptake and transport (*NHX1* and *RCI2A*), have an osmotic function (*P5CS1*), and participate in signaling pathways and control cell growth and leaf function (*SIMKK*, *ZFN*, and *AP2/EREB*). Results showed that the studied population of *M. sativa* and *M. arborea* performed equally well under salt stress, whereas that of *M. sativa* performed better under salt shock. The productivity of the studied population of Alborea exceeded that of its parents under normal conditions. Nevertheless, Alborea was extremely sensitive to all initial salt treatments except the low dose (50 mM NaCl). In addition, significantly higher expression levels of all the studied genes were observed in the population of *M. arborea* under both salt shock and salt stress. On the other hand, in the population of *M. sativa*, *NHX1*, *P5CS1*, and *AP2/EREB* were highly upregulated under salt shock but to a lesser extent under salt stress. Thus, the populations of *M. sativa* and *M. arborea* appear to regulate different components of salt tolerance mechanisms. Knowledge of the different parental mechanisms of salt tolerance could be important when incorporating both mechanisms in Alborea populations.

Keywords: *Medicago*; salinity tolerance; gene expression profile; salt acclimatization

---

1. Introduction

Soil salinity is a major limiting factor in crop productivity in irrigated and non-irrigated areas worldwide [1]. It can be caused by natural processes defined as "primary salinity" and/or by human activities ("secondary salinity") that are mainly due to improper irrigation [2]. The phenomenon of salinization is growing fast in arid and semi-arid areas [3] as a result of the relatively high temperatures and inadequate rainfall that contribute to the accumulation of salts [4]. In this respect, climate change, which is likely to increase temperatures and to affect precipitation will probably increase salinization. From the management point of view, the utilization of salty soils could be enhanced by cultivating crops adapted to salinity.

The exposure of plants to high concentrations of sodium (Na^+) and chloride ions in soil [5] causes salt stress. In experiments, it occurs in one of two forms: (a) when plants are subjected to gradual application of NaCl until a final, prearranged salt concentration is reached, and (b) when plants are exposed to low levels of salinity [6,7]. Salt stress has two main components: a hyperosmotic stress caused by the reduction of water potential, and a hyper-ionic stress resulting from the accumulation of ions to a toxic level for plant growth [8]. Osmotic stress happens instantly when roots come into contact with solutions containing high concentrations of salts, and ionic stress follows the osmotic response.

On the other hand, salt shock is an extreme form of salt stress, where plants are abruptly subjected to high concentrations of NaCl. The main component of the salt shock is the hyperosmotic stress [9,10] due to large differences in osmotic pressure between solutes outside the cell cytoplasm, and solutes inside the cell cytoplasm. Cell plasmolysis and leakage of osmolytes occur under osmotic stress, phenomena that do not appear under ionic stress, indicating that gene expression is different in response to salt stress and salt shock [10].

According to Reference [11], genes that contribute to salt tolerance in plants can be categorized into three functional groups: (1) those that regulate the salt uptake and transport, (2) those with osmotic function, and (3) those that regulate plant growth under saline soil. The latter genes are related to transcription factors and signal transduction proteins. In this regard, the most responsive genes in salt stress experiments were related to transcription and signaling pathways and/or to ion transport, and are expected to be responsive to ionic changes [12–14]. On the other hand, the expression of genes with osmotic function alters rapidly under salt shock, while ionic-responsive genes change to a lesser extent.

The majority of molecular and biochemical research is focused on experiments that are typically performed using high NaCl doses [10,11] although salt shock rarely occurs in nature. This is due to the fact that NaCl concentration rises gradually, either via increasing water levels or by the slow drying of the soil profile due to evaporation and plant uptake. However, imposition of salt stress through gradual application of NaCl rather than a single application of a high concentration of NaCl (salt shock) is preferred for studies regarding salinity tolerance, because this resembles the occurrence of salinity in nature. A comparison of gene expression in parallel experiments using gradual and sudden salt application is the most recommended method to distinguish between plant responses that are specific to salt shock and salt stress [10].

Alfalfa or *Medicago sativa* L. (*M. sativa*) is one of the most important forage legumes for improving soil composition and providing plentiful forage for animals. Even though alfalfa is a moderately salt-tolerant species and less sensitive than other legumes [15], its production and growth rate decreases when soil salt content is between 50 and 200 mM NaCl [16,17]. It is notable that salinity slightly increases the nutritive value of alfalfa [18]. The need to produce legumes with increased salt tolerance and high yield is extensively emphasized given that legumes are second in importance to agriculture and provide 33% of human nitrogen needs [19].

Several studies investigated mainly the physiological and, to a lesser extent, the molecular mechanisms associated with tolerance of *M. sativa* to salt stress/shock [20–23]. *Medicago arborea* L. (*M. arborea*), which is considered the oldest in the genus, is known to be a tolerant species to salt stresses [24–26]. However, there is limited research on the molecular mechanisms related to its salt tolerance. Hybrids between *M. sativa* and *M. arborea* were produced in the United States of America (USA) and Australia, and are named Alborea, a high-yielding hybrid [27,28]. There is evidence that the high-yielding hybrids in some cases are more sensitive to stresses compared to their parents [29]. The aim of the present work was to study the seedling responses of *M. sativa*, *M. arborea*, and their hybrid in terms of transcriptional responses of selected genes under both salt stress and salt shock. The genes were selected based on the aforementioned groups that were proposed by Reference [11].

In particular, *RCI2A* and *NHX1* represent genes encoding ion transporters. Their association with both salt shock and salt stress was reported for several plant species including *Medicago* [30–33]. *P5CS1* controls the de novo synthesis of proline. Several studies associate proline accumulation with plant adaptability to salt stress [34,35]. Finally, *AP2/EREB*, *SIMKK*, and *ZFN* were selected from the group of transcription factors and signal transduction proteins. Their importance in the response of plants to both salt stress and salt shock was highlighted in several reports [12,36,37]. The following questions were addressed: (1) Is there any differentiation in the response of *M. sativa* and *M. arborea* under salt stress and salt shock? (2) What is the response of the hybrid compared to the parental species? (3) Is there any differentiation in the expression of the selected genes under salt stress and salt shock? The results of the present work can be utilized in future Alborea breeding programs.

2. Materials and Methods

2.1. Plant Material

The *M. arborea* parents used to develop Alborea were originally collected from Greece. The *M. sativa* parent population used in this study was a hybrid of the two *M. sativa* parents. Two *M. sativa* parents were used in crosses of *M. sativa* × *M. arborea* to produce 27 initial Alborea hybrids [27]. The Alborea population used in this study was developed from intercrosses of 20 initial hybrids using methods described by Reference [28]. Seeds were obtained from E. Bingham, Agronomy Department, Univ. Wisconsin-Madison, USA.

2.2. Seed Pretreatment

Seeds of *M. sativa*, *M. arborea*, and Alborea were scarified with absolute sulfuric acid for 10 min and then rinsed thoroughly with sterile distilled water. Bleach solution (3%) was added for 1.5–2 min and then rinsed thoroughly. The scarified seeds were placed on petri dishes containing 0.5% agar at 4 °C overnight, and then transferred to 20 °C (dark) for 3–4 days.

2.3. Growth Conditions and Salt Stress Treatments

Seedlings with about 1-cm-long radicle were transplanted individually into pots (8.5 cm in height and 10 cm in diameter) containing a commercial growing medium (Kronos N 50–300 mg/L, P_2O_5 80–300 mg/L, K_2O 80–300 mg/L, pH 5–6.5, salinity <1.75 g/L). The pots were placed in a growth chamber under a 16/8-h day/night regime, 23 °C, 55–65% relative humidity in a completely randomized block (block in the treatments). All pots were randomized within each treatment biweekly. The salt treatments started four days after transfer to pots. The following treatments were implemented: (1) control, no salt; (2) salt stress, with initial treatment of 50 mM NaCl and gradual step acclimatization to 50–100 and 50–100–150 mM final NaCl concentrations; and (3) salt shock, with initial treatment of 100 and 150 mM NaCl concentrations. Thus, the experiment consisted of six sample sets, each comprising control (no salts) and five salt treatments, three of salt stress (50, 50–100 and 50–100–150 mM NaCl), and two of salt shock (100, 150 mM NaCl). Each set had three independent biological replicate pools of four plants. The gradual acclimatization started four days post-transplanting, and the salt concentration was increased in three steps of 10 days from 0 to 50, 100, and 150 mM NaCl or in two steps of 15 days from 0 to 50 and from 50 to 100 mM. Plants were watered every five days when salt treatments started. Once per week, Hoagland solution [38] was added to the salt solution. The growth chamber culture lasted 34 days. Whole shoots, excluding cotyledons and the roots, were transferred in situ into liquid nitrogen in the middle of the light period. When harvested, all plants were in the vegetative stage, and roots did not show nodules. The plants of Alborea at initial salt shock of 100 and 150 mM NaCl were not harvested as they did not survive.

2.4. Growth Characteristics Measurements

The length of the stems from the base to the tip was measured in each seedling at an interval of three days through the duration of the salt treatments. Stem elongation rate (SER) was estimated as SER = (T2 − T1)/t, where T1 and T2 represent the stem length at the beginning and at the end of a time t, respectively. Additionally, the salinity sensitivity index (IS) based on the stem length was estimated according to the formula proposed by Reference [39], IS = (Hs − Ht)/Ht × 100, in which Hs and Ht represent the values of stem length of the salt-stressed and control plants, respectively.

2.5. Determination of Na^+ and K^+ Contents

Leaves and roots from the harvested plants were dried at 65 °C for 48 h. Dried material was ground to powder and the concentrations of Na^+ and K^+ were determined using flame photometry (Corning 410, Sherwood Scientific Ltd., Cambridge, UK). As there was not enough plant material, particularly for the initial treatments of 100 and 150 mM NaCl concentrations, the three replicates were bulked into one. Additionally, there was no plant material of Alborea for these two treatments. The data of Na^+ and K^+ determination were not subject to statistical analysis, and therefore, are considered indicative.

2.6. RNA Isolation and Complementary DNA Synthesis

Total RNA was extracted using the Trizol reagent (Sigma-Aldrich, St. Louis, MO, USA) method according to the manufacturer's protocol. Nucleic acid concentration and quality were assessed using NanoDrop™ ultraviolet (UV) spectrophotometry and by agarose gel electrophoresis. For the complementary DNA (cDNA) synthesis experiment, first-strand cDNA was synthesized from 0.5 μg of total RNA using the Superscript II enzyme (Invitrogen, Carlsbad, CA, USA) following the manufacturer's instructions.

2.7. Real-Time PCR Experiments

Quantitative PCR was performed with a Step One Plus Real-Time PCR system (Applied Biosystems, Foster City, CA, USA) using SYBR Select Master Mix (Applied Biosystems, Foster City, CA, USA). Primer design was accomplished using the Primer blast NCBI otool (Primer Blast, https://www.ncbi.nlm.nih.gov/tools/primer-blast/), using expressed sequence tags (ESTs) of *M. sativa* that were deposited in National Center for Biotechnology Information (NCBI), except for the *AP2/EREB*-like *AINTEGUMENTA* gene. A phylogenetic analysis of the AP2/EREB-like AINTEGUMENTA protein from *M. truncatula* and various AP2/EREB members was performed using the MEGA7 software, Pennsylvania State University, State College, PA, USA) [40]. Best primer pairs were selected utilizing an oligo-analyzer tool. The gene-specific primers which were used are listed in Table 1, and the synthesized cDNA was adjusted to a final concentration of 1.5 ng/μL for qPCR experiments. Serial dilutions of cDNA were used to make a standard curve to optimize amplification efficiency for each primer set. All reactions were performed in triplicate. Melt curves of the reaction products were generated and fluorescence data were collected at a temperature above the melting temperature of non-specific products. Relative expression levels of the studied genes (*NHX1*, *RCI2A*, *P5CS1*, *SIMKK*, *ZFN*-containing CCCH domain, and *AP2/EREB*) were calculated according to the $2^{-\Delta\Delta Ct}$ method [41] (for calculating ΔΔCt value, the control sample of each entry was used). *Actin-2* was used as an internal control for normalization.

Table 1. The real-time PCR primers of Msactin2, MsRCI2A, MsNHX1, MsAP2/EREB, MsSIMKK, MsZFN, and MsP5CS1.

Sequence Name	Genbank Number	Primer Sequence	Amplicon Size
Ms-actin2-F	JQ028730.1	TTCTCACCACACTTCTCGCC	173 bp
Ms-actin2-R		CCAGCCTTCACCATTCCAGT	
Ms-AP2/EREB-F	Not deposited	AATGGGTGGGGAAACGGAAC	95 bp
Ms-AP2/EREB-R		TTTGGTGGTGGAGTGTGGTT	
Ms-NHX1-F	AY513732.1	GCCATGAAATTCACCGACCG	118 bp
Ms-NHX1-R		CTGCCACCAAAAACAGGACG	
Ms-P5CS-F	X98421.1	TTTGCGGTCGGAAGGTGTTA	119 bp
Ms-P5CS-R		CGATTTCCAAGGTGCAAGCC	
Ms-ZFN-F	JX131368.1	CCCAAGCTGCAAGTTTGACC	154 bp
Ms-ZFN-R		TGAGCCCGACTCAACAAGTC	
Ms-SIMKK-F	AJ293274.1	ACCAGAAGCTCCAACGACTG	94 bp
Ms-SIMKK-R		CCTCGAAGCAGTCCATCTCC	
Ms-RCI2-F	JQ665271.1	GTTGTCAGGGGCGTCATTCT	169 bp
Ms-RCI2-R		TCCAAGCAGGACAAAACGGA	

2.8. Statistical Analysis

The repeated-measures ANOVA with the generalized linear model (GLM) was used for detecting the effect of treatment and species on seedlings height and SER. The within-subject factors were the dates and the treatments were the between-subject factor for the species. Additionally, a three-way-ANOVA was performed in order to detect the effect of treatments, species, and organs on gene expression. Tukey's test at the 0.05 probability level was used to detect the differences among means [42]. The IBM SPSS Statistics 23 software (SPSS Inc., Chicago, IL, USA) was used for the statistical analysis. A heat map was generated using Excel 2016, for a visual summary of gene expression.

3. Results

3.1. Growth Parameters and Salinity Sensitivity Index

Significant differences in plant height, stem elongation rate (SER), and salinity sensitivity indexes were recorded among salt treatments, parent species and Alborea (hereafter entries), and dates of treatment (Table 2). However, significant interaction was detected between salt treatments and entries for all the growth parameters, indicating a differentiated response of the entries to the salt treatments (Table 2). Overall, the growth parameters, i.e., seedling height and SER, were significantly reduced under salt treatments, and especially, under salt shock with 100 and 150 mM NaCl initial treatment, where the lowest values were recorded (Figure 1). The studied population of Alborea had significantly higher plant height and SER compared to the others under control and 50 mM NaCl initial treatment (Figure 1). On the other hand, the population of Alborea had the lowest value for all growth parameters at the 100 and 150 mM NaCl initial treatment, and those of the studied population of *M. sativa* were highest for the gradual acclimatizations (Figure 1). Furthermore, the lowest salinity sensitivity index was recorded for the population of *M. sativa* at the gradual acclimatization followed by that of *M. arborea* at 50 mM initial treatment and at 50–100 mM gradual acclimatization (absolute numbers; Figure 2). The highest salinity sensitivity index (absolute number) was detected for the population of Alborea under salt sock at 100 and 150 mM NaCl initial treatments.

Table 2. Statistical significance of F ratios from the analysis of variance for stem height of seedlings, stem elongation rate (SER), and sensitivity index.

Source of Variation	Height	SER	Salinity Sensitivity Index
Salt (A)	$p < 0.05$	$p < 0.05$	$p < 0.05$
Species (B)	$p < 0.05$	$p < 0.05$	$p < 0.05$
Dates (C)	$p < 0.05$	$p < 0.05$	$p < 0.05$
A × B (Interaction)	$p < 0.05$	$p < 0.05$	$p < 0.05$
A × C (Interaction)	$p < 0.05$	ns	ns
B × C (Interaction)	ns *	$p < 0.05$	ns
A × B × C (Interaction)	ns	ns	ns

* ns: not significant at 0.05 level.

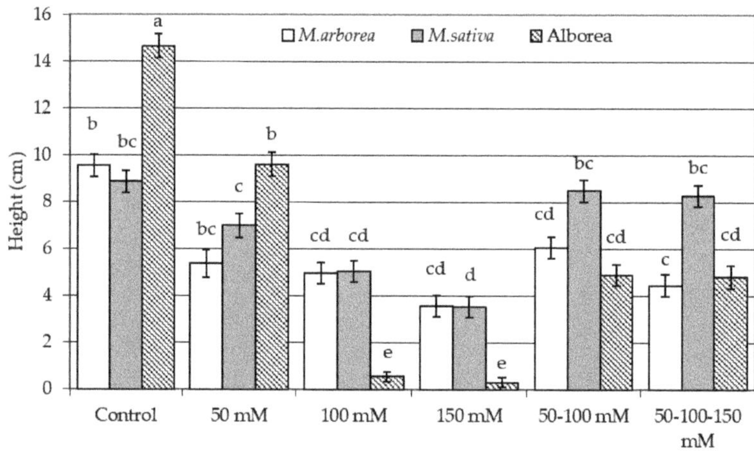

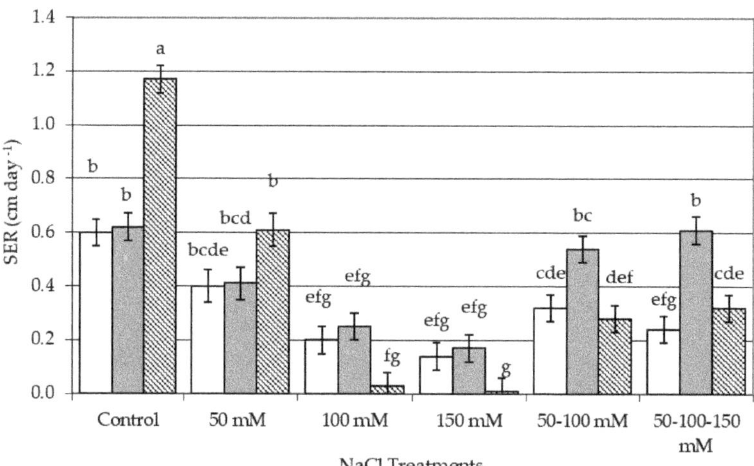

Figure 1. The stem height, and the stem elongation rate (SER) of *M. arborea* (M. ar), *M. sativa* (M. sa), and Alborea (Al) under control, 50 mM, 100 mM, 150 mM, 50–100 mM, and 50–100–150 mM NaCl. The vertical bars indicate the mean ± standard error (SE) of five independent samples. The different letters refer to the significant differences at $p < 0.05$ (Tukey's test).

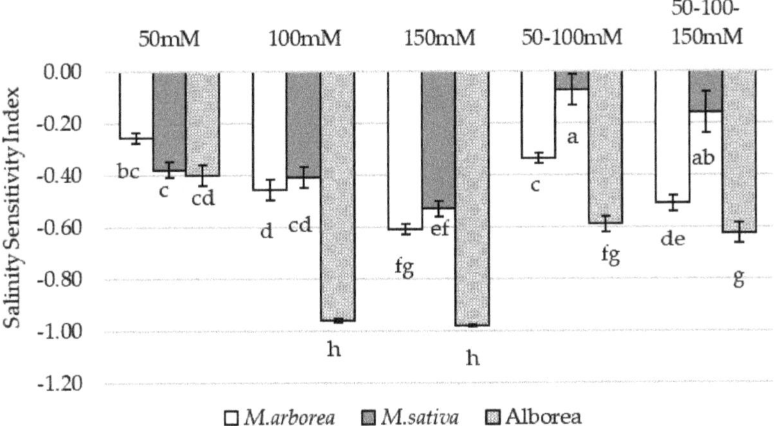

Figure 2. The salinity sensitivity index of *M. arborea*, *M. sativa*, and Alborea under control, 50 mM, 100 mM, 150 mM, 50–100 mM, and 50–100–150 mM NaCl. The vertical bars indicate the mean ± SE of five independent samples. The different letters refer to the significant differences at $p < 0.05$ (Tukey's test).

Significant interaction was detected between salt treatments and dates for seedling heights and between entries and dates for SER (Table 2). Seedling heights gradually increased from the first to the last date of measurements for all the treatments except for those at 100 and 150 mM NaCl initial treatments, where they remained almost unchanged (Figure 3). On the other hand, the SER of the Alborea population remained steady during the experimental period independent of salt treatment, while, for the populations of *M. arborea* and *M. sativa*, it tended to decrease from the first to the final date of treatments (Figure 3).

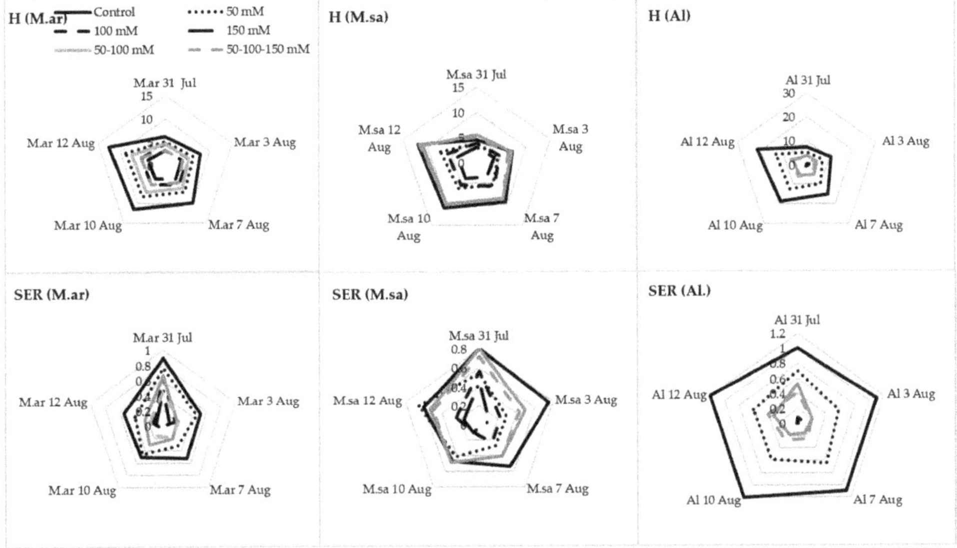

Figure 3. Stem height (H) and stem elongation rate (SER) of *M. arborea* (M. ar), *M. sativa* (M. sa), and Alborea (Al) as affected by control, 50 mM, 100 mM, 150 mM, 50–100 mM, and 50–100–150 mM NaCl during the experimental period.

Regarding the Na^+ and K^+ concentrations in leaves, the results are only indicative, as there was not enough tissue from all the treatments, especially for Alborea. According to the results, the Na^+ concentration was much lower in leaves of the *M. sativa* population under salt shock, compared to those of *M. arborea* and of Alborea (Table 3). The K^+ concentration tended to decrease with the increase in NaCl concentration for the populations of *M. arborea* and Alborea, while it remained at the same level for that of *M. sativa*.

Table 3. Indicative data of sodium and potassium content ($mg \cdot g^{-1}$ dry weight (dw)) of *M. arborea*, *M. sativa*, and Alborea.

		K^+	Na^+
		($mg \cdot g^{-1}$ dw)	($mg \cdot g^{-1}$ dw)
M. arborea	Control	38.5	1.1
	50 mM	24.4	10.6
	50–100 mM	31.8	20.4
	50–100–150 mM	13.8	21.7
M. sativa	Control	22.8	3.5
	50–100 mM	21.5	6.9
	50–100–150 mM	25.1	5.1
Alborea	Control	51.2	2.3
	50 mM	38.1	19.6
	50–100 mM	13.4	23.4

3.2. Gene Expression Levels

The main effects and the interactions between each entry/type of plant organ and different treatments on the expression levels of the studied genes are presented in Table 4. Interactions were statistically significant in most cases. On average, *NHX1* and *AP2/EREB* transcripts were the most highly abundant in all entries (Figure 4). The transcript levels of both genes were highly responsive during salt gradual acclimatization and salt shock, followed by *RCI2A* and *P5CS1*.

In general, the expression levels of all transcripts were higher in the studied population of *M. arborea* followed by that of *M. sativa*. On the other hand, in the studied population of Alborea, the lowest induction of all genes was detected under salt treatments (Figure 5). The induction of all genes in most cases was proportional to the NaCl concentration, and the highest induction levels were detected after 150 mM NaCl or 50–100–150 mM NaCl treatment (Figure 5). The lowest increase in transcript levels for all genes was detected in plants treated with 50 mM NaCl or with 50–100 mM NaCl gradual acclimatization (Figure 5).

Table 4. Statistical significance of F ratios from the analysis of variance for gene expression.

Source of Variation	*RCI2A*	*NHX1*	*AP2/EREB*	*SIMKK*	*ZFN*	*P5CS_1*
Salt (A)	$p < 0.05$	$p < 0.05$	$p < 0.05$	$p \leq 0.05$	$p \leq 0.05$	$p \leq 0.05$
Species (B)	$p < 0.05$	$p < 0.05$	$p < 0.05$	$p \leq 0.05$	$p \leq 0.05$	$p \leq 0.05$
Organs (C)	$p < 0.05$	$p < 0.05$	$p < 0.05$	$p \leq 0.05$	$p \leq 0.05$	$p \leq 0.05$
A × B (Interaction)	$p < 0.05$	$p < 0.05$	$p < 0.05$	$p \leq 0.05$	ns	$p \leq 0.05$
A × C (Interaction)	$p < 0.05$	ns *	$p < 0.05$	$p < 0.05$	$p \leq 0.05$	ns
B × C (Interaction)	$p < 0.05$	$p < 0.05$	$p < 0.05$	ns	ns	$p \leq 0.05$
A × B × C (Interaction)	$p < 0.05$	$p < 0.05$	$p < 0.05$	$p < 0.05$	$p \leq 0.05$	$p \leq 0.05$

* ns: not significant at 0.05 level.

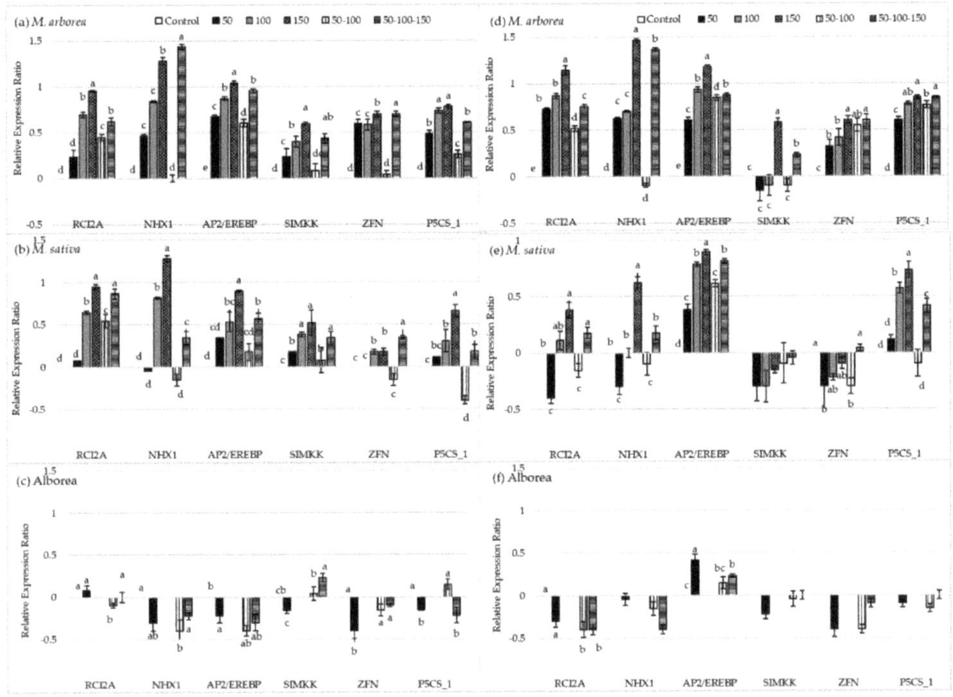

Figure 4. *RCI2A, NHX1, AP2/EREB, SIMKK, ZFN,* and *P5CS* relative expression levels for *M. arborea*, *M. sativa*, and Alborea under control, 50 mM, 100 mM, 150 mM, 50–100 mM, and 50–100–150 mM NaCl for leaves (**a–c**) and roots (**d–f**). The vertical bars indicate the mean ± SE of three independent samples. The different letters refer to the significant differences at $p < 0.05$ (Tukey's test).

All gene transcripts showed the strongest induction in the population of *M. arborea*, and to the same extent in both leaves and roots, apart from *SIMKK* (Figure 4a,d). Moreover, in most cases, gene induction was proportional to the final concentration of NaCl regardless of the type of treatment (stress or shock). On the other hand, in leaves of the *M. sativa* population, *AP2/EREB, NHX1,* and *P5CS1* had higher induction levels in salt shock treatments compared to salt stress ones (Figure 4b,e). Inversely, the transcript accumulation of *RCI2A* was dependent on the final concentration of NaCl regardless of stress or shock treatment. Additionally, the transcript abundance of all genes, except for *AP2/EREB* and *P5CS1*, was higher in leaves of the *M. sativa* population than in roots (Figure 4b,e). Finally, regarding the population of Alborea, a low induction of only *SIMKK* transcripts was detected in leaves and of *AP2/EREB* in roots (Figure 4c,f, respectively). A heat map was constructed in order to visualize possible differences in the relative change of each transcript level among the treatments and the entries (Figure 5). According to the heat map, the population of Alborea exhibited the lowest induction of all genes. Moreover, the most abundant gene transcripts were found for *NHX1* and *AP2/EREB*. The difference in gene induction is evident between leaves and roots in the populations of *M. arborea* and *M. sativa*, as well as the fact that gene induction in the roots of the population of *M. sativa* is much lower compared to leaves.

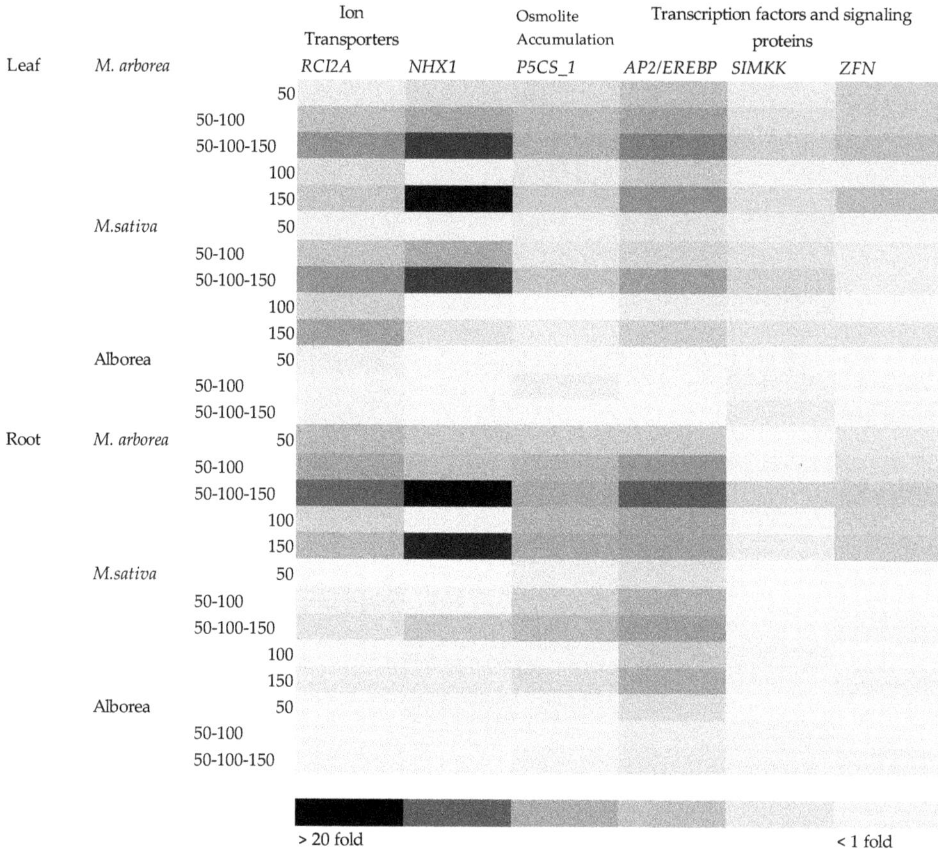

Figure 5. General expression levels of *RCI2A*, *NHX1*, *AP2/EREB*, *SIMKK*, *ZFN*, and *P5CS* under salt shock and salt stress for *M. arborea*, *M. sativa*, and Alborea. Expression levels are black and white-coded to depict the fold change as follows: black (high expression level > 20-fold) to white (low < 1 fold).

4. Discussion

4.1. Growth of Seedlings under Salinity

In the present study, salinity tolerance was tested in populations of two *Medicago* species, *M. sativa* and *M. arborea*, and in their hybrid Alborea. Two salinity regimes were applied: salt stress and salt shock. Plants differed greatly in their tolerance to salinity, as indicated by seedling growth parameters and the sensitivity index. However, the effect was much more severe for salt shock at 100 mM and 150 mM NaCl compared to the gradual acclimatization, reducing the seedling growth parameters and increasing the sensitivity index, on average, by about 70% to 80%, compared to the control. The suppression of plant growth under salinity is generally attributed to the osmotic effects of the salt in the soil in combination with ionic effects of the salt concentration in plant tissues. The plants suffer from a much more intense osmotic stress or plasmolysis under sudden NaCl application than under a gradual one [9]. In this regard, the concentration of shoot Na$^+$ was higher and the growth reduction was greater in wheat plants under salt shock compared to those under gradual application [43]. Similarly, Reference [44] reported that 150–200 mM NaCl salt shock had a lethal effect on rice plants, while a similar effect was not documented for the gradual acclimatization of NaCl to the same level [10].

In our study, seedling growth parameters decreased as the salt level increased for the studied populations of *M. sativa*, *M. arborea*, and their hybrid. The reduced growth of seedlings or more mature plants of different *M. sativa* cultivars under either salt shock [20,45] or salt stress [16,46,47] were broadly reported. A similar reduction in plant growth under salt stress gradually applied was reported for *M. arborea* [24,25,48,49].

Differences in the response of the populations of *M. sativa*, *M. arborea*, and their hybrid to salt shock and salt stress were also observed. The hybrid performed better than its parents under control and low salinity level (50 mM), while the population of *M. sativa* performed better under gradual salt acclimatization. On the other hand, the salt shock of 100 and 150 mM NaCl had a detrimental effect on the hybrid, causing the death of plants seven days after the salt application, whereas its parents continued their growth until the end of the experiment. Additionally, the population of *M. sativa* presented a consistently lower salinity sensitivity index under all salt treatments compared to that of *M. arborea* and the hybrid. According to the results, it seems that the studied population of *M. sativa* is more tolerant to salinity compared to that of *M. arborea* in terms of seedling growth. Other studies conducted with *M. sativa* varieties also featured a correlation between tolerance to salt stress and low salinity sensitivity index, which was also evaluated for stem length [20]. However, the different growth forms of the species could be a factor, in that *M. sativa* is a perennial herbaceous species, whereas *M. arborea* is woody. The response of plants to stresses at the seedling stage is highly related to their growth form [50] and, generally, perennial herbaceous species are characterized by higher relative growth rates in the seedling stage.

4.2. Gene Expression under Salt Treatments

To further explore the response mechanisms of the parent species and their hybrid to salinity, the relative expression ratios of several salt-induced genes were evaluated. The ion transporters *NHX1* and *RCI2A* control the uptake of cations and/or the efflux of anions, and they function as membrane stabilizers. In particular, *NHX1* is a vacuole transporter [51] and its role in the cell is the transport of K^+ or Na^+ into the vacuole exchanging H^+ efflux to the cytosol, maintaining a low Na^+/K^+ cytosolic ratio. Proteins encoded by *RCI2A* act within plasma membranes and they possibly modulate ion transporters or stabilize membrane proteins that affect ion transport [52]. Both gene transcripts were highly accumulated in roots and leaves of the *M. arborea* population and their relative expression was influenced only by the final NaCl concentration. On the other hand, *NHX1* and *RCI2A* upregulation was recorded mainly in leaves of the *M. sativa* population. Additionally, there was a profound upregulation of *NHX1* relative expression under salt shock compared to gradual acclimatization of NaCl mainly in the leaves of the *M. sativa* population. Recent findings by Leidi et al. [31] indicate that proteins encoded by *RCI2A* may be involved in the salt tolerance of *M. sativa* and *M. truncatula*. Moreover, research conducted by Sandhu et al. [23] highlighted the importance of *NHX1* to the salt tolerance of *M. sativa*.

Organic solutes are key players in increasing tolerance of plant tissues to excessive salt concentrations. *P5CS1* controls the de novo synthesis of proline and its accumulation during salt stress [53]. Several studies associate proline abundance with plant adaptability to drought and salt stress [54,55]. Results of the present study indicate that relative transcript levels were highly abundant in leaves and roots of the *M. arborea* population in both salt shock and stress treatments. However, the *P5CS1* in the studied population of *M. sativa* (both roots and leaves) under salt shock was upregulated almost 5.5-fold, while, under salt stress, the induction was lower (0–2-fold).

The *AP2/ERF* transcription factors comprise a large group of genes, which is divided into four subfamilies, namely *AP2*, *ERF*, *DREB*, and *RAV* [56,57]. Moreover, the *AP2* subfamily consists of two subgroups, *AP2* and *AINTEGUMENTA* [58]. Phylogenetic tree analysis revealed that the *M. sativa* AINTEGUMENTA-like transcription factor was identical to the *M. truncatula* homologous protein and very similar to the *Cicer arietium* protein, as well as to *Arachis ipaensis*, *Vigna anqularis*, and *Phaseolus vulgaris* homologous proteins (Supplementary Materials). Interestingly, not many studies demonstrated the contribution of *AINTEGUMENTA* subfamily members to abiotic tolerance. Meng et al. [59] reported that *AINTEGUMENTA* genes regulate salt tolerance in *Arabidopsis*. The findings of the present study highlight the importance of an *AP2/EREB* transcription factor belonging to the *AINTEGUMENTA* subgroup during the response of the studied populations of *M. sativa* and *M. arborea* to salt treatments. The trend of *AINTEGUMENTA* upregulation in most cases was the same as described for the aforementioned genes. It is noteworthy that *AINTEGUMENTA* gene was the only gene that was highly expressed in roots of *M. sativa's* population under both salt stress and salt shock.

MAP kinases are represented by multigene families in plants that perceive and transmit various signals including specific stimuli from abiotic stresses [60]. Specifically, the studied *SIMKK* gene was previously found to upregulate the downstream *SIMKK* gene and induce salt tolerance in alfalfa [61]. Moreover, transcription factors such as zinc finger proteins (ZFPs) regulate gene expression under many abiotic stresses including salt [62]. The expression level of *ZFN* was highly induced after salt treatment in alfalfa tolerant varieties [36,63]. In the present study, there was a fivefold upregulation of *ZFN* under salt shock (150 mM NaCl) and salt stress of 50–100–150 gradual acclimatization for the studied population of *M. arborea*. However, in that of *M. sativa*, no profound differences were observed between control and salt-treated plants. In general, a similar expression pattern was detected for the *SIMKK* gene.

Findings in the present study indicate that differences occur between expression ratios of salt shock (initial salt treatments of 100 and 150 mM) and salt stress (initial salt treatments 50 mM and gradual 50–100 and 50–100–150 mM) depending on the type of gene, the species, and the plant part. Among the studied genes, it seems that the ion transporters *NHX1* and *RCI2A*, the transcription factor of the *AINTEGUMENTA* subgroup, and the organic solute *P5CS1* play critical roles in the response of studied populations of *M. sativa* and *M. arborea* to both salt shock and salt stress. In experiments conducted in *Lotus japonicus* [14] using gradual salt stress treatment (up to 150 mM NaCl), it was reported that the most responsive genes were clearly related to transcription and signaling pathways and were likely to be involved in ionic changes and not involved in osmoregulation [10]. Moreover, some studies demonstrated clear differences between the genes that were upregulated between salt shock and stress in parallel experiments [64,65]. In our case, the same genes are key players to both salt shock and stress. This is probably attributed to the fact that, in salt-shock-treated plants, the ionic phase occurs early [10] and, as a result, expression changes of genes related to ionic stress responses can be detected especially after long exposure to salt shock (about four weeks in our case).

The population of *M. arborea* constitutively expresses all studied genes even under control conditions. Moreover, it upregulates all studied genes regardless of the salt regime (salt stress or salt shock). This strategy of *M. arborea* appears to have a negative impact on its growth rate probably due to the large energy cost. For example, as mentioned by many authors, proline synthesis is a metabolically expensive strategy [66,67]. On the other hand, a clear difference in gene expression levels under the two salt regimes was detected in the population of *M. sativa*. It seems that some responses are not necessary at low salt concentrations and are activated only when a threshold is reached [14]. The key genes according to the present study were moderately or hardly upregulated under salt stress, and highly under salt shock, indicating that *M. sativa* "orchestrates" a fine-tuning of gene induction up to the point it combines both salinity tolerance and the lowest growth reduction.

According to Sanches et al. [14], sensitive genotypes can compensate under low salt stresses, which is also the case with Alborea. The hybrid performed better than its parents under low concentrations of NaCl. This could be due to tolerance to low concentrations. In any case, it cannot efficiently respond under moderate-to-high salt concentrations.

We can conclude that the hybrid of *M. sativa* and *M. arborea* was much more productive compared to its parents under normal conditions and low salt treatments, but extremely sensitive to salt stress. On the other hand, the studied populations of *M. sativa* and *M. arborea* were relatively tolerant to all salt treatments. Furthermore, the population of *M. sativa* performed better than that of *M. arborea* under both salt stress and salt shock at the seedling stage, probably by activating a more cost-efficient strategy.

The two parental species of Alborea appear to regulate different components of the salt tolerance mechanism. There was no selection for salt tolerance in the development of the Alborea population used in our experiments; thus, Alborea plants could be genetically deficient for both mechanisms. Fortunately, the Alborea population was developed from several initial hybrids. Hence, all the genes for both mechanisms for salt tolerance should be present in the population per se, and potentially could be pyramided in individual plants by cycles of selection for salt tolerance.

Further investigation of the mechanisms of *M. sativa* and *M. arborea* through transcriptomic and metabolomic analysis, as well as under field conditions, could contribute to a better understanding of salt stress tolerance. Specific attention should be given to further analyze the function of *AINTEGUMENTA* under both salt shock and stress. This could be valuable when breeding salt tolerance into the highly productive hybrid.

Supplementary Materials: The following are available online at http://www.mdpi.com/2073-4395/8/10/231/s1, Figure S1: Phylogeny analysis of *M. sativa* AINTEGUMENTA protein with other AP2/EREB transcription factors. The GenBank accession numbers are as follows: AP2 like *Phaseolus vulgaris* (ESW30355.1), AIL6 *Vigna angularis* var. *angularis* (BAT99446.1), AIL6 *Arachis ipaensis* (XP_016201070.1), AIL like *Medicago truncatula* (XM_003612982.1), AIL6 *Cicer arietinum* (XP_004509770), AIL6 *Lupinus angustifolius* (XP_019421574.1), AIL6 *Cucurbita moschata* (XP_022942874), AIL6 *Gossypium raimondii* (KJB18166.1), AP2 like *Citrus clementina* (ESR39494.1), AIL6 *Nicotiana attenuate* (O1T03126.1) and AP2 like *Coffea canephora* (CDP01573.1).

Author Contributions: Conceptualization, E.T. and E.M.A. Methodology, E.T. and E.M.A. Formal analysis, E.M.A. Investigation, E.S., M.G., G.A., and M.P. Resources, E.B. Data curation, E.S. Writing—original draft preparation, E.T. and E.M.A. Writing—review and editing, E.B. Visualization, E.M.A. Supervision, E.T. and E.M.A. Project administration, E.T. and E.M.A. Funding acquisition, G.N.S.

Funding: This research received no external funding.

Acknowledgments: We would like to thank Antonia Tzampazy and Joseph Pentheroudakis (native English speakers) for carefully editing our manuscript.

Conflicts of Interest: The authors declare no conflict of interest.

References

1. Pitman, M.G.; Läuchli, A. Global impact of salinity and agricultural ecosystems. In *Salinity: Environment—Plants—Molecules*; Läuchli, A., Luttge, U., Eds.; Springer: Dordrecht, The Netherlands, 2002; pp. 3–20.

2. Rengasamy, P. Soil chemistry factors confounding crop salinity tolerance—A review. *Agronomy* **2016**, *6*, 53. [CrossRef]

3. Young, J.; Udeigwe, T.K.; Weindorf, D.C.; Kandakji, T.; Gautam, P.; Mahmoud, M.A. Evaluating management-induced soil salinization in golf courses in semi-arid landscapes. *Solid Earth* **2015**, *6*, 393–402. [CrossRef]

4. Pariente, S. Soluble salts dynamics in the soil under different climatic conditions. *Catena* **2001**, *43*, 307–321. [CrossRef]

5. Ispail, A.; Takeda, S.; Nick, P. Life and death under salt stress: Same players, different timing? *J. Exp. Bot.* **2014**, *65*, 2963–2979.

6. Munns, R.; James, R.A. Screening methods for salinity tolerance: A case study with tetraploid wheat. *Plant. Soil* **2003**, *253*, 201–218. [CrossRef]

7. Shavrukov, Y.; Langridge, P.; Tester, M. Salinity tolerance and sodium exclusion in genus *Triticum*. *Breed. Sci.* **2009**, *59*, 671–678. [CrossRef]

8. Apse, M.P.; Aharon, G.S.; Snedden, W.A.; Blumwald, E. Salt tolerance conferred by overexpression of a vacuolar Na^+/H^+ antiport in Arabidopsis. *Science* **1999**, *285*, 1256–1258. [CrossRef] [PubMed]

9. Munns, R. Comparative physiology of salt and water stress. *Plant Cell Environ.* **2002**, *25*, 239–250. [CrossRef] [PubMed]

10. Shavrukov, Y. Salt stress or salt shock: Which genes are we studying? *J. Exp. Bot.* **2013**, *64*, 119–127. [CrossRef] [PubMed]

11. Munns, R. Genes and salt tolerance: Bringing them together. *New Phytol.* **2005**, *167*, 645–663. [CrossRef] [PubMed]

12. Chakraborty, K.; Sairam, R.K.; Bhattacharya, R.C. Differential expression of salt overly sensitive pathway genes determines salinity stress tolerance in Brassica genotypes. *Plant Physiol. Biochem.* **2012**, *51*, 90–101. [CrossRef] [PubMed]

13. Roshandel, P.; Flowers, T. The ionic effects of NaCl on physiology and gene expression in rice genotypes differing in salt tolerance. *Plant Soil* **2009**, *315*, 135–147. [CrossRef]

14. Sanchez, D.H.; Lippold, F.; Redestig, H.; Hannah, M.A.; Erban, A.; Krämer, U.; Kopka, J.; Udvardi, M.K. Integrative functional genomics of salt acclimatization in the model legume *Lotus japonicus*. *Plant J.* **2008**, *53*, 973–987. [CrossRef] [PubMed]

15. Munns, R.; Tester, M. Mechanisms of salinity tolerance. *Annu. Rev. Plant Biol.* **2008**, *59*, 651–681. [CrossRef] [PubMed]

16. Li, R.; Shi, F.; Fukuda, K.; Yang, Y. Effects of salt and alkali stresses on germination, growth, photosynthesis and ion accumulation in alfalfa (*Medicago sativa* L.). *Soil Sci. Plant. Nutr.* **2010**, *56*, 725–733. [CrossRef]

17. Yang, Q.C.; Wu, M.S.; Wang, P.Q.; Kang, J.M.; Zhou, X.L. Cloning and expression analysis of a vacuolar Na$^+$/H$^+$ antiporter gene from alfalfa. *DNA Seq.* **2005**, *16*, 352–357. [CrossRef] [PubMed]

18. Ferreira, J.F.S.; Cornacchione, M.V.; Liu, X.; Suarez, D.L. Nutrient composition, forage parameters, and antioxidant capacity of alfalfa (*Medicago sativa*, L.) in response to saline irrigation water. *Agriculture* **2015**, *5*, 577–597. [CrossRef]

19. Graham, P.H.; Vance, C.P. Legumes: Importance and constraints to greater use. *Plant Physiol.* **2003**, *131*, 872–877. [CrossRef] [PubMed]

20. Campanelli, A.; Ruta, C.; Morone-Fortunato, I.; De Mastro, G. Alfalfa (*Medicago sativa* L.) clones tolerant to salt stress: In vitro selection. *Cent. Eur. J. Biol.* **2013**, *8*, 765–776. [CrossRef]

21. Postnikova, O.A.; Shao, J.; Nemchinov, L.G. Analysis of the alfalfa root transcriptome in response to salinity stress. *Plant Cell Physiol.* **2013**, *54*, 1041–1055. [CrossRef] [PubMed]

22. Quan, W.L.; Liu, X.; Wang, H.Q.; Chan, Z.L. Physiological and transcriptional responses of contrasting alfalfa (*Medicago sativa* L.) varieties to salt stress. *Plant Cell Tissue Organ Cult.* **2016**, *126*, 105–115. [CrossRef]

23. Sandhu, D.; Cornacchione, M.V.; Ferreira, J.F.S.; Suarez, D.L. Variable salinity responses of 12 alfalfa genotypes and comparative expression analyses of salt-response genes. *Sci. Rep.* **2017**, *7*, 42958. [CrossRef] [PubMed]

24. Boughalleb, F.; Denden, M.; Ben Tiba, B. Photosystem II photochemistry and physiological parameters of three fodder shrubs, *Nitraria retusa*, *Atriplex halimus* and *Medicago arborea* under salt stress. *Acta Physiol. Plant.* **2009**, *31*, 463–476. [CrossRef]

25. Boughalleb, F.; Denden, M.; Tiba, B.B. Anatomical changes induced by increasing NaCl salinity in three fodder shrubs, *Nitraria retusa*, *Atriplex halimus* and *Medicago arborea*. *Acta Physiol. Plant.* **2009**, *31*, 947–960. [CrossRef]

26. Lefi, E.; Conesa, M.À.; Cifre, J.; Gulías, J.; Medrano, H. Dry matter allocation in *Medicago arborea* and *Medicago citrina* in response to drought and defoliation. *Crop Pasture Sci.* **2012**, *63*, 179–189. [CrossRef]

27. Bingham, E.T.; Armour, D.; Irwin, J.A.G. The hybridization barrier between herbaceous *Medicago sativa* and woody *M. arborea* is weakened by selection of seed parents. *Plants* **2013**, *2*, 343–353. [CrossRef] [PubMed]

28. Irwin, J.A.G.; Sewell, J.C.; Woodfield, D.R.; Bingham, E.T. Restructuring lucerne (*Medicago sativa*) through introgression of the *Medicago arborea* genome. *Agric. Sci.* **2016**, *28*, 40–46.

29. Clifton-Brown, J.C.; Lewandowski, I. Water use efficiency and biomass partitioning of three different *Miscanthus* genotypes with limited and unlimited water supply. *Ann. Bot.* **2000**, *86*, 191–200. [CrossRef]

30. Lei, Y.; Xu, Y.; Hettenhausen, C.; Lu, C.; Shen, G.; Zhang, C.H.; Li, J.G.; Song, J.; Lin, H.; Wu, J. Comparative analysis of alfalfa (*Medicago sativa* L.) leaf transcriptomes reveals genotype-specific salt tolerance mechanisms. *BMC Plant Biol.* **2018**, *18*, 35. [CrossRef] [PubMed]

31. Leidi, E.O.; Barragán, V.; Rubio, L.; El-Hamdaoui, A.; Ruiz, M.T.; Cubero, B.; Fernández, J.A.; Bressan, R.A.; Hasegawa, P.M.; Quintero, F.J.; et al. The AtNHX1 exchanger mediates potassium compartmentation in vacuoles of transgenic tomato. *Plant J.* **2010**, *61*, 495–506. [CrossRef] [PubMed]

32. Long, R.C.; Zhang, F.; Li, Z.Y.; Li, M.N.; Cong, L.L.; Kang, J.M.; Zhang, T.J.; Zhao, Z.X.; Sun, Y.; Yang, Q.C. Isolation and functional characterization of salt-stress induced RCI2-like genes from *Medicago sativa* and *Medicago truncatula*. *J. Plant Res.* **2015**, *128*, 697–707. [CrossRef] [PubMed]

33. Mitsuya, S.; Taniguchi, M.; Miyake, H.; Takabe, T. Overexpression of RCI2A decreases Na$^+$ uptake and mitigates salinity-induced damages in *Arabidopsis thaliana* plants. *Physiol. Plant.* **2006**, *128*, 95–102. [CrossRef]

34. Ashraf, M.; Foolad, M.R. Roles of glycine betaine and proline in improving plant abiotic stress resistance. *Environ. Exp. Bot.* **2007**, *59*, 206–216. [CrossRef]

35. Kaur, G.; Asthir, B. Proline: A key player in plant abiotic stress tolerance. *Biol. Plant.* **2015**, *59*, 609–619. [CrossRef]

36. Wang, X.H.; Gao, B.W.; Liu, X.; Dong, X.J.; Zhang, Z.X.; Fan, H.Y.; Zhang, L.; Wang, J.; Shi, S.P.; Tu, P.F. Salinity stress induces the production of 2-(2-phenylethyl)chromones and regulates novel classes of responsive genes involved in signal transduction in *Aquilaria sinensis calli*. *BMC Plant Biol.* **2016**, *16*, 119. [CrossRef] [PubMed]

37. Yang, Y.; Guo, Y. Elucidating the molecular mechanisms mediating plant salt-stress responses. *New Phytol.* **2018**, *217*, 523–539. [CrossRef] [PubMed]

38. Suh, A. Evaluation of Bioactivity of Phytotoxins from Pathogenic Fungi of *Orobanche* sp. Ph.D. Thesis, Agricultural University of Athens, Athina, Greece, 2011.

39. Hamrouni, L.; Ben Abdallah, F.; Abdelly, C.; Ghorbel, A. La culture in vitro: Un moyen rapide et efficace pour sélectionner des génotypes de vigne tolérant la salinité. *C. R. Biol.* **2008**, *331*, 152–163. [CrossRef] [PubMed]

40. Kumar, S.; Stecher, G.; Tamura, K. MEGA7: Molecular Evolutionary Genetics Analysis Version 7.0 for Bigger Datasets. *Mol. Biol. Evol.* **2016**, *33*, 1870–1874. [CrossRef] [PubMed]

41. Livak, K.J.; Schmittgen, T.D. Analysis of relative gene expression data using realtime quantitative PCR and the 2$^{-\Delta\Delta Ct}$ Method. *Methods* **2001**, *25*, 402–408. [CrossRef] [PubMed]

42. Beyer, H.T.; John, W. Exploratory Data Analysis. Addison-Wesley Publishing Company Reading, Mass.—Menlo Park, Cal., London, Amsterdam, Don Mills, Ontario, Sydney 1977, XVI, 688 S. *Biom. J.* **1981**, *23*, 413–414. [CrossRef]

43. Almansouri, M.; Kinet, J.-M.; Lutts, S. Effect of salt and osmotic stresses on germination in durum wheat (*Triticum durum* Desf.). *Plant Soil* **2001**, *231*, 243–254. [CrossRef]

44. Zou, J.; Liu, C.F.; Liu, A.L.; Zou, D.; Chen, X.B. Overexpression of OsHsp17.0 and OsHsp23.7 enhances drought and salt tolerance in rice. *J. Plant Physiol.* **2012**, *169*, 628–635. [CrossRef] [PubMed]

45. Scasta, J.D.; Trostle, C.L.; Foster, M.A. Evaluating alfalfa (*Medicago sativa* L.) cultivars for salt tolerance using laboratory, greenhouse and field methods. *J. Agric. Sci.* **2012**, *4*, 90. [CrossRef]

46. Cornacchione, M.V.; Suarez, D.L. Evaluation of alfalfa (*Medicago sativa* L.) populations' response to salinity stress. *Crop Sci.* **2017**, *57*, 137–150. [CrossRef]

47. Farissi, M.; Bouizgaren, A.; Faghire, M.; Bargaz, A.; Ghoulam, C. Agro-physiological responses of Moroccan alfalfa (*Medicago sativa* L.) populations to salt stress during germination and early seedling stages. *Seed Sci. Technol.* **2011**, *39*, 389–401. [CrossRef]

48. Boughalleb, F.; Hajlaoui, H.; Mhamdi, M.; Dendon, M. Possible involvement of organic compounds and the antioxidant defense system in salt tolerance of *Medicago arborea* (L.). *Agric. J.* **2011**, *6*, 353–365. [CrossRef]

49. Sibole, J.V.; Cabot, C.; Poschenrieder, C.; Barcelo, J. Ion allocation in two different salt-tolerant Mediterranean Medicago species. *J. Plant Physiol.* **2003**, *160*, 1361–1365. [CrossRef] [PubMed]

50. Galmes, J.; Cifre, J.; Medrano, H.; Flexas, J. Modulation of relative growth rate and its components by water stress in Mediterranean species with different growth forms. *Oecologia* **2005**, *145*, 21–31. [CrossRef] [PubMed]

51. Teakle, N.L.; Amtmann, A.; Real, D.; Colmer, T.D. *Lotus tenuis* tolerates combined salinity and waterlogging: Maintaining O$_2$ transport to roots and expression of an NHX1-like gene contribute to regulation of Na$^+$ transport. *Physiol. Plant.* **2010**, *139*, 358–374. [CrossRef] [PubMed]

52. Rocha, P.S.C.F. Plant abiotic stress-related RCI2/PMP3s: Multigenes for multiple roles. *Planta* **2016**, *244*, 287. [CrossRef] [PubMed]

53. Sairam, R.K.; Tyagi, A. Physiology and molecular biology of salinity stress tolerance in plants. *Curr. Sci.* **2004**, *86*, 407–421.

54. Gharsallah, C.; Fakhfakh, H.; Grubb, D.; Gorsane, F. Effect of salt stress on ion concentration, proline content, antioxidant enzyme activities and gene expression in tomato cultivars. *AOB Plants* **2016**, *8*, plw055. [CrossRef] [PubMed]

55. Silva-Ortega, C.O.; Ochoa-Alfaro, A.E.; Reyes-Agüero, J.A.; Aguado-Santacruz, G.A.; Jiménez-Bremont, J.F. Salt stress increases the expression of P5CS gene and induces proline accumulation in cactus pear. *Plant Physiol. Biochem.* **2008**, *46*, 82–92. [CrossRef] [PubMed]

56. Mizoi, J.; Shinozaki, K.; Yamaguchi-Shinozaki, K. AP2/ERF family transcription factors in plant abiotic stress responses. *Biochim. Biophys. Acta* **2012**, *1819*, 86–96. [CrossRef] [PubMed]
57. Nakano, T.; Suzuki, K.; Fujimura, T.; Shinshi, H. Genome-wide analysis of the ERF gene family in Arabidopsis and rice. *Plant Physiol.* **2006**, *140*, 411–432. [CrossRef] [PubMed]
58. Mizukami, Y.; Fischer, R.L. Plant organ size control: AINTEGUMENTA regulates growth and cell numbers during organogenesis. *Proc. Natl. Acad. Sci. USA* **2000**, *97*, 942–947. [CrossRef] [PubMed]
59. Meng, L.S.; Wang, Y.B.; Yao, S.Q.; Liu, A.Z. Arabidopsis AINTEGUMENTA mediates salt tolerance by trans-repressing SCABP8. *J. Cell Sci.* **2015**, *128*, 2919–2927. [CrossRef] [PubMed]
60. Sinha, A.K.; Jaggi, M.; Raghuram, B.; Tuteja, N. Mitogen-activated protein kinase signaling in plants under abiotic stress. *Plant Signal. Behav.* **2011**, *6*, 196–203. [CrossRef] [PubMed]
61. Kiegerl, S.; Cardinale, F.; Siligan, C.; Gross, A.; Baudouin, E.; Liwosz, A.; Eklof, S.; Till, S.; Bogre, L.; Hirt, H.; et al. SIMKK, a mitogen-activated protein kinase (MAPK) kinase, is a specific activator of the salt stress-induced MAPK, SIMK. *Plant Cell* **2000**, *12*, 2247–2258. [CrossRef] [PubMed]
62. Tang, L.L.; Cai, H.; Ji, W.; Luo, X.; Wang, Z.Y.; Wu, J.; Wang, X.D.; Cui, L.; Wang, Y.; Zhu, Y.M.; et al. Overexpression of GsZFP1 enhances salt and drought tolerance in transgenic alfalfa (*Medicago sativa* L.). *Plant Physiol. Biochem.* **2013**, *71*, 22–30. [CrossRef] [PubMed]
63. Chao, Y.; Kang, J.; Sun, Y.; Yang, Q.; Wang, P.; Wu, M.; Li, Y.; Long, R.; Qin, Z. Molecular cloning and characterization of a novel gene encoding zinc finger protein from *Medicago sativa* L. *Mol. Biol. Rep.* **2009**, *36*, 2315. [CrossRef] [PubMed]
64. Kawasaki, S.; Borchert, C.; Deyholos, M.; Wang, H.; Brazille, S.; Kawai, K.; Galbraith, D.; Bohnert, H.J. Gene expression profiles during the initial phase of salt stress in rice. *Plant Cell* **2001**, *13*, 889–905. [CrossRef] [PubMed]
65. Lisa, A.L.; Elias, S.; Rahman, M.; Shahid, S.; Iwasaki, T.; Hasan, A.K.M.M.; Kosuge, K.; Fukami, Y.; Seraj, Z. Physiology and gene expression of the rice landrace Horkuch under salt stress. *Funct. Plant Biol.* **2011**, *38*, 282–292. [CrossRef]
66. Bojórquez-Quintal, E.; Velarde-Buendía, A.; Ku-González, Á.; Carillo-Pech, M.; Ortega-Camacho, D.; Echevarría-Machado, I.; Pottosin, I.; Martínez-Estévez, M. Mechanisms of salt tolerance in habanero pepper plants (*Capsicum chinense* Jacq.): Proline accumulation, ions dynamics and sodium root-shoot partition and compartmentation. *Front. Plant Sci.* **2014**, *5*, 605. [PubMed]
67. Shabala, S. Learning from halophytes: Physiological basis and strategies to improve abiotic stress tolerance in crops. *Ann. Bot.* **2013**, *112*, 1209–1221. [CrossRef] [PubMed]

MDPI
St. Alban-Anlage 66
4052 Basel
Switzerland
Tel. +41 61 683 77 34
Fax +41 61 302 89 18
www.mdpi.com

Agronomy Editorial Office
E-mail: agronomy@mdpi.com
www.mdpi.com/journal/agronomy